OXYGEN TRANSFER
FROM ATMOSPHERE
TO TISSUES

ADVANCES IN EXPERIMENTAL MEDICINE AND BIOLOGY

Recent Volumes in this Series

A Continuation Order Plan is available for this series. A continuation order will bring delivery of each new volume immediately upon publication. Volumes are billed only upon actual shipment. For further information please contact the publisher.

OXYGEN TRANSFER FROM ATMOSPHERE TO TISSUES

Edited by

Norberto C. Gonzalez

University of Kansas Medical Center
Kansas City, Kansas

and

M. Roger Fedde

College of Veterinary Medicine
Kansas State University
Manhattan, Kansas

PLENUM PRESS • NEW YORK AND LONDON

Library of Congress Cataloging in Publication Data

Oxygen transfer from atmosphere to tissues.

(Advances in experimental medicine and biology; v. 227)
Proceedings of a symposium held Mar. 27–28, 1987, in Kansas City, Kan. as a result
of the celebration of the centennial of American Physiological Society.
Includes bibliographies and index.
1. Oxygen transport (Physiology) — Congresses. 2. Tissue respiration — Congresses.
3. Respiration — Congresses. I. Gonzalez, Norberto C. II. Fedde, M. Roger. III.
American Physiological Society (1887–   ) IV. Series. [DNLM: 1. Biological Transport
— congresses. 2. Oxygen — blood — congresses. 3. Oxygen — metabolism — congresses. 4.
Oxygen Consumption — congresses. W1 AD559 v.227/QV 312 0973 1987]
QP99.3.090934   1988   599′.0124   88-2531
ISBN 0-306-42825-3

Proceedings of a symposium on Oxygen Transfer from Atmosphere
to Tissues, held March 27–28, 1987, in Kansas City, Kansas

© 1988 Plenum Press, New York
A Division of Plenum Publishing Corporation
233 Spring Street, New York, N.Y. 10013

PREFACE

This volume summarizes the Proceedings of a meeting held at the University of Kansas Medical Center, Kansas City, Kansas, on March 27 and 28, 1987. The meeting developed as a result of the celebration of the Centennial of the American Physiological Society. We took the opportunity offered by the presence, in the United States, of several European scientists who were guests of the American Physiological Society. Most of the symposium participants also attended the FASEB meeting which took place immediately afterwards in Washington, D.C.

We wish to express our appreciation to the contributors to this book and to all those who attended the meeting. The symposium and this book would not have been possible without the generous support of the following:

The Parker B. Francis III Foundation, Kansas City, Missouri

The Hans Rudolph Company, Kansas City, Missouri

The School of Medicine, University of Kansas Medical Center, Kansas City, Kansas

The Department of Physiology, University of Kansas Medical Center, Kansas City, Kansas

The College of Veterinary Medicine, Kansas State University, Manhattan, Kansas

We also want to thank Gail Widener, Antoinette Bray and Janet Pauly, of the Department of Physiology, University of Kansas Medical Center, for their invaluable help during the symposium, and Mrs. Linda Carr, who had the arduous task of typing the manuscripts.

Norberto C. Gonzalez

M. Roger Fedde

Kansas City

Kjell Johansen, VIKING and Physiologist

One of the distinguished scientists invited by the American Physiological Society to participate in its Centennial Celebration was Dr. Kjell Johansen, founder and Chairman of the Department of Zoophysiology of the University of Aarhus, Denmark. Dr. Johansen died on March 4, 1987, while vacationing in France, shortly before he was due to depart for his visit to the United States.

Dr. Johansen had accepted our invitation to attend this symposium and give the keynote lecture on the session of Oxygen Transfer by the Circulation. The title of his lecture was to be "Blood Circulation and Respiratory Gas Transport: A Comparative Assessment", certainly a subject to which he had made many important contributions.

Dr. Johansen left a large legacy. His research is well known to all interested in the physiological adaptation of animals to the environment. The laboratory that he headed is, largely thanks to his efforts, one of the leading laboratories in the world in the area of comparative physiology. His passing will be mourned by students, friends and colleagues around the world.

CONTENTS

OXYGEN TRANSFER IN THE TISSUES

LIMITATIONS TO OXYGEN TRANSFER

HIGH POINTS IN THE PHYSIOLOGY OF EXTREME ALTITUDE[1]

John B. West

Department of Medicine
University of California, San Diego
La Jolla, CA 92093

The history of the physiology of extreme altitude is one of the most colorful in the whole area of physiology. Man seems to have a fundamental urge to climb higher and higher, and reaching the summit of Mt. Everest is still often referred to as one of the basic human aspirations. An interesting feature of the history is that physiologists through the years have repeatedly been astonished by some new altitude record but in spite of this have confidently predicted on each occasion that man can go no higher!

In this brief survey, I shall touch on some of the most important events in the gradual improvement of our understanding of the physiology of extreme altitude. Space limitations preclude me from covering many important topics and I hope readers will excuse the personal nature of the selection.

One of the first descriptions of the debilitating effects of high altitude on the human body was given by the Jesuit missionary, Jose de Acosta who accompanied the early Spanish conquistadors to Peru in the 16th century. He described how as he traveled over a high mountain, he "was suddenly surprised with so mortall and strange a pang, that I was ready to fall from the top to the ground." He went on to add "I was surprised with such pangs of straining and casting, as I thought to cast up my heart too; for having cast up meate, fleugme, and choller, both yellow and greene; in the end I cast up blood, with the straining of my stomacke." And finally "I therefore perswade myselfe, that the element of the aire is there so subtile and delicate, as it is not proportional with the breathing of man, which requires a more grosse and temprate aire" (Acosta, 1590). The reference to the thinness of the air was an inspired guess because it was not until 1648 that Blaise Pascal arranged to have a mercury barometer taken to the top of the Puy de Dome in central France and showed that the pressure fell (Pascal, 1648). Incidentally, Acosta's dramatic description of vomiting at high altitude is not typical of acute mountain sickness.

Acosta's book which also contains much valuable information about the Inca civilization, was widely read in both Spanish and English versions and influenced Robert Boyle and others who took part in the great surge of scientific learning in the 17th century. However, it seems that Chinese

---

[1]Lecture given at the Symposium Banquet, March 27, 1987

Fig. 1. Professor De Saussure and guides during his ascent of Mont Blanc in 1787. The guides are carrying scientific equipment for measurements of temperature, snow conditions, etc., on the summit. From a contemporary painting.

travelers were aware of mountain sickness at least 1600 years before Acosta gave his historic description. In the classical Chinese history of the period preceeding the Han dynasty, the Qian Han Shu written about 30 B.C., we read that the route from the Western regions of China to the Hindu Kush (now in Afghanistan) crosses "the Great Headache Mountain, the Little Headache Mountain" and that "men's bodies become feverish, they lose colour, and are attacked with headache and vomiting; the asses and cattle being all in like condition" (Wylie, 1881).

In the 18th century, climbers in the European Alps reported a variety of disagreeable sensations at high altitude which now seem to us greatly exaggerated. An important event was the first ascent of Mont Blanc (4807 m, 15,782 ft.) in 1786. Twenty-six years earlier, Professor De Sassure, a physicist in Geneva, had offered a generous reward to anyone who could find a way to the summit in the hope that he could first use it himself. To his chagrin, the first ascent was made by Dr. Michel-Gabriel Paccard with the guide Jacques Balmat in 1786. Paccard took a mercury barometer and scratched the glass with a diamond in the village of Chamonix where they began, and at various stages of the climb. They had great difficulties circumventing the Jonction, a meeting point of two glaciers where there were many crevasses, and they apparently only saved themselves on four occasions by falling flat on their faces when they felt the snow giving way beneath them. These desperate measures apparently introduced a bubble of air into the barometer because heights above 3350 m registered about 150 m too high. The two men suffered frostbite and snow blindness but gave little account of the physiological problems of the great altitude. Subsequent accounts of the climb disagreed considerably about the state of the two men near the summit.

De Sassure made the ascent the following year (Fig. 1) and he was eloquent on the physiological problems. "When I began this ascent" he reported "I was quite out of breath from the rarity of the air.... The kind of fatigue which results from the rarity of the air is absolutely uncon-

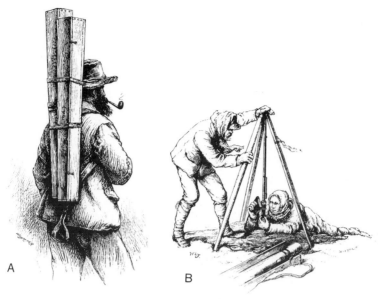

A

B

JEAN-ANTOINE AND THE BABIES.

'LOWER IT WOULD NOT GO.

Fig. 2. A. Jean-Antoine Carrel, one of Whymper's guides in the first ascent of Chimborazo in 1879. Carrel was entrusted with the care of the two Fortin barometers ("babies"). B. Measurement of barometric pressure on the summit of Chimborazo using one of the barometers. The height of the mercury was 14.100 inches at a temperature of 21°F. From Whymper (1891).

querable; when it is at its height, the most terrible danger would not make you take a single step further." When he was near the summit he complained of extreme exhaustion. "The need of rest was absolutely unconquerable; if I tried to overcome it, my legs refused to move, I felt the beginning of a faint, and was seized by dizziness.." On the summit itself he reported "When I had to get to work to set out the instruments and observe them, I was constantly forced to interrupt my work and devote myself to breathing". (De Sassure, 1786-1790). All these complaints at the comparatively modest altitude of 4807 m (15,782 ft.) or less reflect a combination of an almost complete lack of acclimatization and the fear of the unknown.

In the 19th century, numerous ascents were made of higher mountains including those in the South American Andes and there were abundant accounts of the distressing effects of extreme altitude. In 1879, Whymper made the first ascent of Chimborazo in Ecuador and described how at an altitude of 16,664 ft. (5079 m) he was incapacitated by the thin air - "In about an hour I found myself lying on my back, along with the Carrels (guides), placed hors de combat, and incapable of making the least exertion... We were unable to satisfy our desire for air, except by breathing with open mouths... Besides having our normal rate of breathing largely accelerated, we found it impossible to sustain life without every now and then giving spasmodic gulps, just like fishes when taken out of water (Whymper, 1891). However, Whymper and his two guides gradually recovered their strength and eventually attained the summit (6268 m, 20563 ft.). The climb was interesting because of the elaborate precautions to carry two Fortin mercury barometers to determine altitudes. Fig. 2 shows sketches of Jean-Antoine Carrel, the guide entrusted with the care of the two barometers, and the measurement of barometric pressure on the summit. Another interesting facet was that Whymper was aware of the beneficial effects of high altitude acclimatization.

In the latter part of the nineteenth century, there was considerable interest in the highest altitude that could be tolerated by man. Thomas W. Hinchliff, President of the (British) Alpine Club, wrote an account of his travelings around the world and described his feelings as he looked at the view from Santiago in Chile. "Lover of mountains as I am, and familiar with such summits as those of Mont Blanc, Monte Rosa and other Alpine heights, I could not repress a strange feeling as I looked at Tupungato and Aconcagua, and reflected that endless successions of men must in all probability be forever debarred from their lofty crests... Those who, like Major Godwin Austen, have had all the advantages of experience and acclimatization to aid them in attacks upon the higher Himalayas, agree that 21,500 ft. [6553 m] is near the limit at which man ceases to be capable of slightest further exertion" (Hinchliff, 1876).

Of course men had been to much greater altitudes in hydrogen balloons but these were purely passive ascents which did not require exertion on their part. The history of early ballooning is a fascinating story in its own right and there is not space to do it justice here. It is difficult for us to appreciate the drama of the first hot-air balloon ascents. Until the Montgolfier brothers astounded the citizens of Paris with their first ascent in November 1783, the only other way to reach high altitude was by climbing a mountain. The ability to leave the surface of the earth for the first time understandably created enormous excitement.

One of the most colorful balloon ascents was that made by Glaisher and Coxwell in 1862. James Glaisher was the chief meterologist at the Royal Observatory in Greenwich, and he wanted to obtain samples of air from the upper atmosphere for analysis. After several unsuccessful attempts to persuade younger colleagues to do this, he made the first balloon ascents himself at the age of 53. Henry Coxwell, a dentist who had become a professional balloonist, was in charge and the two made a series of scien- tific flights. In one, the balloon rapidly rose to an altitude estimated at 7 miles (over 11,000 m), considerably higher than the summit of Mt. Everest. High in the flight, Glaisher collapsed and Coxwell tried to vent hydrogen from the balloon, but because his hands were paralyzed by the hypoxia, he had to seize the cord with his teeth and dip his head two or three times. Just as remarkable was the fact that when the ballooon eventu- ally landed, Glaisher felt "no inconvenience" and walked over 7 miles to the nearest village because they came down in a remote country area (Glaisher et al., 1871).

Perhaps the most famous balloon ascent in this period was that of the balloon "Zenith" in 1875 with the three French balloonists, Tissandier, Croce-Spinelli, and Sivel. Their balloon reached an altitude estimated to be 8600 m (28,000 ft.) and when it returned to the ground Croce-Spinelli and Sivel had succumbed to the acute hypoxia. Tragically, although oxygen had been provided for the aeronauts, there was not enough, and further it was difficult to inhale because it was bubbled through bottles of water to remove an unpleasant odor. The disaster caused a sensation in France. This was an age of competitive ballooning with each balloonist trying to exceed the altitude of the other with little notion of the hazards. For example, in the funeral oration that followed the "Zenith" tragedy, Pastor Coquerel remarked that "they say that an Englishman could live and make observations above 8000 meters: the flag we carry must float higher yet!" (Bert, 1878).

The great French physiologist, Paul Bert, had a link with the Zenith tragedy because he recommended the balloonists to take more oxygen. However his warning letter arrived too late. Bert's work was pivotal in the devel- opment of our understanding of the physiological effects of high altitude. At his laboratory at the Faculte des Sciences in Paris, he had a number of both low and high pressure chambers, and he was able to show conclusively that the deleterious effects of low pressure were caused by the reduced PO_2. He exposed animals to air in low pressure chambers, and compared their responses with those of animals exposed to low oxygen concentrations at normal sea level pressures. By doing this, he was able to demonstrate convincingly that the partial pressure of oxygen was the important factor (Bert, 1878). Interestingly, he made measurements on Croce-Spinelli and Sivel when they visited his laboratory a year before their fateful balloon ascent. Bert decompressed them in his chamber to a pressure of 304 mmHg where they duly noted impairment of vision and hearing, and mental dullness. Oxygen administration caused a marked improvement. For anyone interested in the history of high altitude, Bert's great work "La Pression Barometrique" is essential reading, and a good English translation is available (Bert, 1878).

Although most physiologists were convinced by Bert's demonstrations that hypoxia was the crucial factor at high altitude, some were not. Among the latter was Angelo Mosso, Professor of Physiology at the University of Turin, Italy. Mosso built the first high altitude station for physiological research, completing it in 1894 (Fig. 3). The funds for the building were provided by Queen Margherita of Italy, an ardent mountaineer herself, and it was placed at an altitude of 4570 m (15,000 ft.) on the Punta Gnifetti, one of the peaks of the Monte Rosa in the Italian Alps. Many important observations were made in this laboratory which provided the first opportunity to expose man to high altitude for prolonged periods under relatively comfortable conditions. Mosso was the first to make extensive studies of the periodic breathing which occurs at high altitude. He also coined the term "acapnia", the reduction in PCO_2 which occurs at high altitude as the result of hyperventilation, and he thought that many of the symptoms of mountain sickness could be ascribed to this (Mosso, 1898).

The Capanna Margherita as it was called had an important influence on the development of the physiology of high altitude. However, the station was rather inaccessible and the eminent English physiologist Joseph Barcroft thought the conditions "too rigorous"; he remarked that "the difficulty of transport greatly restricts the possibilities both of research and gastronomy" (Barcroft, 1925). The Margherita station is still in use today but still can only be reached after a strenuous walk over the snow.

Fig. 3. Capanna Margherita on the Punta Gnifetti on Monte Rosa at an altitude of 4750 m (15,000 ft.) This was the first high altitude laboratory for physiology. From Barcroft (1925).

Mosso also had a low pressure chamber in his laboratories in Turin and he carried out some decompression experiments which rivaled the enterprises of the early balloonists. In one experiment, he decompressed his faithful technician, Georgio Mondo, to a barometric pressure of 246 torr, equivalent (as Mosso pointed out) to the summit of Mt. Everest (actually somewhat higher) though some additional oxygen had apparently been added to the chamber. The composition of the expired gas showed a PO_2 of about 40 and PCO_2 of about 5 mmHg! (Mosso, 1898).

The Capanna Margherita was a meeting place for many of the great high altitude physiologists of the day. Perhaps this is where the idea of the first high altitude expedition for physiological studies was generated by the German physiologist, Nathan Zuntz (Fig. 4). Previous expeditions to high regions had included some physiological observations, for example the geographical explorations of Alexander von Humboldt, but Zuntz's expedition was apparently the first to have physiology as its primary objective. The expedition was to Mount Tenerife in the Canary Islands in 1910 and the highest station was the Alta Vista Hut at about 3350 m (11,000 ft.). The international expedition included two Englishmen, C.G. Douglas and Joseph Barcroft, and the latter made an interesting observation on mountain sickness. Although Bert had claimed that the deleterious effects of high altitude were caused by the low PO_2, Mosso was unable to confirm this finding and suggested instead that it was the low blood PCO_2 (Mosso, 1898). However, at the Alta Vista Hut, Barcroft had an almost normal alveolar PCO_2 (38 torr) but was completely incapacitated by the altitude, whereas Douglas whose PCO_2 was only 32 torr was "perfectly free from all symptoms" (Barcroft, 1925). Thus, hypoxia (which was more severe in Barcroft because he did not increase his ventilation) was implicated as the culprit.

Fig. 4. Nathan Zuntz, German physiologist who organized the first high altitude physiology expedition to Mount Tenerife in the Canary Islands in 1910.

We saw earlier that Hinchliff made the gloomy prediction in 1876 that about 21,500 ft (6553 m) was near the limit of high altitude for a mountain climber. Both mountaineers and physiologists alike were, therefore, astonished when the aristocratic Italian climber, the Duke of the Abruzzi, reached the remarkable altitude of 7500 m (24,600 ft.) during an attempted ascent of K2 in the Karakorum Mountains in 1909. According to the Duke's biographer, one of the reasons given for this expedition was to determine "the greatest height to which man may attain in mountain climbing" (Filippi, 1912), and certainly the climb had a dramatic effect on both the mountaineering and medical communities interested in high altitude tolerance. In contrast to the florid accounts of paralyzing fatigue and breathlessness given by De Sassure, Whymper and others at much lower altitudes, the Duke made light of the physiological problems associated with this great altitude.

The Duke's climb prompted several physiologists to sharpen their pencils and try to calculate how the body could survive such hypoxic conditions. Douglas, Haldane and their co-workers estimated from the reported barometric pressure of 312 torr that the alveolar PO_2 must have been only 30 torr, and they concluded that adequate oxygenation of the arterial blood would be impossible under these conditions without active oxygen secretion by the lung (Douglas et al., 1913). However, this conclusion was disputed by Marie Krogh who argued that Douglas and his colleagues had markedly underestimated the diffusing capacity of the lung (Krogh, 1915). Incidentally, we now know that Douglas et al's. estimate of an alveolar PO_2 of 30 torr for a barometric pressure of 312 torr was much too low; the actual value is approximately 35 torr (Gill et al., 1962) in a region of the oxygen dissociation curve where an increase of 5 torr makes a world of difference.

Douglas and Haldane tested their hypothesis of oxygen secretion on the Anglo-American Pike's Peak Expedition in 1911. Pike's Peak near Colorado Springs has a summit observatory at an altitude of 4300 m (14,110 ft.) with the great advantage of easy accessibility by means of a cog railway to the summit. Douglas, Haldane and their co-workers measured the arterial PO_2 using a complicated carbon monoxide rebreathing technique, and reported that the arterial PO_2 could exceed the alveolar value by as much as 30 torr (Douglas et al., 1913). To this day it is uncertain where the measurement errors were made. An interesting facet of this expedition was the involvement of Miss Mabel FitzGerald. The story goes that it would have been improper for her to join the men in the Pike's Peak laboratory because she was unchaperoned. Instead she set off alone to tour the boisterous mining towns of Colorado collecting alveolar gas and blood specimens with which she was able to show the gradual decline in PCO_2, and increase in hemoglobin concentration with increasing altitude (FitzGerald, 1913).

At about this time, important contributions were made by a man whose work has been almost completely overlooked. This was Alexander M. Kellas (1868-1921), a lecturer in chemistry at the Middlesex Hospital Medical School in London (Fig. 5). In spite of his full-time faculty position, Kellas managed to make eight expeditions to the Himalayas in the first two decades of the century, and probably spent more time above 20,000 ft. (6100 m) than anyone else. In 1919 he wrote an extensive paper entitled "A consideration of the possibility of ascending Mt. Everest" which unfortunately was never properly published, appearing only in French in the proceedings of an obscure congress (Kellas, 1921). In this paper, he analyzed the physiology of a climber near the Everest summit including a discussion of the summit altitude, barometric pressure, alveolar PO_2, arterial oxygen saturation, maximal oxygen consumption, and maximal ascent rate. On the basis of his study he concluded that "Mt. Everest could be ascended by a man of excellent physical and mental constitution in first-rate training without adventitious aids [supplementary oxygen] if the physical difficulties

Fig. 5. Alexander M. Kellas (1868-1921)
He carried out the first analysis of the
limiting factors on the summit of Mt.
Everest but his contributions have been
almost completely overlooked. Archives
of the Royal Geographical Society.

of the mountain are not too great." The importance of this study was not so much that he reached the correct conclusion; he had so few data that many of his calculations were inevitably incorrect. However, Kellas asked all the right questions and he can claim the distinction of being the first physiologist to seriously analyze the limiting factors at the highest point on earth. It was not until almost 60 years later that his prediction was fulfilled.

Kellas collaborated with Haldane in an interesting set of experiments in which two subjects (Haldane and himself) spent four consecutive days in a low pressure chamber at altitudes equivalent to 11,600, 16,000, 21,000 and 25,000 ft. The studies showed that as little as 3 days of acclimatization apparently increased tolerance to an altitude of 25,000 ft. (7620 m). The full description (Haldane et al., 1919-20) which was published in the Journal of Physiology (London) makes good reading and contains a number of vignettes which modern day editors disdain. For example, on the last day in the chamber which was the climax of the experiment, Haldane apparently cut the experiment short and "came out about 4:00 p.m., as it was necessary to catch a train". Presumably, this was the last train to Oxford. Haldane's son, J.B.S. Haldane who later became an eminent biologist, also took part in some of the acute experiments. On one occasion when the pressure was 330 mmHg, outside observers noted that he could not stand properly and looked very blue and shaky. The paper records: "The emergency tap was therefore opened so as to raise the pressure. There is a correspondingly indignant and just legible note 'some bastard has turned the tap,' after which the notes become quite legible again as the pressure rose."

Kellas was a member of the first official reconnaisance expedition to Everest in 1921. Tragically, he died during the approach march just as the expedition had its first view of the mountain they came to climb. Three years later, E.F. Norton, who was a member of the third Everest expedition reached a height of about 8580 m (28,150 ft.) on the north side of Everest without supplementary oxygen. He was accompanied to just below this altitude by Dr. T.H. Somervell, who collected alveolar gas samples at an altitude of 7010 m (23,000 ft.) though unfortunately these were stored in rubber bladders through which the carbon dioxide rapidly diffused (Somervell, 1925). As a result the alveolar PCO_2 was given as 8 torr, which is much too low for this altitude; the correct value is approximately double this. It is interesting to note that although the calculated respiratory exchange ratios were clearly erroneously low, being between 0.31 and 0.44, no less a physiologist than Barcroft used the data in his predictions of the physiological changes at high altitudes (Barcroft, 1925). Somervell referred to the extreme breathlessness at those great altitudes stating that "for every step forward and upward, 7 to 10 complete respirations were required." The early Everest expeditions make vivid reading; an eyewitness account reports how Norton had part of his frostbitten ear snipped off in one of the camps by another climber using a pair of scissors (Fig. 6).

Barcroft led an important expedition to Cerro de Pasco in Peru in the winter of 1921-22. Careful measurements were made to test the theory of oxygen secretion which had been championed by Haldane, but no evidence for this was found. In fact, Barcroft showed that the arterial oxygen saturation fell during exercise at high altitude and he argued that this could be explained by the failure of equilbration of PO_2 between alveolar gas and pulmonary capillary blood (Barcroft et al., 1923). This was one of the first direct demonstrations of diffusion limitation at high altitude, a finding that has now been reproduced many times.

Fig. 6. Colonel E.F. Norton, who reached an altitude of about 8580 m (28,150 ft.) on Mt. Everest in 1924 without supplemental oxygen. The upper part of his right ear was removed with a pair of scissors by an expedition member after it became frostbitten.

One of the most significant outcomes of this expedition was the interest that it aroused in the physiology of permanent residents of high altitudes. The expedition parties were enormously impressed by the capacity of the residents of Cerro de Pasco for physical work, and they were astonished at the popularity of energetic sports, such as soccer. Nevertheless, Barcroft concluded that "all dwellers at high altitudes [are] persons of impaired physical and mental powers", a sentiment that incensed the Peruvian physician Carlos Monge. He and his pupil, Alberto Hurtado, subsequently embarked on an extensive study of these high altitude natives. It could be argued that the development of this influential Peruvian school of high altitude physiology dated from Barcroft's expedition, and possibly from this unguarded remark.

Monge and Hurtado went on to analyze how these high altitude dwellers had adapted to the "climatic aggression" (as they called it) of high altitude. They pointed out that one of the major differences between the Spanish conquistadors and the resident native population in the 16th and 17th centuries was that the Spanish could reproduce only by going to low altitudes. Monge also unearthed some interesting facts about the Incas, who had developed their sophisticated civilization in the high Andes in the 14th century. He noted, for example, that the Incas recognized the advantages of high altitude acclimatization to such an extent that they kept two armies; one operated only at high altitudes while the other fought only on the lower plains (Monge, 1948).

An important international high altitude expedition took place in 1935 under the scientific leadership of D.B. Dill. The site was Aucanquilcha in the Andes of north Chile where there is a sulfur mine which was said to be at an altitude of 5800 m (19,000 ft.). The primary purpose of the expedition was to obtain measurements on the expedition members, but some data were also obtained on the local residents. The main thrust of the research program was a study of the changes in blood chemistry at high altitude, and the results form the basis of our present knowledge in this area (Dill et al., 1937).

Fig. 7. Right: Justo Copa, who had been residing at an altitude of 5950m (19,500 ft.) for 2 years when this photograph was taken. Left: Dr. Raimundo Santolaya, Chilean physiologist.

The Aucanquilcha miners described by the expedition (Keys, 1936) have frequently been referred to as the highest inhabitants of the world. The expedition reported that although they worked at the mine, they lived at an altitude of 5340 m (17,500 ft.) preferring to climb 500 m on foot every day, which took about an hour and a half. The implication was that indefinite residence above an altitude of about 5340 m was impossible. However, a recent visit to the mine resulted in the surprising finding that a small group of miners live indefinetely at the mine itself, which is actually at an altitude of 5950 m (19,500 ft.) (West, 1986). One man had been living there for two years (Fig. 7) but most of the caretakers reside for shorter periods. This is hardly surprising given the primitive accommodation (Fig. 8). It was also clear from conversations with the mine manager that only a small subset of miners are able to tolerate the altitude. Nevertheless, it is apparent that some people can live at this altitude for indefinite periods of time, which makes them easily the highest inhabitants of the world.

It was the end of an era when the summit of Mt. Everest was finally attained by Hillary and Tensing in 1953. However, the fact that the two climbers used supplementary oxygen still did not answer the question of whether it was possible to reach the highest point on earth breathing air. Hillary did remove his oxygen mask on the summit for about 10 minutes and at the end of the time reported "I realized that I was becoming rather clumsy-fingered and slow-moving, so I quickly replaced my oxygen set and experienced once more the stimulating effect of even a few liters of oxygen" (Hunt, 1953). Nevertheless, the fact that he could survive for a few minutes without additional oxygen came as a surprise to some physiologists who had predicted that he would lose consciousness.

Fig. 8. Hut where the caretakers of the Aucanquilcha mine live at an altitude of 5950 m (19,500 ft.). Dr. Santolaya is holding a barometer which reads 373.6 torr.

The British expedition of 1953, which first reached the Everest summit, included the physiologist, Dr. L.G.C.E. Pugh. Naturally, opportunities for research on this expedition were very limited, but Pugh organized another expedition in 1960-1961 which made important additional contributions to our knowledge of the physiology of extreme altitude. The Himalayan Scientific and Mountaineering Expedition (or Silver Hut expedition as it was also called) spent almost nine months in the region near Mt. Everest and included a period of some four months when a group of eight physiologists wintered at an altitude of 5800 m (19,000 ft.) in a prefabricated hut where the barometric pressure averaged about 380 mm Hg. In the subsequent spring, the party moved across to Mt. Makalu (8475 m, 27,790 ft.), the world's fifth highest mountain, where measurements of maximal oxygen consumption were made using a bicycle ergometer (Fig. 9) up to an altitude of 7440 m (24,400 ft.) (Pugh et al., 1964). In addition, alveolar gas samples were obtained as high as 7830 m (25,700 ft.) (Gill et al., 1962). The measurements of maximal oxygen consumption gave tantalizing predictions on whether man could hope to climb Mt. Everest without supplementary oxygen. Extrapolation of the line relating maximal oxygen uptake to barometric pressure to the pressure of 250 torr on the Everest summit suggested that almost all the oxygen available would be required for basal oxygen uptake (West and Wagner, 1980). Thus, these results strongly suggested that if man could reach the Everest summit without supplementary oxygen, he would be very near the limit of human tolerance.

Perhaps the ultimate climbing achievement occurred when Reinhold Messner and Peter Habeler reached the summit of Everest without supplementary oxygen in May 1978. Messner's account (Messner, 1979) makes it clear that he had very little in reserve. "After every few steps, we huddle over our ice axes, mouths agape, struggling for sufficient breath... As we get higher it becomes necessary to lie down to recover our breath... Breathing becomes such a strenuous business that we scarcely have strength to go on."

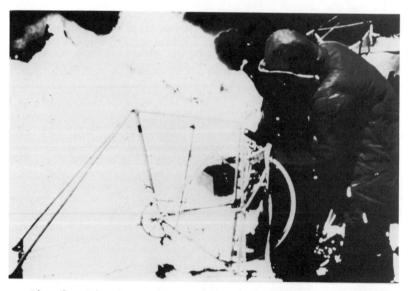

Fig. 9. Bicycle ergometer being assembled on the Makalu Col (7440 m, 24,400 ft.) during the Himalayan Scientific and Mountaineering Expedition of 1960-1961. These experiments gave the highest measurements of maximal oxygen uptake to date. In the background is Everest on the right and Lhotse on the left.

And when he eventually reached the summit, "in my state of spiritual abstraction, I no longer belong to myself and to my eyesight. I am nothing more than a single, narrow gasping lung, floating over the mists and the summits."

This dramatic description confirmed the prediction that man is very near the limits of tolerance at these great altitudes. This is further underlined when we recall that Norton and Somervell ascended to within 300 m of the Everest summit as early as 1924, but it was not until 1978 that climbers reached the top without supplementary oxygen. Thus, the last 300 m took 54 years!

Partly stimulated by the remarkable ascent of Messner and Habeler in 1978, the American Medical Research Expedition to Everest in 1981 had as one of its major objectives the gathering of data at altitudes over 8000 m (26,200 ft.). Dr. Christopher Pizzo obtained the first alveolar gas samples on the summit and also made the first direct measurement of barometric pressure (Fig. 10). Both results provided some surprises. Pizzo's alveolar PCO_2 was only 7.5 torr on the summit indicating the astonishingly high level of hyperventilation which successful climbers have to develop in order to defend their alveolar PO_2 (West et al., 1983a). The measured barometric pressure of 253 torr was surprising to those physiologists who had been using the standard altitude-pressure tables which gave a barometric pressure of only 246 torr on the summit. The difference can be explained by the fact that the barometric pressure at these altitudes is latitude-dependent because of the large mass of very cold air in the stratosphere above the equator. The result of this "climatic idosyncrasy" is that it is just possible for man to reach the highest point on earth without supplementary oxygen (West et al., 1983b)

Fig. 10. Dr. Christopher Pizzo on the summit of Mt. Everest during the American Medical Research Expedition to Everest, 1981. He is about to obtain the world's high altitude Frisbee record.

A little over a year ago a very successful simulation of an Everest ascent was carried out over a period of 40 days and nights in a low pressure chamber by Charles S. Houston, John R. Sutton and their colleagues. Many valuable new data were obtained, especially on the cardiovascular changes which occur at extremely low barometric pressures. In addition, a surprising degree of impairment of pulmonary gas exchange was found, particulary after rapid increases in simulated altitude. Although not all the data have yet been analyzed, it is clear that many fascinating questions about human physiology at extreme altitude remain to be answered.

REFERENCES

Acosta, I. de (1590). Historia natural y moral de las Indias... Seville: Iuan de Leon, Lib. 3, Cap. 9. [English translation of pertinent pages in: High Altitude Physiology edited by J.B. West, Hutchinson Ross Publishing Company, Stroudsburg, PA, 1981).

Barcroft, J., C.A. Binger, A.V. Bock, J.H. Doggart, H.S. Forbes, G. Harrop, J.C. Meakins, and A.C. Redfield (1923). Observations upon the effect of high altitude on the physiological processes of the human body, carried out in the Peruvian Andes, chiefly at Cerro de Pasco. Phil. Trans. Royal Soc., Ser. B, 211:351-480.

Barcroft, J. (1925). The respiratory functions of the blood, Part I: Lessons from high altitudes. Cambridge, Cambridge University Press.

Bert, P. (1978). La pression barometrique. Masson, Paris. (English translation by M.A. Hitchcock and F.A. Hitchcock. College Book Company, Columbus, OH, 1943).

Brown, T.G. and G. De Beer (1957) The first ascent of Mont Blanc. London, Oxford University Press.

De Sassure, H.B. (1786-1798). Voyages dans les alpes. 4 Vol.: Neuchatel: Louis Fauche-Borel.

Dill, D.B., J.H. Talbott and W.V. Consolazio (1937). Blood as a physico-chemical system XII. Man at high altitudes. J. Biol. Chem. 188:649-666.

Douglas, C.G., J.S. Haldane, Y. Henderson, and E.C. Schneider. (1913). Physiological observations made on Pike's Peak, Colorado, with special reference to adaptations to low barometric pressures. Phil. Trans. Royal Soc. Ser. B, 203:185-318.

Filippi, F.D. (1912). Karakorum and Western Himalaya. London: Constable.

FitzGerald, M.P. (1913). The changes in the breathing and the blood of various altitudes. Phil. Trans. Royal Soc. London Ser. B, 203:351-371.

Gill, M.G., J.S. Milledge, L.G.C.E. Pugh, and J.B. West (1962). Alveolar gas composition at 21,000 to 25,700 ft. (6400-7830 m). J. Physiol. London 163:373-377.

Glaisher, J., C. Flammarion, E. De Fonvielle and G. Tissandier (1871). Ascents from Wolverhampton. In: Travels in the Air, edited by J. Glaisher. Philadelphia, PA: Lippincott, pp. 50-58.

Haldane, J.S., A.M. Kellas and E.L. Kennaway (1919-1920). Experiments on acclimatization to reduced atmospheric pressure. J. Physiol. London 53:181-206.

Hinchliff, T.W. (1876). Over the sea and far away. London: Longmans, Green.

Hunt, J. (1953). The ascent of Everest. London: Hodder and Stoughton.

Kellas, A.M. (1921). Sur les possibilites de faire l'ascension du Mount Everest. Congres de l'Alpinisme, Monaco, 1920. Comptes Rendus des Seances, Paris 1:451-521.

Keys, A. (1936). The physiology of life at high altitudes: the International High Altitude Expedition to Chile 1935. Sci. Mon. 43:289-312.

Krogh, M. (1915). The diffusion of gases through the lungs of man. J. Physiol. London 49:271-296.

Messner, T. (1979). *Everest: expedition to the ultimate*. London: Kaye and
    Ward.
Monge, C. (1948). *Acclimatization in the Andes: historical confirmations
    of "climatic aggression" in the Andean man*. Baltimore, Johns Hopkins
    Press.
Mosso, A. (1898). *Life of man on the high alps*. London: Fisher Unwin.
Pascal, B. (1648). Story of the great experiment on the equilibrium of
    fluids. (English translation in: *High Altitude Physiology* edited by
    J.B. West, Hutchinson Ross Publishing Company, Stroudsburg, PA, 1981).
Pugh, L.G.C.E., M.B. Gill, S. Lahiri, J.S. Milledge, M.P. Ward, and J.B.
    West (1964). Muscular exercise at great altitudes. J. Appl. Physiol.
    19:431-440.
Somervell, T.H. (1925). Note on the composition of alveolar air at extreme
    heights. J. Physiol. London, 60:282-285.
West, J.B. (1986). Highest inhabitants of the world. Nature 324:517.
West, J.B., P.H. Hackett, K.H. Maret, J.S. Milledge, R.M. Peters, Jr., C.J.
    Pizzo and R.M. Winslow (1983a). Pulmonary gas exchange on the summit
    of Mount Everest. J. Appl. Physiol. 55:678-687.
West, J.B., S. Lahiri, K.H. Maret, R.M. Peters, Jr and C.J. Pizzo (1983b).
    Barometric pressures at extreme altitudes on Mt. Everest: Physiologi-
    cal significance. J. Appl. Physiol. 54:1188-1199.
West, J.B. and P.D. Wagner (1980). Predicted gas exchange on the summit of
    Mt. Everest. Respir. Physiol. 42:1-16.
Whymper, E. (1891). *Travels amongst the great Andes of the equator*.
    London: Murray.
Wylie, A. (1881). Notes on the Western Regions (translated from the Qian
    Han Shu, Book 96, Part 1). J. Anthrop. Inst. 10:20-73.

OXYGEN TRANSFER AND GAS EXCHANGE IN THE LUNGS

PULMONARY DIFFUSING CAPACITY AND ALVEOLAR-CAPILLARY EQUILIBRATION

Johannes Piiper

Max Planck Institute for Experimental Medicine
Göttingen
Federal Republic of Germany

In this short overview, theoretical and experimental work by the author and his co-workers pertaining to alveolar-capillary equilibration in mammalian lungs is summarized. The report is subdivided into four sections, each dealing with a different aspect of the topic: I. Diffusing capacity and diffusion/perfusion model. II. Determination of pulmonary diffusing capacity by rebreathing. III. O_2 kinetics of red blood cells measured by stopped-flow. IV. Alveolar-capillary CO_2 equilibration. The topics of the sections I, II and IV have been previously reviewed in more detail (Piiper and Scheid, 1980).

I. Diffusing capacity and diffusion/perfusion model

In its simplest model the pulmonary diffusing capacity, D, is the diffusive conductance of a uniform flat sheet forming a barrier between alveolar gas and pulmonary capillary blood. According to Fick's diffusion law, the diffusion conductance, defined as transfer rate (M) divided by partial pressure difference across the barrier is equal to the surface area/thickness ratio (F/x) and to the Krogh diffusion constant (K):

$$D = K \cdot \frac{F}{x} \qquad\qquad (1)$$

and K is given by the product of diffusion coefficient (d) and solubility (α):

$$K = d \cdot \alpha \qquad\qquad (2)$$

The next step in simple model simulation of alveolar gas transfer is to introduce blood flow (flow rate $\dot{Q}$) in the form of a flowing blood sheet (Fig. 1). Equal in importance to blood flow is the capacitance coefficient or effective solubility, β (Piiper et al., 1971). For inert gases, β is equal to physical solubility. For gases chemically bound by blood (CO_2, O_2, CO) β is the slope of the dissociation curve plotted as concentration against partial pressure.

Assuming that diffusion occurs perpendicular to flow and that there is no diffusive or convective mixing in the direction of flow and furthermore assuming that the effective solubility (β) in blood is constant, the following relationship is easily derived (Piiper and Scheid, 1981):

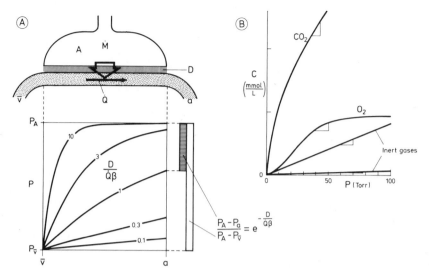

Fig. 1. Model for diffusion limitation in capillary-alveolar gas transfer. A. Model with diffusing capacity, D, and blood flow, $\dot{Q}$, and the partial pressure (P) profiles in the capillary, from the mixed venous ($\bar{v}$) to the arterial (end-capillary) values (a). The deficit in diffusive equilibration or diffusion limitation $(P_A-P_a)/(P_A-P_{\bar{v}})$ increases with decreasing $D/(\dot{Q}\beta)$. For $D/(\dot{Q}\beta) = 1$ it is represented by columns. For gases eliminated by lungs P increases bottomwards. B. Blood concentration-partial pressure relationships (dissociation curves) are shown for CO_2, O_2, a highly soluble gas and a gas of lower solubility. The slopes, $\Delta C/\Delta P$, are capacitance coefficients, β. After Piiper and Scheid, 1981.

$$\frac{P_A - P_a}{P_A - P_{\bar{v}}} = \exp[-D/(\dot{Q}\beta)] \qquad (3)$$

This relationship shows that the incompleteness of gas/blood equilibration, expressed as the alveolar-arterial difference relative to the alveolar-mixed-venous difference, is determined by the ratio of diffusion conductance (diffusing capacity D) and perfusion conductance ($\dot{Q}\beta$).

With $D/(\dot{Q}\beta) > 3$, gas transfer is predominantly limited by perfusion; with $3 > D/(\dot{Q}\beta) > 0.1$, there is combined perfusion and diffusion limitation, the latter becoming progressively more important with decreasing $D/(\dot{Q}\beta)$, and with $D/(\dot{Q}\beta) < 0.1$ gas transfer is diffusion-limited, being practically independent of perfusion.

When different gas species are compared, it is appropriate to combine eqs. (1), (2) and (3) and to group the variables into those independent of gas species (F, x and $\dot{Q}$) and the variables describing the properties of the gases (d, α, β):

$$\frac{D}{\dot{Q}\beta} = \frac{d \cdot \alpha}{\beta} \cdot \frac{F}{x \cdot \dot{Q}} \qquad (4)$$

The diffusion coefficient (d), both in gaseous and liquid media, is roughly inversely proportional to the square root of the molecular weight (Graham's law). For He and SF_6, gas species of very different molecular weight, the d ratio is about 7. The blood/tissue solubility ratio β/α can

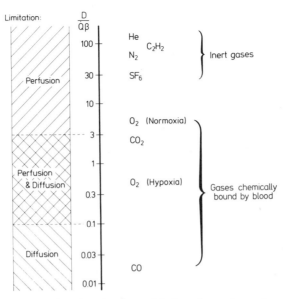

Fig. 2. Diffusion/perfusion limitation in alveolar-
capillary transfer of various gases. Ordinate: $D/(\dot{Q}\beta)$
on logarithmic scale. The position of O_2 in hypoxia is
estimated from Do_2, those of other gases are placed
relative to O_2 in hypoxia according to eq. (4). In
contrast to inert gases, gases chemically bound to blood
are diffusion-limited due to high β. From Piiper and
Scheid (1981).

vary even more. It is close to unity for inert gases, but much higher for
the gases which are chemically bound. For human blood the following values
are estimated for the factorial increase of β over the physical solubility
and thus represent estimates for the β/α ratio (Piiper and Scheid, 1981):
CO_2: 10; O_2 in normoxia, 30; O_2 in hypoxia, 170; CO in hypoxia, 40,000.

$D/(\dot{Q}\beta)$ values for various gases in alveolar-capillary transfer in
human lungs are pictured in Fig. 2. Thus all inert gases are expected to be
practically perfusion-limited. O_2 (and CO_2) should be limited by both
perfusion and diffusion, and CO, by diffusion only (except at very low $\dot{Q}$).

The case of O_2 is of particular interest. In normoxia (and more so in
hyperoxia) the exchange should be predominantly flow-limited, but in
hypoxia additional diffusion limitation is expected to develop. The
following values have been estimated (Piiper and Scheid, 1981) from
measurements at 5350 m altitude (PB = 390 Torr, effective PIo_2 = 72 Torr)
performed by Cerretelli (1976) on man at rest and during maximum O_2 uptake
exercise (resting value/exercise value): $D/(\dot{Q}\beta)$ = 2.1/0.44, diffusion
limitation, 12%/64%; alveolar-endcapillary Po_2 difference, 1.7 Torr/27
Torr. The alveolar-endcapillary Po_2 difference values due to diffusion
limitation are in order-of-magnitude agreement with recent measurements by
Torre-Bueno et al. (1985), Wagner et al. (1986) and Hammond et al. (1986).

Defining the total O_2 conductance, g_{tot}, as O_2 uptake divided by the
alveolar-to-mixed-venous Po_2 difference,

$$G_{tot} = \frac{\dot{M}}{P_A - P_{\bar{V}}}$$

(5)

21

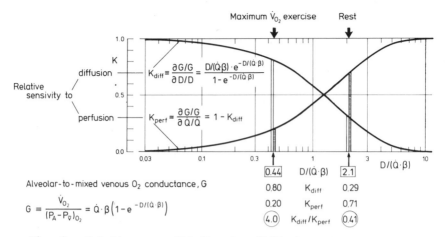

Fig. 3. Relative sensitivity to diffusion, K_{diff}, and to perfusion, K_{perf}, as function of $D/(\dot{Q}\beta)$. Arrows mark values calculated for hypoxia after measurements by Cerretelli (1976). Of particular interest is $K_{diff}/K_{perf} < 1$ in maximum O_2 uptake exercise, meaning that O_2 uptake is more limited by pulmonary diffusing capacity than by cardiac output. After Piiper and Scheid (1983).

diffusion/perfusion limitation relationships can be expressed in terms of relative increment of the total conductance, $\Delta G/G$, brought about by a relative increment in diffusing capacity, $\Delta D/D$, or in pulmonary blood flow, $\Delta\dot{Q}/\dot{Q}$ (Piiper and Scheid, 1983). One obtains the relative sensitivity to diffusion, $K_{diff} = (\Delta G/G)/(\Delta D/D)$, and to perfusion, $K_{perf} = (\Delta G/G)/(\Delta\dot{Q}/\dot{Q})$. These relationships are shown in Fig. 3, along with values for O_2 transfer according to the above-mentioned data obtained in high altitude hypoxia. Whereas at rest the K_{diff}/K_{perf} ratio is < 1.0, meaning less limitation by diffusion as compared to blood flow, at maximum O_2 uptake the ratio is > 1.0. This means that the O_2 uptake is more limited by diffusion than by perfusion and Do_2 becomes the major variable limiting aerobic performance.

Conclusions

1. The role of diffusion limitation in alveolar-capillary equilibration can be assessed using a simple model. The decisive parameter is the $D/(\dot{Q}\beta)$ ratio (D, pulmonary diffusing capacity; $\dot{Q}$, pulmonary capillary blood flow; β, solubility or slope of the blood dissociation curve).

2. Alveolar-capillary transfer of inert gases is limited by perfusion, that of CO, by diffusion; exchange of O_2 and CO_2 is mainly limited by perfusion, but diffusion limitation may become important in certain conditions (exercise; O_2 exchange in hypoxia).

II. Determination of pulmonary diffusing capacity by rebreathing

The main difficulties in determining the pulmonary diffusing capacity (D) are due to functional inhomogeneities, i.e. unequal distribution of alveolar ventilation ($\dot{V}_A$), diffusing capacity and blood flow ($\dot{Q}$) within the lungs. Generally, D is underestimated with the steady-state and breath-holding (single breath) methods when the functional inhomogeneities are not properly taken into account. Such corrections are difficult because the kind and the extent of inhomogeneity must be known and the procedures may be complex. It may easily happen that inhomogeneity effects are over-

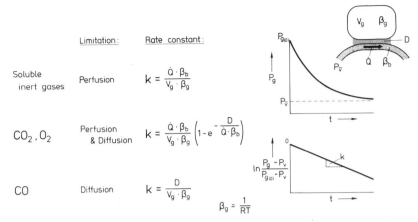

Fig. 4. Rebreathing equilibration for determination of pulmonary blood flow ($\dot{Q}$) and diffusing capacity (D). Model I: complete mixing between lung and rebreathing bag assumed (volume of both, V_g). From the rate constants, k, for the approach of the partial pressure in gas (P_g) to that in mixed venous blood ($P_{\bar{v}}$), $\dot{Q}$ and D are calculated.

corrected, leading to overestimation of D (Chinet et al., 1971). In most cases the uncorrected D values are higher for CO than for O_2. Such a finding may be considered as characteristic for inhomogeneities effects which in many cases reduce the apparent Do_2 more than the apparent Dco (Savoy et al., 1980; Geiser et al., 1983).

The rebreathing methods are probably much less influenced by functional inhomogeneities, because the local differences in alveolar gas composition produced by functional inhomogeneities should be much reduced by the rebreathing procedure (fast and deep breathing in closed system). Ideally, the whole lung-rebreathing bag system is completely homogenized by rebreathing (Fig. 4). From the gas to mixed-venous blood equilibration kinetics the parameters limiting the process can be derived (Adaro et al., 1973):

1. from soluble inert gases, perfusion, $\dot{Q}$ (i.e. pulmonary capillary blood flow);

2. from CO, diffusing capacity for CO, Dco.

3. from O_2 in hypoxia, Do_2 (when $\dot{Q}$ is known, e.g. from inert soluble gas equilibration).

Unfortunately, in many cases a high enough effective ventilation ($\dot{V}eff$) between lung and rebreathing bag cannot be achieved, and a two-compartment model, consisting of rebreathing bag (mean volume VR) and lung (mean volume VA) should be used (Fig. 5) (Adaro et al., 1973; Cerretelli et al., 1974). In this model the equilibration is biexponential, and the calculations are based on the second rate constant, k, of the slower component (Fig. 5).

The following equations may be used (s.i.g., soluble inert gas):

$$k_{s.i.g.}: \dot{Q} = k_{s.i.g.} \cdot \frac{\beta_g}{\beta_{s.i.g.}} \cdot V_A' \left(1 + \frac{V_R/V_A'}{1 - k_{s.i.g.} \cdot \dot{V}_R/\dot{V}_{eff}}\right) \qquad (6)$$

23

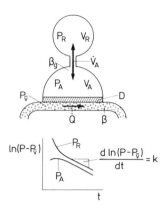

Fig. 5. Rebreathing equilibration for determination of pulmonary blood flow (Q̇) and diffusing capacity (D). Model II: with ventilation limitation (V̇A = V̇eff < ∞). In comparison with model I, the additional variables V̇eff, V̇A and VR (= Vg - VA) are required. Calculations according to eqs. (6)-(8). After Adaro et al. (1973).

$$k_{O_2}: \quad D_{O_2} = -\dot{Q} \cdot \beta_{O_2} \cdot \ln \left[1 - \frac{\dot{V}_A \cdot \beta_g}{\dot{Q} \cdot \beta_{O_2}} \cdot k_{O_2} \left(1 + \frac{V_R/V_A'}{1 - k_{O_2} \cdot V_R/\dot{V}_{eff}} \right) \right] \quad (7)$$

$$k_{CO}: \quad D_{CO} = k \cdot \beta_g \cdot \dot{V}_A \left(1 + \frac{V_R/V_A}{1 - k \cdot V_R/\dot{V}_{eff}} \right) \quad (8)$$

The value V_A', to be used with gases of high tissue solubility, takes into account the lung tissue volume (including pulmonary capillary volume) which equilibrates rapidly with alveolar gas (Teichmann et al., 1974).

V_A and $\dot{V}_{eff}$ are determined from insoluble gas, i.g. (e.g. He) equilibration (which ideally is monoexponential):

$$V_A = V_R \cdot (P_{i.g.(initial)}/P_{i.g.(final)} - 1) \quad (9)$$

$$\dot{V}_{eff} = \frac{k_{i.g.}}{1/V_A + 1/V_R} \quad (10)$$

The rebreathing equilibration of O_2 may be measured as approach to mixed-venous P_{O_2}, but it is preferable to use a (non-radioactive) isotope, like 18-O-16-O or 18-O-18-O (Meyer et al., 1981). The advantages are:

(1) mixed-venous P is zero

(2) effective β_b for the rare isotope is equal to the ratio $C_{\bar{v}}/P_{\bar{v}}$ for the abundant isotope if the ratio P(isotopic O_2)/P(abundant O_2) is very small and the abundant O_2 is in blood/gas equilibrium.

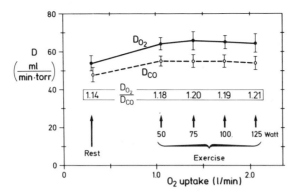

Fig. 6. Pulmonary diffusing capacities for O_2 (Do_2) and CO (Dco) in hypoxia at rest and during exercise. Mean values ± SD in 6 normal, young males. There is only a minor increase in Do_2 and Dco with increasing O_2 uptake during exercise. The Do_2/Dco ratio is close to constant, averaging 1.2. After Meyer et al. (1981).

The mean values (± SE) of Do_2 and Dco, determined simultaneously by rebreathing, at rest and during exercise in normal human subjects are plotted in Fig. 6 against the O_2 uptake (Meyer et al., 1981). The following features are of interest.

1) At rest the values are much higher than the conventional values. This may be due to underestimation by steady-state and single breath methods or to recruitment of pulmonary capillaries by the hyperpnea of rebreathing.

2) There is only a slight increase in exercise, the values levelling off at O_2 uptakes above 1 l/min.

3) The ratio Do_2/Dco is close to 1.2, which is the O_2/CO Krogh diffusion constant ratio. This may mean that the uptake of both O_2 and CO is mainly limited by diffusion. It must be remembered that the values are determined in hypoxia, with pulmonary capillary Po_2 around 25-40 Torr; in normoxia and in hyperoxia CO uptake is clearly reaction-limited, according to Roughton and Forster (1957).

Conclusions

1. The pulmonary diffusing capacity (D) for O_2 and CO can be reliably determined by rebreathing techniques using stable isotopes.

2. The experimental Do_2/Dco ratio of about 1.2 in hypoxia is in agreement with diffusion (and not chemical reaction) being the process limiting alveolar-capillary equilibration of both O_2 and CO.

III. O_2 kinetics of red blood cells measured by stopped-flow

Since Roughton and Forster (1957) it is customary to subdivide the total CO uptake resistance into a 'membrane component' (1/Dm) and into a blood or red blood cell component ($1/(\theta \cdot V_c)$). Later, the equation has also been applied to O_2 uptake (Staub et al., 1962):

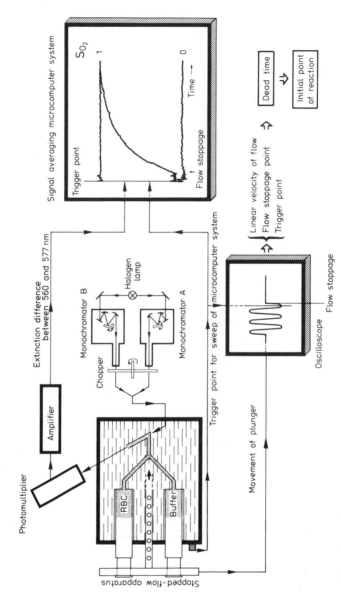

Fig. 7. The stopped flow method for measurement of O_2 (and CO) kinetics of red blood cells (RBC). A suspension of RBC is rapidly mixed with a buffer of different Po_2, and after a sudden flow stop the O_2 saturation change is measured by a dual-wavelength spectrophotometer. The record with low noise level is obtained by repeated measurements with electronic averaging.

$$\frac{1}{D} = \frac{1}{D_M} + \frac{1}{\theta \cdot V_c} \qquad\qquad (11)$$

We redetermined θ for O_2 using the stopped-flow principle as modified by Holland et al. (1985) (Fig. 7). A suspension of RBC is rapidly mixed with a buffer, of higher or lower Po_2, and after sudden flow stop the rapid change in Hb O_2 saturation is measured. O_2 release from RBC is usually measured by mixing the RBC suspension with an O_2-absorbing dithionite solution.

The obtained oxygenation (and deoxygenation) kinetics curves were evaluated on the basis of the Bohr integration for co-current RBC and medium flow (Fig. 8). The value calculated was the specific RBC conductance for O_2 (or specific RBC diffusing capacity for O_2), go_2. The parameter θo_2, the specific O_2 conductance of blood, is obtained from the relationship, $\theta o_2 = g$ x hematocrit.

With human RBC at $37^{\circ}C$, $Pco_2 = 40$ Torr and medium pH = 7.4, the following results were obtained (Yamaguchi et al., 1985).

1) In measurements of O_2 uptake g decreases rapidly. This effect is evidently due to pericellular O_2 depletion with lengthening of diffusion path length. The initial g value decreases with increasing albumin concentration in the medium, by which the O_2 diffusion coefficient is reduced and the Reynolds number is decreased (meaning decreased convective mixing). Similar results and conclusions have been obtained in a number of studies (Coin and Olson, 1979; Rice, 1980; Huxley and Kutchai, 1981, 1983; Vandegriff and Olson, 1984a, b, c).

2) Identical g values were calculated for O_2 uptake and O_2 release (without dithionite) taking place in the same O_2 saturation range (Fig. 9). This appears to indicate that diffusion (in the red cells and in the medium) is the main limiting process in both O_2 uptake and release.

3) O_2 release from red cells was accelerated by increasing dithionite concentration, a ceiling being reached at 40 mmol/l dithionite in the medium. Apparently at this concentration no appreciable limiting extracellular Po_2 gradient develops. Therefore, this highest g value, 8.7 ml/(min·Torr·ml RBC) corresponding to $\theta = 3.9$ ml/(min·Torr·ml blood) was taken as best approximation to diffusion conductance of red blood cells.

This θo_2 value is higher than all previous values which range from 0.9 (Mochizuki, 1966) to 2.7 ml/(min·Torr·ml blood) (Staub et al., 1962), apparently in part due to the fact that pericellular diffusion resistance has not been taken into account.

Use of this relatively high θo_2 value in the Roughton-Forster equation, assuming $V_c = 100$ ml and $Do_2 = 54$ ml/(min·Torr), leads to the result that only 12% of the O_2 uptake resistance is due to red cells. The main diffusive resistance appears to be outside the red cells, in plasma and in capillary-alveolar wall.

Conclusions

1. Stopped-flow measurements of O_2 transfer between red cells and medium are importantly limited by diffusion in the pericellular medium layer.

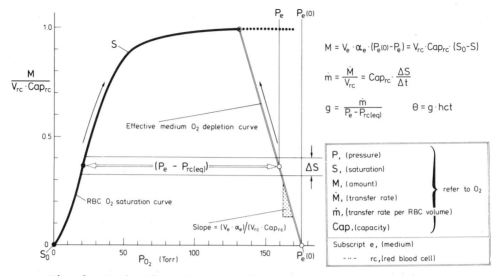

Fig. 8. Evaluation of stopped-flow measurements. The specific O_2 conductance of RBC, g, is defined as specific O_2 uptake ($\dot{m}$) divided by the difference between medium Po_2 (P_e) and the RBC hemoglobin equilibrium Po_2 ($P_{rc(eq)}$). The Po_2 difference is visualized in the diagram. The curve S corresponds to the O_2 dissociation curve in terms of O_2 saturation (S), the P_e value is on the effective medium O_2 depletion line, obtained from O_2 mass balance considerations (top equation). The intersection of the S curve and the medium O_2 depletion line is the final equilibrium point. The specific O_2 uptake ($\dot{m}$) is calculated from the measured rate of change of O_2 saturation ($\Delta S/\Delta t$) and the RBC O_2 capacity (Cap_{rc}).

2. Attempted corrections for diffusion limitation yield higher values for specific O_2 conductance of RBC than generally accepted: θo_2 = 3.9 ml/(min·Torr·ml blood) for human blood at $37^{\circ}C$, hematocrit 45%.

3. There is no indication of reaction limitation in the O_2 saturation range 10% to 80%. (Unfortunately, in the physiologically important saturation range > 80% no reliable measurements could be performed.)

4. Combination with measured Do_2 values indicates only a small fraction of O_2 uptake resistance (in hypoxia) to reside in the red cells.

IV. Alveolar-capillary CO_2 equilibration

Using 13-CO_2 as test gas, we succeeded in determining an effective pulmonary diffusing capacity for CO_2 in normal subjects by similar rebreathing techniques used for measurement of Do_2 and Dco_2 (Piiper et al., 1980). Particular difficulties were the separation of the 13-CO_2 signal from that of 12-CO_2 in mass spectrometry and the accurate assessment of the lung tissue distribution space for 13-CO_2. From the 'effective solubility' of 13-CO_2 and the physical solubility of CO_2, an apparent pH = 7.0 could be estimated for lung tissue.

As diffusing capacity for CO_2, the following mean values were estimated: rest, 180; exercise 75 Watt, 300 ml/(min·Torr). Comparison with Do_2 determined on the same subjects (see above) yielded for the ratio Dco_2/Do_2 values around 4. If diffusion were the limiting process, the ratio should

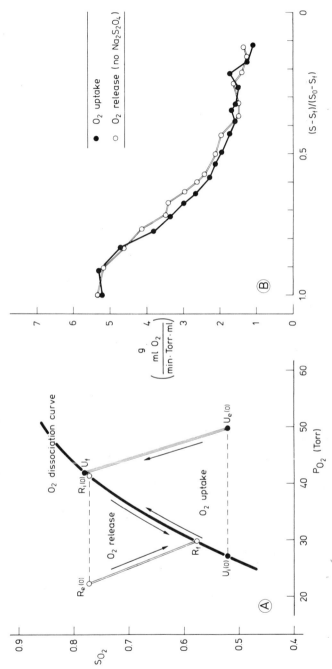

Fig. 9. Measurement of specific O_2 conductance of RBC from O_2 uptake and O_2 release in 'symmetrical' conditions. Left hand side: O_2 dissociation curve and medium O_2 depletion (enrichment) lines (see Fig. 8) for O_2 uptake (initial values, $U_{i(0)}$ and $U_{e(0)}$; final equilibration value, U_f) and for O_2 release by RBC (initial values, $R_{i(0)}$ and $R_{e(0)}$; final equilibrium value, R_f). Right hand side: the kinetics of O_2 uptake and release are very similar. After Yamaguchi et al. (1985).

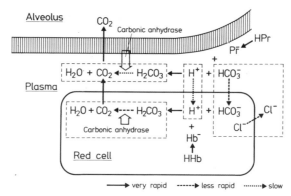

Fig. 10. Limiting processes in blood-gas CO_2 transfer. Simplified schema (Bohr effect and carbamate are neglected). Equilibration of molecular CO_2 probably is rapidly completed. The equilibration of the $H^+/HCO_3^-/CO_2$ systems in red cells and plasma is probably limited by slow exchange HCO_3^-/Cl^- across the red cell membrane and the slowness of dehydration of H_2CO_3 in plasma (in spite of the presence of some carbonic anhydrase activity on the capillary wall).

be close to the Krogh diffusion coefficient ratio, which is about 20. What is the explanation of the ratio being substantially lower?

The complex array of processes involved in alveolar-plasma-RBC CO_2 transfer in lungs is schematically represented in Fig. 10. The primary equilibration of CO_2 is expected to be rapid. But the subsequent equilibration of the $H^+/HCO_3^-/CO_2$ system in RBC and plasma appears to be relatively slow due to (1) limited rate of HCO_3^-/Cl^- exchange between RBC and plasma and (2) to slowness of the CO_2/HCO_3^- equilibration in RBC and, particularly, in plasma.

The equilibration process is qualitatively depicted in Fig. 11. Pco_2 equilibrium is expected to be achieved very rapidly. This has recently been shown by Schuster (1985, 1987) who used oxygen-labeled CO_2. The calculated Dco_2 was very high, amounting to 900–1,100 ml/(min·Torr). These values are in fact about 20 times higher than Do_2, as expected on the basis of Krogh diffusion constant ratio CO_2/O_2. But the equilibration of the system $H^+/HCO_3^-/CO_2$ within red cells and within plasma, and between red cells and plasma, may be much slower. The net result of this slowness is a reduced efficiency of alveolar-capillary CO_2 exchange. Thus our Dco_2 value should preferably be designated as 'equilibration capacity' rather than 'diffusing capacity'.

On the basis of the Dco_2 values, the alveolar-arterial Pco_2 difference due to incomplete equilibration was estimated: rest, 0.2 Torr; medium exercise, 1 Torr; maximum O_2 uptake exercise (extrapolated), 7 Torr.

Conclusions

1. The effective 'diffusing capacity for CO_2', Dco_2, could be estimated in man from rebreathing kinetics of $13\text{-}CO_2$.

2. As Dco_2 was only about 4 times higher than Do_2, capillary/alveolar CO_2 transfer seemed to be limited by reequilibration of the $H^+/HCO_3^-/CO_2$ system in blood rather than by CO_2 diffusion.

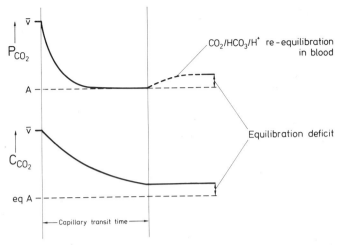

Fig. 11. Hypothetical time course of capillary-alveolar
CO_2 equilibration. CO_2 equilibrates rapidly so that
complete Pco_2 equality is reached during the capillary
transit time. But no equilibrium is reached for the total
CO_2, particularly not for bicarbonate. Reequilibration in
blood after leaving the pulmonary capillaries leads to
increase of Pco_2, and to a positive arterial-to-alveolar
Pco_2 difference.

3. Calculations based on $D/(\dot{Q}\beta)$ indicate that capillary-alveolar equili-
    bration of CO_2 is close to complete in resting conditions, but may
    become limiting in strenuous exercise.

REFERENCES

Adaro, F., P. Scheid, J. Teichmann and J. Piiper (1973). A rebreathing
    method for estimating pulmonary Do_2: theory and measurements in dog
    lungs. Respir. Physiol. 18:43–63.
Cerretelli, P. (1976). Limiting factors to oxygen transport on Mount
    Everest. J. Appl. Physiol. 40:658–667.
Cerretelli, P., A. Veicsteinas, J. Teichmann, H. Magnussen and J. Piiper
    (1974). Estimation by a rebreathing method of pulmonary O_2 diffusing
    capacity in man. J. Appl. Physiol. 37:526–532.
Chinet, A., J. L. Micheli and P. Haab (1971). Inhomogeneity effects on O_2
    and CO pulmonary diffusing capacity estimates by steady-state methods.
    Theory. Respir. Physiol. 13:1–22.
Coin, J.T. and J.S. Olson (1979). The rate of oxygen uptake by human red
    blood cells. J. Biol. Chem. 254:1178–1190.
Geiser, J., H. Schibli and P. Haab (1983). Pulmonary O_2 diffusing capacity
    estimates from assumed lognormal $\dot{V}_A$, $\dot{Q}$ and D_L distributions. Respir.
    Physiol. 52:53–67.
Hammond, M.D., G.E. Gale, K.S. Kapitan, A. Ries and P.D. Wagner (1986).
    Pulmonary gas exchange in humans during normobaric hypoxic exercise.
    J. Appl. Physiol. 61:1749–1757.
Holland, R.A.B., H. Shibata, P. Scheid and J. Piiper (1985). Kinetics of O_2
    uptake and release by red cells in stopped-flow apparatus: effects of
    unstirred layer. Respir. Physiol. 59:71–91.
Huxley, B.H. and H. Kutchai (1981). The effect of the red cell membrane and
    a diffusion boundary layer on the rate of oxygen uptake by human
    erythrocytes. J. Physiol. London 316:75–88.

Huxley, V.H. and H. Kutchai (1983). Effect of diffusion boundary layers on the initial uptake of O_2 by red cells. Theory versus experiments. Microvasc. Res. 26:89-107.

Meyer, M., P. Scheid, G. Riepl, H.J. Wagner and J. Piiper (1981). Pulmonary diffusing capacities for O_2 and CO_2 measured by a rebreathing technique. J. Appl. Physiol. 51:1643-1650.

Mochizuki, M. (1966). Study on the oxygenation velocity of the human red cell. Jpn. J. Physiol. 16:635-648.

Piiper, J., P. Dejours, P. Haab and H. Rahn (1971). Concepts and basic quantities in gas exchange physiology. Respir. Physiol. 13:292-304.

Piiper, J., M. Meyer, C. Marconi and P. Scheid (1980). Alveolar-capillary equilibration kinetics of 13-CO_2 in human lungs studied by rebreathing. Respir. Physiol. 42:29-41.

Piiper, J. and P. Scheid (1980). Blood-gas equilibration in lungs. In: Pulmonary Gas Exchange. Volume I. Ventilation, Blood Flow and Diffusion, ed. by J.B. West. New York, Academic Press, pp. 131-171.

Piiper, J. and P. Scheid (1981). Model for capillary-alveolar equilibration with special reference to O_2 uptake in hypoxia. Respir. Physiol. 46:193-208.

Piiper, J. and P. Scheid (1983). Comparison of diffusion and perfusion limitations in alveolar gas exchange. Respir. Physiol. 51:287-290.

Rice, S. (1980). Hydrodynamic and diffusion considerations of rapid-mix experiments with red blood cells. Biophys. J. 29:65-78.

Roughton, F.J.W. and R.E. Forster (1957). Relative importance of diffusion and chemical reaction rates in determining rate of exchange of gases in the human lung, with special reference to true diffusing capacity of pulmonary membrane and volume of blood in the lung capillaries. J. Appl. Physiol. 11:290-302.

Savoy, J., M.C. Michoud, M. Robert, J. Geiser, P. Haab and J. Piiper (1980). Comparison of steady state pulmonary diffusing capacity estimates for O_2 and CO in dogs. Respir. Physiol. 42:43-59.

Schuster, K.D. (1985). Kinetics of pulmonary CO_2 transfer studied by using labeled carbon dioxide C-16-O-18-O. Respir. Physiol. 60:21-37.

Schuster, K.D. (1987). Diffusion limitation and limitation by chemical reactions during alveolar-capillary transfer of oxygen-labeled CO_2. Respir. Physiol. 67:13-22.

Staub, N.C., J.M. Bishop and R.E. Forster (1962). Importance of diffusion and chemical reaction rates in O_2 uptake in the lung. J. Appl. Physiol. 17:21-27.

Teichmann, J., F. Adaro, A. Veicsteinas, P. Cerretelli and J. Piiper (1974). Determination of pulmonary blood flow by rebreathing of soluble inert gases. Respiration 31:296-309.

Torre-Bueno, J., P.D. Wagner, H.A. Saltzman, G.E. Gale and R.E. Moon (1985). Diffusion limitation in normal humans during exercise at sea level and simulated altitude. J. Appl. Physiol. 58:989-995.

Vandegriff, K.D. and J.S. Olson (1984a). The kinetics of O_2 release by human red blood cells in the presence of external sodium dithionite. J. Biol. Chem. 259:12609-12618.

Vandegriff, K.D. and J.S. Olson (1984b). Morphological and physiological factors affecting oxygen uptake and release by red blood cells. J. Biol. Chem. 259:12619-12627.

Vandegriff, K.D. and J.S. Olson (1984c). A quantitative description in three dimensions of oxygen uptake by human red blood cells. Biophys. J. 45:825-835.

Wagner, P.D., G.E. Gale, R.E. Moon, J.R. Torre-Bueno, B.W. Stolp and H.A. Saltzman (1986). Pulmonary gas exchange in humans exercising at sea level and simulated altitude. J. Appl. Physiol. 61:260-270.

Yamaguchi, K., D. Nguyen-Phu, P. Scheid and J. Piiper (1985). Kinetics of O_2 uptake and release by human red blood cells studied by a stopped-flow technique. J. Appl. Physiol. 58:1215-1224.

INFLUENCE OF GAS PHYSICAL PROPERTIES ON PULMONARY GAS EXCHANGE

Michael P. Hlastala, David D. Ralph and Albert L. Babb

Departments of Chemical Engineering, Medicine
Nuclear Engineering and Physiology and Biophysics
University of Washington
Seattle, WA 98195

INTRODUCTION

The gas exchange properties of the lung depend on the relative uniformity of the matching of ventilation and perfusion in the individual acinar units. The exchange of inert gases in a homogeneous acinar unit is dependent on the physical properties of those inert gases. This presentation deals with three aspects of the inert gas physical properties governing their exchange. The primary determinant of inert gas exchange is the solubility of the gas in blood. Second, gas exchange is weakly dependent on the molecular weight of the gas. Finally, the exchange of the very soluble inert gases depends on interaction of gas with the airways as air passes through the airways during inspiration and expiration.

Blood-Gas Partition Coefficient

A quantitative analysis of the influence of the blood solubility of an inert gas on its steady-state alveolar gas exchange was first put forward by Farhi (1967) who used the idea to describe a method for analyzing the exchange of two gases of differing solubility using a lung with two compartments, each having a different $\dot{V}A/\dot{Q}$ ratio (Farhi and Yokoyama, 1967). If an inert gas dissolved in mixed venous blood enters the alveolus, there is a rapid exchange of that gas between the blood and the gas in the alveolus. Calculations suggest that equilibration will take place within a small fraction of the time available for exchange. In a small gas exchange unit, alveolar partial pressure (PA) is equal to the end-capillary partial pressure (Pa). The fraction of gas retained in the blood (Pa/Pv or R) is a function of the Ostwald partition coefficient (λ) of that gas between the gas phase and blood and the relative amounts of ventilation ($\dot{V}A$) and of perfusion ($\dot{Q}$) to the alveolus:

$$R = \frac{\lambda}{\lambda + \dot{V}A/\dot{Q}}$$

For an alveolar unit with a fixed $\dot{V}A$ and $\dot{Q}$, retention of a gas is a function of its solubility in blood. If a gas with a $\lambda = 1$ is infused into the venous circulation, it will be eliminated by an alveolus with a $\dot{V}A/\dot{Q}$ of 1.0 with a retention of 0.5 ($\lambda = \dot{V}A/\dot{Q}$). A gas with greater λ (say 10.0) will

33

have a greater retention. A gas with lower λ (say 0.1) will have a lower retention. Thus the alveolus acts as a high pass filter retaining gas with high λ and eliminating gas with low λ.

Retention of the gas also depends on the $\dot{V}A/\dot{Q}$ of the alveolus. If a gas with $\lambda = 1$ passes through alveoli of differing $\dot{V}A/\dot{Q}$, there will be a greater retention in the low $\dot{V}A/\dot{Q}$ alveolus and a low retention in the high $\dot{V}A/\dot{Q}$ alveolus. Different alveoli process a gas differently depending on the gas λ and the alveolus $\dot{V}A/\dot{Q}$ ratio. Thus the exchange of gas in the lung is highly sensitive to the $\dot{V}A/\dot{Q}$ ratio and the gas λ.

This approach has been used to advantage in the multiple inert gas elimination technique (MIGET) to provide a method for evaluating the heterogeneity of $\dot{V}A/\dot{Q}$ in the lung. Wagner et al., (1974) have used the analysis of the elimination of six inert gases with varying solubility to assess the multicompartment $\dot{V}A/\dot{Q}$ distribution in the lung. The method provides information about the dead space, shunt and general description of the relative heterogeneity of $\dot{V}A/\dot{Q}$ distribution (Hlastala, 1984). Because MIGET provides such useful information, it has been used with success by a number of investigators to deal with physiological questions about the lung.

Molecular Weight

The assumption of diffusional equilibration for inert gases has often been incorporated in analytical models. From analysis of inert gas exchange data, there has generally been no substantial evidence of diffusional impairment. However Scheid et al., (1981) pointed out that the particular approach used in MIGET with gases distributed over a wide range of partition coefficients would be relatively insensitive to small impairments in diffusional equilibrium. An alternative approach is to examine the elimination of inert gases with similar partition coefficients in blood and different molecular weights (Adaro and Farhi, 1971; Hlastala et al, 1982). By the assumptions of the homogenous alveolus model, the retentions and excretions of those different gases should be nearly identical. Robertson et al (1986a) compared the elimination of three inert gases with similar partition coefficients but with molecular weights ranging from 26 to 184.5 and demonstrated a consistent impairment of exchange in the higher molecular weight gases. The difference observed was small, but similar in magnitude to that proposed by Scheid et al., (1981) and Hlastala et al., (1981). The latter papers demonstrated that this degree of molecular weight effect would be perceived by MIGET as representing additional ventilation-perfusion heterogeneity. The importance of these observations does not relate to the resultant relatively small errors induced in the ventilation-perfusion distributions, but rather does establish that there is some form of diffusion impairment that can be measured for the inert gases in dog lungs. In other species, diffusion impairment may be more prominent (Truog et al., 1979).

The original assumption made in the consideration of the inert gas molecular weight effect was that this effect observed during normal gas exchange was related to the molecular weight-dependent separation observed with single breath studies involving boli of insoluble gases with different molecular weights. This effect has been described as a gas phase diffusion abnormality (also called stratified inhomogeneity; Scheid and Piiper, 1980), or attributed to more complex interactions between the convective and diffusive movements of gases (Paiva and Engel, 1982). Recent studies of tidal gas concentrations by Robertson et al., (1986b) appear to confirm that measurable molecular weight differences in elimination of infused inert gases can be manifested as different Fowler dead spaces in the gases. The magnitude of the effect in these two preliminary reports, however, was not as large as that reported by Robertson et al., (1986a), and it appears likely that other mechanisms may also be involved.

Soluble Gas Deposition in the Airways

Finally, other mechanisms exist which may play a role under certain circumstances. Classically, the airways of the lung have been considered as a volume of air which simply conducts and dilutes exhaled gas and has no other influence on gas exchange. However, new information regarding the physiology of the airways has highlighted four mechanisms within the conducting airways which influence overall gas exchange: (a) the mixing effect caused by reinspiration of series dead space results in an effective homogenization of the $\dot{V}A/\dot{Q}$ distribution of the lung (Ross and Farhi, 1960); (b) the airways take up soluble gases and some particles that enter the nasopharynx in the inspired gas (Aharonson et al., 1974); (c) the airways participate in the warming and humidification of inspired air (McFadden, 1983). This energy exchange process results in a change in airway tissue temperature during the breathing cycle (Gard et al., 1984), and (d) the bronchial circulation can exchange gas with air passing through the airways (Swenson et al., 1986). These result in a complex interaction of dynamic processes which influence the exchange of soluble gases.

Often the study of $\dot{V}A/\dot{Q}$ distribution with the multiple inert gas elimination technique is complicated by unexplicably high or low excretion values of the very soluble inert gases during exercise or other conditions associated with large minute ventilations, such as high-frequency ventilation. These changes, which can vary a great deal from individual to individual, may well be related to interactions between airway cooling, airway drying, and changes in bronchial blood flow. High $\dot{V}A/\dot{Q}$ regions have been characterized following application of positive airway pressure. It was originally postulated that the regions of lung in a zone I condition actually received intermittent perfusion from the pulmonary circuit to explain the gas exchange observations. It may well be that these high $\dot{V}A/\dot{Q}$ zones arise as a result of bronchial perfusion of zones of lung whose alveolar pressure prevent any pulmonary artery flow. This is a potential explanation for the observations of Dueck et al., (1977) and Coffey et al., (1983) that the inert gas dead space measured over a wide range of PEEP applications only increases as much as can be accounted for by the small fraction of airway distension. The much higher CO_2 physiologic dead space produced by high levels of PEEP might reflect the effective loss of pulmonary artery perfusion to zones of lung served only by the bronchial artery perfusion.

A common finding in many inert gas elimination technique outputs is the computed existence of a small fraction of ventilation distributed to very high $\dot{V}A/\dot{Q}$ regions. No alveolar zone of such high $\dot{V}A/\dot{Q}$ ratio seems physiologically likely, and gas exchange by the airways may provide the best explanation for this finding. Finally, the one inert gas exchange phenomenon where airways are thought to influence the exchange of soluble inert gases is with high frequency ventilation (HFV). While current theories explaining this phenomenon only incorporate the interaction of mucosal liquid with the high degree of airway mixing (McEvoy et al., 1982), bronchial blood flow and airway temperature also exert an influence. Robertson et al., (1982) measured tracheal temperature during HFV and found tracheal temperature to be three to five degrees above body temperature, a possible explanation for the observation that acetone dead spaces were smaller than those reported in other HFV inert gas studies (McEvoy et al., 1982). High airway temperatures would increase the partial pressure of inert gases delivered to the airway mucosa, facilitating their transport. In addition, the gradual heating of the airways during HFV might cause a transient elevation in exhaled soluble gas as the increasing mucosal temperature and decreasing mucosal inert gas solubility drives off previously dissolved soluble gas into the exhaled air.

Exhaled partial pressure of highly soluble gases such as ethyl alcohol is affected by airway interactions. The λ for ethyl alcohol has been measured to be 1756 (Jones, 1983a) for human blood at 37°C. Single breath testing for estimation of blood alcohol concentration is a method widely used in both scientific and legal environments. Since the early description of breath alcohol testing, it has been assumed that the concentration of alcohol in end-exhaled air is constant, and the same as the alcohol concentration in the alveolar air. However, several recent studies (Wright, 1962; Slemeyer, 1981; Jones, 1982a; Jones, 1983b) have shown that alcohol has considerable interaction with the airway surface during exhalation, which precludes an equilibrium between end-tidal and alveolar gas concentration. The magnitude of this interaction depends principally upon tissue solubility of the gas and ambient temperature and humidity (Jones, 1982b). In addition, the breathing pattern prior to the delivery of the breath sample influences the breath alcohol concentration. Hyperventilation reduces alcohol concentration in the breath, while a breathhold period prior to the delivery of the sample breath increases breath alcohol concentration (Jones, 1982c). Breath alcohol concentration increases with exhaled volume, and a flat alcohol plateau, predicted in the absence of airway interactions, is never reached if the exhaled flow rate is maintained (Jones, 1982a; Slemeyer, 1981).

Airway Heat Exchange

It has been widely accepted that inspired air is simultaneously heated and humidified such that the gas is fully saturated with water vapor at body temperature by the time it reaches the delicate gas exchange surfaces of the alveoli (Verzar, 1953). As the air is warmed, its water capacity increases and the air is humidified by similar mechanisms. As water vapor is formed, additional heat exchange occurs to supply the latent heat of vaporization. Calculations have shown that the exchange of water accounts for over 85% of the total energy exchanged from the airway surface during inspiration (Tsu et al., 1987). The net effect on inspiration is to cool the airway mucosa to a degree dependent on position along the airway, temperature and humidity of the inspired air, and breathing pattern. Recent measurements using thermistors placed within the airways in humans (McFadden, 1983) indicated a tracheal temperature of 32.0°C and a subsegmental bronchus temperature of 35.5°C breathing room air at rest. Hyperventilation decreased the tracheal temperature to 20.6°C and the subsegmental bronchus temperature to 31.6°C. These results suggest that thermal exchange occurs all the way down the tracheobronchial tree to airways of 2 mm in diameter or less (McFadden, 1983). Furthermore, it is likely that the airway temperature, and therefore inert gas solubility, is changing during the respiratory cycle as inhalation and exhalation alternately cool and warm the airway tissue. These data suggest that interactive heat and gas exchange with the airways may play a role in affecting the analysis of gas exchange.

SUMMARY

Overall, the exchange of gas by the lung is strongly dependent on the blood-gas partition coefficient of that gas and weakly dependent on the molecular weight of the gas. The exchange of very soluble inert gases is dependent on interaction with the airway surface during inspiration and expiration.

REFERENCES

Adaro, F. and L.E. Farhi (1971). Effects of intralobular gas diffusion on alveolar gas exchange. Fed. Proc. 30:437.

Aharonson, E.F., H. Menkes, G.Gurtner, D.L. Swift and D.F. Proctor (1974). Effect of respiratory airflow rate on removal of soluble vapors by the nose. J. Appl. Physiol. 37:654-657.

Coffey, R.L., R.K. Albert and H.T. Robertson (1983). Mechanisms of physiological dead space response to PEEP after acute oleic acid lung injury. J. Appl. Physiol. 55:1550-1557.

Dueck, R., P.D. Wagner and J.B. West (1977). Effects of positive end-expiratory pressure on gas exchange in dogs with normal and edematous lungs. Anesthesiology 47:359-366.

Farhi, L.E. (1967). Elimination of inert gas by the lung. Respir. Physiol. 3:1-11.

Farhi, L.E. and T. Yokoyama (1967). Effect of ventilation-perfusion inequality on elimination of inert gases. Respir. Physiol. 3:12-20.

Gard, R.G., M.P. Hlastala and A.L. Babb (1984). Heat and water exchange in an airway tissue model. Physiologist 27:259.

Hlastala, M.P. (1984). Multiple inert gas elimination technique. J. Appl. Physiol. 56:1-7.

Hlastala, M.P., P. Scheid and J. Piiper (1981). Interpretation of inert gas retention and excretion in the presence of stratified inhomogeneity. Respir. Physiol. 46:247-259.

Hlastala, M.P., H.P. McKenna, M. Middaugh and H.T. Robertson (1982). Role of diffusion dependent inhomogeneity in gas exchange in the dog. Bull Eur Physiolpathol. Resp. 18:373-380.

Jones, A.W. (1982a). Quantitative measurements of the alcohol concentration and the temperature of breath during a prolonged exhalation. Acta Physiol. Scand. 114:407-412.

Jones, A.W. (1982b). Effects of temperature and humidity of inhaled air on the concentration of ethanol in a man's exhaled breath. Clin. Sci. 63:441-445.

Jones, A.W. (1982c). How breathing technique can influence the results of breath-alcohol analysis. Med. Sci. Law 22:275-280.

Jones, A.W. (1983a). Determination of liquid/air partition coefficients for dilute solutions of ethanol in water, whole blood, and plasma. J. Anal. Toxicol. 7:193-197.

Jones, A.W. (1983b). Role of rebreathing in determination of the blood-breath ratio of expired ethanol. J. Appl. Physiol.: Respir. Environ. Exercise Physiol. 55:1237-1241.

McEvoy, R.D., N.J.H. Davies, F.L. Mannino, R.J. Prutow, P.T. Schumacker, P.D. Wagner and J.B. West (1982). Pulmonary gas exchange during high-frequency ventilation. J. Appl. Physiol. 52:1278-1287.

McFadden, E.R., Jr. (1983). Respiratory heat and water exchange: physiological and clinical implications. J. Appl. Physiol. 54:331-336.

Paiva, M. and L.A. Engel (1982). Influence of bronchial asymmetry on cardiogenic gas mixing in the lung. Respir. Physiol. 49:325-338.

Robertson, H.T., R.L. Coffey, T.A. Standaert and W.E. Truog (1982). Respiratory and inert gas exchange during high-frequency ventilation. J. Appl. Physiol. 52:683-689.

Robertson, H.T., J. Whitehead and M.P. Hlastala (1986a). Diffusion-related differences in the elimination of inert gases from the lung. J. Appl. Physiol. 61:1162-1172.

Robertson, H.T., D.D. Ralph and M.P. Hlastala (1986b). Diffusion dependent differences in the Fowler dead spaces of intravenously infused inert gases. Fed. Proc. 45:394.

Ross, B.B. and L.E. Farhi (1960). Dead space ventilation as a determinant in the ventilation-perfusion concept. J. Appl. Physiol. 15:363-371.

Scheid, P. and J. Piiper (1980). Intrapulmonary gas mixing and stratification. In: Pulmonary Gas Exchange. Ed: J.B. West, Vol. 1, pp 87-130.

Scheid, P., M.P. Hlastala and J. Piiper (1981). Inert gas elimination from lungs with stratified inhomogeneity: theory. Respir. Physiol. 44:299-309.

Slemeyer, A. (1981). An analytical model describing the exchange processes of alcohol in the respiratory system. _Alcohol, Drugs and Traffic Safety_. Ed: L. Goldberg. Almqvist and Wiksell International, pp. 456-468.

Swenson, E.R., H.T.R. Robertson and M.P. Hlastala (1986). Gas exchange in the _in situ_ isolated trachea. Amer. Rev. Resp. Dis. 133:A27.

Truog, W.E., M.P. Hlastala, T.A. Standaert, H.P. McKenna and W.A. Hodson (1979). Oxygen induced alteration of ventilation-perfusion relationships in rats. J. Appl. Physiol: Respirat. Environ. Exercise Physiol. 47:1112-1117.

Tsu, M.E., A.L. Babb, D.D. Ralph and M.P. Hlastala (1987). Analysis of heat, water and soluble gas exchange in an airway. Fed. Proc. 46:1427.

Verzar, F. (1953). Temperature und Feuchtigkeit der Luft in den Atemwegen. Pflügers Archiv. 257:400-416.

Wagner, P.D., H.A. Saltzman and J.B. West (1974). Measurement of continuous distribution of ventilation-perfusion ratio: theory. J. Appl. Physiol. 36:588-599.

CONVECTION, DIFFUSION AND THEIR INTERACTION IN THE BRONCHIAL TREE

H. K. Chang

Department of Biomedical Engineering
University of Southern California
Los Angeles, CA   90089-1451

INTRODUCTION

Oxygen transfer from the ambiance to the gas exchange units of the lung takes place in the bronchial tree. In this asymmetrically multi-generation conduit, the inspired gas meets the resident gas and mixing occurs.

To understand this process physiologists have investigated various physical mechanisms of gas transport. The two basic mechanisms of gas transport in the lung are bulk convection and molecular diffusion. However, for a given set of geometric and mechanical properties of the lung, these two mechanisms may interact to generate other modes of transport.

In this communication, I shall describe briefly these gas transport mechanisms and discuss their role in intrapulmonary gas mixing. A more detailed account of gas transport in the conducting airways can be found in a recent review by Ultman (1985).

Because the bronchial tree is a highly complex structure and breathing is a time-varying, cyclic process, the exact nature of gas transport is difficult to understand. By necessity and also by design, much of the discussion here will be focused on simpler cases. For example, steady state gas transport will be assumed wherever and whenever feasible; gas transport during high frequency ventilation, in which unsteady effects must be taken into account, has been reviewed elsewhere (Chang, 1984).

The study of gas transport in the bronchial tree has often been theoretical. This is because direct measurement in the lung is very difficult if not impossible to make. One of the few experiments that have yielded much useful information has been the single-breath washout test. A typical but schematic single-breath washout curve is shown in Fig. 1. The concentration of nitrogen measured at the mouth during the expiration immediately following a large inspiration of an inert gas or oxygen is plotted against the expired volume. This curve has three distinct phases. The initial flat portion or phase I represents the dead space gas. The frontal portion or phase II reflects the degree of mixing between the inspired gas and the resident gas in the bronchial tree. The gently rising plateau or phase III originates from the alveolar space. Both phase II and phase III give indications of the gas transport processes in the bronchial tree. They have been extensively studied but still remain the focus of much controversy.

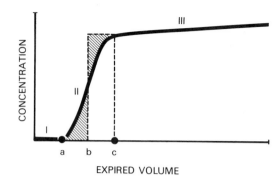

Fig. 1. Single-breath washout curve. The initial flat portion or phase I represents dead space gas; the frontal portion or phase II reflects the degree of mixing in the conducting airways; the gently rising plateau or phase III originates from alveolar space.

I hope that the following passages will be helpful in achieving a better interpretation of this frequently used test.

Bulk Convection

Bulk convection is the consequence of hydrodynamic pressure gradients. As a mode of transport, it moves all species in a mixture regardless of the concentrations of the individual species in this mixture. Bulk convection through the airways occurs whenever there is a pressure difference between the alveolar pressure and the airway opening pressure, the direction of the convective flow depending on the sign of the pressure difference. During inspiration the bulk flow moves from the mouth toward the alveoli, accelerating from zero flow until a peak flow is reached. It then decelerates to zero flow again where a post-inspiratory pause usually ensues. During expiration, the bulk flow occurs in qualitatively the same manner as during inspiration except that the flow is opposite in direction. If the breathing rate is high, the convective flow accelerates and decelerates rapidly and inertia plays an important role in the convective flow. In such cases, the convective flow is said to be unsteady. If, on the other hand, the breathing rate is low, convective flows accelerate and decelerate moderately, inertia is probably insignificant, and the flow can be considered quasi-steady.

If the convective flow is quasi-steady, the dimensionless parameter which characterizes the flow is Reynolds number, defined as:

$$Re = \frac{ud}{\nu} \tag{1}$$

where u is the mean convective velocity in a conduit such as a tube, d is the diameter if the conduit is a tube, and ν is the kinematic viscosity coefficient of the gas mixture being convected. Reynolds number is the ratio of inertial forces of the convective flow to the viscous forces tending to oppose the flow. A large Reynolds number indicates strong inertial forces relative to the viscous forces; a small Reynolds number signifies strong viscous effects. In the latter case, the flow is likely to be laminar. At large Reynolds numbers the flow is more likely to be turbulent.

Since the bronchial tree consists of many generations of airways with varying diameters, there is usually a wide range of Reynolds numbers present in the lung at any given instant of the respiratory cycle. Table I gives

Table I. Geometric and Dynamic Characteristics of the Bronchial Tree
(Based on a flowrate of 500 $cm^3 \cdot s^{-1}$)

Generation Number	Diameter (cm)	Length (cm)	Velocity ($cm \cdot s^{-1}$)	Reynolds Number	Residence Time (s)	Cumulative Res. Time (s)
Trachea	1.8	12.0	197	2325	0.06	0.06
1	1.22	4.76	215	1719	0.02	0.08
2	0.83	1.90	235	1281	0.006	0.009
3	0.56	0.76	250	921	0.006	0.10
4	0.45	1.27	202	594	0.006	0.10
5	0.35	1.07	161	369	0.006	0.10
10	0.130	0.46	38	32	0.01	0.15
15	0.066	0.20	4.4	1.9	0.04	0.28
20	0.045	0.083	0.3	0.1	0.28	1.0

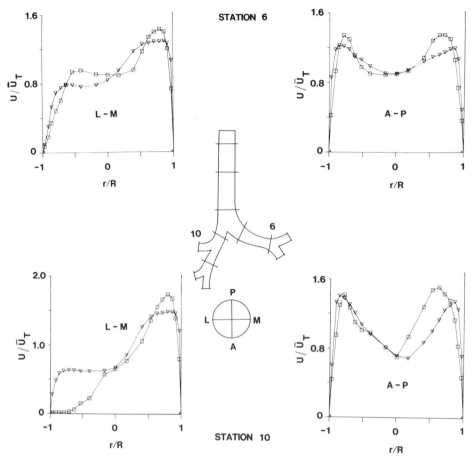

Fig. 2. Two sets of steady flow axial velocity profiles in the
left main bronchus (Station 6) and the right upper bronchus
(Station 10) of a model airway. Velocity profiles across the
diameters in the laterial-medial (L-M) plane are plotted in
the panels on the left; those across the diameters in the
anterior-posterior (A-P) plane are on the right. In each
panel, the middle position 0 refers to the center of the
airway; the left position -1, the lateral or anterior wall;
and the right position 1, the medial or posterior wall. All
velocity magnitudes are normalized by the tracheal velocity
$\bar{V}_T$. $\bar{V}$, high tracheal flow rate, Re = 8846; $\square$, low tracheal
flow rate, Re = 2123.

the typical dimensions of some airways and the corresponding Reynolds
numbers based on the mean convective velocities, which are also shown.

Due to the curvatures and the bifurcations of the airways, the velocity
profiles within each airway are highly nonuniform (Chang and El Masry, 1982;
Menon et al., 1984). The velocity profiles measured in the left main
bronchus and right upper lobar bronchus of a 3:1 scale model airway (Chang
and El Masry, 1982) are shown in Fig. 2. In this figure, the velocity
profiles in the left main bronchus (Station 6) are given in the two upper
plots; those in the right upper bronchus (Station 10) are found in the two
lower plots. The plots on the left give velocity profiles across the
diameter in the frontal or lateral-medial (L-M) plane and those on the
right, in the sagittal or anterior-posterior (A-P) plane. The middle

position (marked 0) in each plot refers to the center of the airway, the left position (marked -1) refers to the lateral or anterior wall, the right position (marked 1) refers to the medial or posterior wall. All the velocity profiles in Fig. 2 are normalized by the mean velocity in the trachea ($\bar{U}T$). The profiles joined by the triangles (∇) and the squares ($\square$) pertain to two different tracheal flow rates. It is evident from this figure that airflow dynamics in the conducting airways is quite complex. More detailed information on air flow dynamics in the human airways may be found in a recent review by Chang and Menon (1985).

Following the varying time course of an inspiration, through the highly asymmetric and complex structure known as the bronchial tree, bulk convection brings the inspired gas to the peripheral part of the lung where the mechanism of molecular diffusion is important.

Molecular Diffusion

Molecular diffusion is a transport mechanism arising from random thermal oscillation of the molecule. So long as the molecules in a mixture have a temperature above absolute zero, molecular diffusion always occurs. In respiratory physiology it is recognized that molecular diffusion is not only responsible for gas exchange across the alveolo-capillary membrane but also contributes to the transport of O_2 and CO_2 in the gas phase near the membrane (Scheid and Piiper, 1980).

The physical law governing molecular diffusion in a binary mixture has the following form:

$$J_1 = -D_{12} \nabla x_1 \qquad (2)$$

$$J_2 = -D_{21} C \nabla x_2 \qquad (3)$$

and is called Fick's law,

In the above equations J_1 and J_2 are fluxes of components 1 and 2, x_1 and x_2 are the fractions of components 1 and 2 in the mixture whose total molar concentration is C, the symbol "∇" denotes gradient and D_{12} and D_{21} are diffusion coefficients. If the net flux of the system, i.e., $J_1 + J_2$, with respect to a fixed frame of reference is zero, then $J_1 = J_2$ and $D_{12} = D_{21}$.

Since the respiratory gas mixtures in the lung are never truly binary but multicomponent, it may be useful to recall that the laws governing multicomponent diffusion are different from those governing binary diffusion. Therefore, the behavior of the gas species in the lung may deviate from that predicted from Fick's law. The effect of multicomponent diffusion in the lung may be considerable if a carrier gas whose density is drastically different from that of oxygen (e.g., He) is involved. The relevant issues on multicomponent diffusion have been reviewed by Chang (1980), who also computed the effective diffusion coefficients for O_2 and CO_2 in some ternary mixtures. Table II gives the values of these effective diffusion coefficients as well as the corresponding binary diffusion coefficients.

A pertinent dimensionless parameter for diffusion is the Schmidt number, defined as

$$Sc = \frac{\nu}{D} \qquad (4)$$

Table II. Computed Effective Diffusion Coefficients

Mixture	Diffusing Gas	$D'(cm^2 s^{-1})$	$D(cm^2 s^{-1})$	D'/D
O_2-CO_2-He	O_2	0.5303	0.7909	0.67
	CO_2	0.4380	0.6345	0.69
$O_2-CO_2-N_2$	O_2	0.2402	0.2187	1.10
	CO_2	0.1768	0.1663	1.06
$O_2-CO_2-SF_6$	O_2	0.1431	0.1001	1.43
	CO_2	0.0961	0.0707	1.46

Computed effective coefficients (D') of O_2 and CO_2 in three different gas mixtures. The binary diffusion coefficients (D) and their ratios (D'/D) are also given.

where ν is the kinematic viscosity of the binary mixture and D is the molecular diffusivity of this pair. Physically, the Schmidt number is the ratio of the viscous force to the momentum of diffusive flux. Obviously, a large Schmidt number indicates weak diffusion.

Although molecular diffusion takes place at all times during a breath, its effect may be overshadowed by the velocity of convective flow if the latter is high compared to the speed of diffusion flux. From Table I one sees that this would indeed be the case in the first 20 generations of the bronchial tree if either inspiratory or expiratory flow were constant at a flowrate of 500 $cm^3 s^{-1}$. However, both flows diminish quickly to zero and are usually followed by a post-inspiratory or post-expiratory pause. During these periods, molecular diffusion becomes the predominant mechanism of gas transport in the lung.

Coupled Longitudinal Convection and Diffusion

The complexity of the spatial and temporal variations of gas transport in the bronchial tree defies detailed mathematical modeling and, consequently, precise mental understanding. As a means to gain a basic understanding of this complex process, simplified cases have been considered. The most useful simplification thus far has been the so-called "trumpet lung". In this idealized lung model, as depicted in Fig. 3, the bronchial tree is regarded as an axisymmetric conduit shaped like a trumpet or a thumb tack. The cross sectional area A at any axial position x is equal to the cumulative cross section of all the airways in a given generation of the Weibel lung geometry (Weibel, 1963). The shaded area represents the volume of the alveoli.

When convection and diffusion in the longitudinal or axial direction are considered simultaneously, the governing equation is of the form

$$\frac{\partial C}{\partial t} + U \frac{\partial C}{\partial x} = \frac{1}{A} \frac{\partial}{\partial x} \left(A D \frac{\partial C}{\partial x} \right) \qquad (5)$$

where C, t and x are concentration, time and the spatial variable, respectively, A is the cross sectional area of the trumpet at a given x, U is the convective velocity at x, and D is the molecular diffusion coefficient.

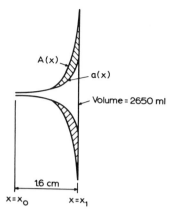

A(x)

a(x)

Volume = 2650 ml

1.6 cm

$x = x_0$ $x = x_1$

Fig. 3. The "trumpet lung" model studied by Paiva (1973).

Numerical solutions of Eq. (5) have been obtained by a number of authors for a variety of geometric and physiological conditions. These have been recently reviewed by Paiva (1985). The most instructive of this type of study was made by Paiva (1973) whose main results are illustrated in Fig. 4. On the abssisa is plotted distance (x) from the terminal alveolar sacs; equivalently, the generation numbers in the Weibel model of the bronchial tree are given in the upper axis. The ordinate represents the fractional concentration of oxygen. Each curve in the figure gives an instantaneous longitudinal profile of oxygen concentration in the "trumpet lung". For a sinusoidal breath beginning with inspiration and lasting 4.0 s, these profiles reveal several remarkable features of intrapulmonary gas mixing due to coupled convection and longitudinal diffusion. First, Fig. 4 shows that an interface or a front is established between the inspired oxygen and the resident gas. Even though oxygen penetrates rapidly (t = 0.8s) into the peripheral region of the lung, the front is held more or less stable over most of the inspiratory period (t = 0.8 - 2.0 s). Second, due mainly to molecular diffusion during the second half of the inspiration (t = 0.8 - 2.0 s), oxygen fraction at the terminal surface of the trumpet (or alveolar membrane) rises gradually. Third, shortly after expiration begins (t = 2.4 s) the oxygen front is quickly abolished by the mouthward bulk convection. This means that in a symmetric lung the slope of phase III in a single-breath washout curve must be zero; the observed slope of the alveolar plateau in such a washout curve must be associated with asymmetry of the bronchial tree and dynamic factors.

When bulk convection and molecular diffusion are considered together, a dimensionless parameter is often used. Known as the Peclet number, it is in fact the product of the Reynolds number and the Schmidt number, i.e.,

$$Pe = ReSc = \frac{ud}{D} \tag{6}$$

The Peclet number is the ratio of inertial forces to the momentum of diffusive flux. If bulk convection dominates over molecular diffusion, the Peclet number tends to be large. Conversely, a small Peclet number signifies the importance of molecular diffusion. Since the Schmidt number for gas mixtures is roughly of the order one, the Peclet number and the Reynolds number have roughly the same numerical value throughout the bronchial tree. At the level of respiratory bronchioles, the Peclet number assumes the value of unity (Table I) indicating that in this region bulk convection and molecular diffusion are of similar importance.

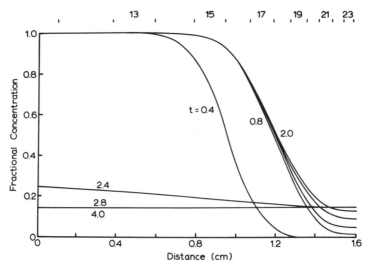

Fig. 4. Simulated oxygen concentration profiles in a "trumpet lung" model during a single cycle of oxygen breathing. Respiratory flow is constant at 0.25 L s^{-1} and tidal volume is 0.5L. Distance is measured from the terminal alveolar sacs, with generation number indicated on upper axis. The profiles are separated by 0.4s intervals.

The existence of a quasi-steady front between inspired gas and resident gas suggests that the local Peclet number should be of order one. If either bulk convection or molecular diffusion predominates, the front must be moving in one direction or the other. In this regard it is interesting to note that in Fig. 4 the center of the quasi-stationary front lies around the 18th generation.

Taylor-Dispersion - Coupled Axial Convection and Radial Diffusion

Although its importance in intrapulmonary gas mixing under normal physiological conditions is questionable, Taylor dispersion has been extensively discussed in the literature of respiratory physiology. Unfortunately, there is some confusion about how and where this mechanism functions. Thus, it may be useful to introduce the principles of this mechanism at some length; a more mathematical description based on the formulations of Taylor (1953) is given in the Appendix. The example used here will be laminar flow in a straight circular tube.

As shown in Fig. 5A, a parabolic velocity profile exists everywhere along a straight tube, with maximum velocity, u_m, occurring at the center of the tube. The mean velocity moving in this tube is U or 0.5 u_m. If at time t_0 a nondiffusible material is introduced in the tube to the left of x_0, then this material will be transported downstream by the moving fluid. As a result of the parabolic velocity distribution across the tube, the tracer material near the center will always move much faster than that near the tube wall. At that time t_1, the concentration profile of this material will be like that depicted in Fig. 5B, in the shape of a "spike" as Henderson et al. (1915) called it. If we can define a mean concentration value C of this material over a cross section of the tube, the longitudinal profile of this mean concentration is as shown in Fig. 5C. Now if an observer travels downstream with the mean velocity U and always measures mean concentration C, he will see a longitudinal spreading or dispersion of this material both upstream and downstream from him.

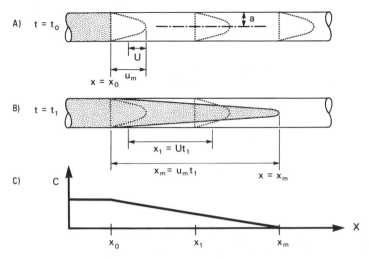

Fig. 5. A: parabolic velocity profile everywhere along tube
at all instants. At $t = t_0$, a new material is introduced to
left of point $x = x_0$.
B: this material is dispersed downstream by axial convective
velocity to form a long narrow "spike". Mean velocity in tube
is U, whereas peak velocity at center is U_m. Distance x_1 is
traveled by moving frame of reference at speed U after a time
period t_1, and x_m is distance traveled by material in center of
tube in same period of time.
C: longitudinal profile of mean cross-sectional concentration
of this material.

If the material has a finite molecular diffusivity, the longitudinal
spreading or dispersion will be impeded because radial diffusion will occur
as a result of the radial concentration gradient. Material in the central
core will diffuse toward the periphery, tending to move the tracer material
to the slow moving zone. In other words, once a molecule of this material
diffuses out of the central core, it will be retarded by the local fluid
velocity and cannot continue to move forward as rapidly as those molecules
remaining in the central core. Thus for the observer who travels with the
mean velocity U, a highly diffusible material disperses less rapidly in the
longitudinal direction than one that has a low molecular diffusivity.
Taylor (1953) showed mathematically that this dispersion process, which is
the result of interaction between axial convection and radial or lateral
diffusion, may be viewed as a virtual diffusion problem with a moving frame
of reference. Namely, Fick's law will be obeyed to a plane that moves with
velocity U

$$M = -DA \frac{dC}{dx'} \tag{7}$$

where M is the mass flux relative to the moving frame of reference whose
longitudinal coordinate is $x' = x - Ut$, dC/dx', the longitudinal concentration gradient, A, the cross-sectional area of the tube, and D, the virtual
diffusion coefficient or dispersion coefficient. For the case considered
above

$$D_{lam} \propto \frac{a^2 U^2}{D_{mol}} \tag{8}$$

47

namely, this coefficient varies with the square of mean axial velocity and tube radius a but is inversely proportional to the molecular diffusivity, D_{mol}.

When the flow is turbulent (tur) in a straight tube, the above-described concept still holds, but now the radial or lateral transport is provided by the random convective eddies that are found in turbulent flows rather than molecular diffusion as in the case of laminar (lam) dispersion. The mixing between the central core and the peripheral fluid is much faster in turbulent flow, and Taylor (1954) showed that in this case the dispersion coefficient

$$D_{tur} \propto Ua \tag{9}$$

Because radial molecular diffusion plays only a negligible role in this case, D_{mol} is not featured in Eq. 9. The radius a is a characteristic length in which lateral mixing occurs; hence the dispersion coefficient is proportional to the product of this mixing length and the mean convective velocity.

The concept of Taylor dispersion was first applied to the problem of intrapulmonary gas transport by Wilson and Lin (1970). These authors took the lung geometry given by Weibel (1963) and calculated the region of significant Taylor dispersion according to the two criteria set forth by Taylor (see Appendix). They modified the two criteria to be

$$\frac{aU}{D} \left(\frac{a}{L} \right) < 10 \text{ and } \frac{aU}{D} > 10 \tag{10}$$

and took L to be the length of a given airway generation. Using an unspecified tidal volume and a diffusivity presumably for the nitrogen-oxygen pair, Wilson and Lin concluded that

(i) between 0th and 7th generation: $\frac{aU}{D} \left(\frac{a}{L} \right) > 10$, therefore convection dominates;

(ii) between the 8th and 11th generation, $\frac{aU}{D} \cdot \frac{a}{L} < 10$,

and $\frac{aU}{D} > 10$, therefore Taylor dispersion dominates;

(iii) between 12th and 18th generation, $\frac{aU}{D} < 10$, therefore longitudinal diffusion is significant.

The above predictions provide explicit information about the region in which Taylor dispersion is effective. However, this information is difficult to verify experimentally.

Obviously, the bronchial tree has a drastically different geometry from a straight tube and the velocity profiles are most likely non-parabolic. Additionally, there are secondary swirls in the airways (Isabey and Chang, 1982) and cardiogenic oscillations (Engel, 1985). These factors definitely modify but do not eliminate the above-described dispersion process. We can call the modified Taylor-dispersion process Taylor-type dispersion in the lung.

Several investigators have reported that, contrary to previous belief, breathing a light gas mixture increases the alveolar-arterial difference of oxygen partial pressure, while breathing a dense gas mixture reduces the (A-a)DO_2 (Wood et al., 1976; Christopherson and Hlastala, 1982). In view of the original description of Taylor laminar dispersion in which the degree of dispersion is inversely related to molecular diffusivity, it is tempting to attribute these experimental findings to the influence of Taylor dispersion; several authors have in fact done so. The remaining questions are: (1) whether Taylor-type dispersion in the lung is really responsible for significantly altering gas exchange; (2) whether valid alternative explanations exist? Based on the current understanding of gas transport processes in the lung, the answer to the first question is likely to be negative. Unfortunately, the answer to the second question is still elusive.

SUMMARY

The bronchial tree is an asymmetric multi-generation conduit for air to pass through. Breathing is a time-varying, cyclic process. Therefore, gas transport in the bronchial tree during breathing is a problem with both spatial and temporal complexities. However, it is possible to gain a fair understanding of this complex process by examining the roles of bulk convection, molecular diffusion and the various modes of their interaction.

Convection brings the inspired gas into the peripheral region of the lung and removes the mixed inspired-resident gas from the lung. During the post-inspiratory and post-expiratory pauses and in the very distal regions of the lung, molecular diffusion alone is responsible for the mixing process.

Convection and diffusion together are responsible for establishing a quasisteady front of the inspired gas in the peripheral lung region. The position of the front depends on lung volume, tidal volume, inspiratory flowrate as well as diffusivity of the inspired gas. Taylor-type dispersion, while certainly existing in the lung, and despite its gas mixing potentials, is of questionable effectiveness in the overall gas transport process in the bronchial tree.

The various modes of interaction between convection and diffusion are likely to improve gas mixing in the bronchial tree and have an effect on the phase II of the single-breath washout curve.

REFERENCES

Chang, H.K. (1980). Multicomponent diffusion in the lung. Fed. Proc. 39:2759-2764.
Chang, H.K. and O.A. El Masry (1982). A model study of flow dynamics in human central airways. I. Axial velocity profiles. Respir. Physiol. 49:75-95.
Chang, H.K. (1984). Mechanisms of gas transport during ventilation by high-frequency oscillation. J. Appl. Physiol.: Respirat. Environ. Exercise Physiol. 56:553-563.
Chang, H.K. and A.S. Menon (1985). Air flow dynamics in the airways. In: Aerosol in Medicine, eds. F. Moren, M.T. Newhouse and M.E. Dolovich, Elsevier, Amsterdam (pp. 77-122).
Christopherson, S.K. and M.P. Hlastala (1982). Pulmonary gas exchange during altered density gas breathing. J. Appl. Physiol. Respir. Environ. Exercise Physiol. 52(1):221-225.

Engel, L.A. (1985). Intraregional gas mixing and distribution. In: <u>Gas Mixing and Distribution in the Lung</u>, eds. L.A. Engel and M. Paiva, Marcel Dekker, Inc., New York, pp. 287-358.

Henderson, Y., F.P. Chillingworth and J.L. Whitney (1915). The respiratory dead space. Am. J. Physiol. 38:1-19.

Isabey, D. and H.K. Chang (1982). A model study of flow dynamics in human central airways. II. Secondary flow velocities. Respir. Physiol. 49:97-113.

Menon, A.S., M.E. Weber and H.K. Chang (1984). Model study of flow dynamics in human central airways. Part III. Oscillatory velocity profiles. Respir. Physiol. 55:255-275.

Paiva, M. (1973). Gas transportation in the human lung. J. Appl. Physiol. 35:401-410.

Paiva, M. (1985). Theoretical studies of gas mixing in the lung. In: <u>Gas Mixing and Distribution in the Lung</u>, eds. L.A. Engel and M. Paiva, Marcel Dekker, Inc., New York, pp. 221-286.

Scheid, P. and J. Piiper (1980). Intrapulmonary gas mixing and stratification. In: <u>Pulmonary Gas Exchange</u>, Vol. I, Ventilation, Blood Flow and Diffusion, ed. J.B. West, Academic, New York, pp. 87-130.

Taylor, G.I. (1953). Dispersion of soluble matter in solvent flowing slowly through a tube. Proc. R. Soc. London 219:186-203.

Taylor, G.I. (1954). The dispersion of matter in turbulent flow through a pipe. Proc. R. Soc. London 223:446-468.

Ultman, J.S. (1985). Gas transport in the conducting airways. In: <u>Gas Mixing and Distribution in the Lung</u>, eds. L.A. Engel and M. Paiva, Marcel Dekker, Inc., New York (pp. 63-136).

Weibel, E.R. (1963). Morphology of the Lung. New York, Academic Press, Chapt. 11.

Wilson, T.A. and K. Lin (1970). Convection and diffusion in the airways and the design of the bronchial tree. In: Airway Dynamics, ed. by A. Bohuys. Springfield, Ill., Charles C. Thomas, pp. 5-19.

Wood, L.D. H., A.C. Bryan, S.K. Bau, T.R. Weng and H. Levison (1976). Effect of increased gas density on pulmonary gas exchange in man. J. Appl. Physiol. 41:206-210.

APPENDIX: A BRIEF DESCRIPTION OF TAYLOR LAMINAR DISPERSION

When a soluble substance is introduced into a fluid flowing slowly through a circular tube, it spreads out under the combined action of molecular diffusion and the variation of velocity over the cross section. This problem was first considered by Taylor (1953) who showed analytically that "the distribution produced this way is centered on a point which moves with the mean speed of flow and is symmetric about it in spite of the asymmetry of the flow".

Referring to Fig. 5, the incoming substance in the tube is assumed to have a parabolic velocity profile, i.e.,

$$u = u_m \left(1 - \frac{r^2}{a^2}\right) \tag{A1}$$

where u_m is the maximum velocity in the center of the tube, a is the radius of the tube and r denotes various radial positions. The concentration of the injected substance varies both longitudinally and radially, i.e.,

$$C = f(x,r)$$

and the distribution of concentrations is influenced by radial molecular diffusion, which tends to diminish the concentration gradient established by the non-uniform velocity profile across the tube, and by longitudinal convection, which tends to send the material in the central portion of the tube far ahead of that in the peripheral regions.

If longitudinal convection is very rapid and radial diffusion is slow, the incoming substance will be dispersed in the tube like an ever-lengthening "spike" with its sharp point located in the center and leaving the peripheral regions free of this substance. If, on the other hand, longitudinal convection is slow and radial diffusion is relatively rapid, then the injected substance will move out of the paraboloid to the slow-moving peripheral regions of the tube and longitudinal dispersion of this substance will thus be curtailed. It is this latter situation which is known as Taylor dispersion.

Taylor showed that by defining a mean concentration over a cross-section,

$$\bar{C} = \frac{2}{a^2} \int_0^a Crdr, \tag{A2}$$

and by adopting a frame of reference moving with the speed of mean flow, U, i.e.

$$x' = x_0 - Ut_0 = x_0 - \frac{1}{2} u_m t, \tag{A3}$$

the longitudinal dispersion of the injected substance may be described by a simple equation resembling that of one-dimensional diffusion equation, namely,

$$\frac{\partial \bar{C}}{\partial t} = k \frac{\partial^2 \bar{C}}{\partial x'^2} \tag{A4}$$

Interpreted physically, Eq. (A4) means that to an observer moving with the mean flow, the cross-sectional mean concentration of the injected

51

substance appears to spread out symmetrically about him in the same manner as ordinary molecular diffusion would spread out a concentrated substance with respect to a fixed frame of reference. The speed with which the injected matter spreads depends on the coefficient k in Eq. (A4), and Taylor found that

$$k = \frac{a^2 U^2}{48D} \tag{A5}$$

One notes that the Taylor dispersion coefficient k is inversely proportional to D, the molecular diffusion coefficient between the injected substance and the fluid originally flowing in the tube. This is to be expected. For a given tube and a given mean flow, a more diffusible substance will have more effective radial diffusion, and, thereby may be less effectively dispersed in the longitudinal direction.

In Taylor dispersion the mass flux with respect to any moving cross-section having the speed of mean flow can be expressed in terms of a quasi-Fick's equation,

$$J_{x_1} = - \pi a^2 \left(\frac{a^2 U^2}{48D}\right) \frac{\partial C}{\partial x_1} = - \frac{\pi a^4 U^2}{48D} \frac{\partial C}{\partial x_1} \tag{A6}$$

For eqs. (A4), (A5) and (A6) to be valid, two conditions must be met. First, longitudinal convection must be slow compared to radial diffusion. Second, longitudinal diffusion is negligible.

The first condition means that the time necessary for radial variation of concentration to die down must be small compared to the time necessary for appreciable change in concentration to appear owing to longitudinal convective transport. As derived by Taylor (1953), this condition can be expressed by the following inequality,

$$\frac{L}{U} >> \frac{a^2}{7.2D} \tag{A7}$$

where L is the length of tube over which convective transport makes an appreciable change in concentration.

In order that the longitudinal molecular diffusion may be negligible compared with the dispersive effect represented by the Taylor dispersion coefficient k, it is necessary that

$$D << \frac{a^2 U^2}{48D}$$

or

$$\frac{AU}{D} >> 6.9 \tag{A8}$$

Recalling the definition of the Peclet number in Eq. (6), Eqs. (A7) and (A8) may be rearranged and combined to be approximately

$$14(L/a) >> Pe >> 14 \tag{A9}$$

which states the conditions under which dispersion of a soluble matter flowing laminarly in a straight circular tube may be described by Taylor's formulations.

COMPARATIVE PHYSIOLOGY OF OXYGEN TRANSFER IN LUNGS

Frank L. Powell and Steven C. Hempleman

M-023A, Department of Medicine
University of California, San Diego
La Jolla, CA 92093

INTRODUCTION

There is a strong tradition of comparative physiology in the field of respiration. Over the years, physiologists have been fascinated by the differences in respiratory organs which evolved for the common problem of oxygen transfer. This led to a rich experimental literature about possible advantages that one type of structure may have over another, but it also led to many speculations.

In 1975, a paper by Piiper and Scheid entitled "Gas transport efficacy of gills, lungs and skin: Theory and experimental data" was published. Four basic models of gas exchange were considered: (1) counter-current, applicable to fish gills, (2) cross-current, applicable to parabronchial avian lungs, (3) ventilated pool, applicable to alveolar mammalian lungs and the uni-, pauci- and multicameral lungs of amphibians and reptiles, and (4) infinite pool, applicable to cutaneous gas exchange. This study offered the first systematic theoretical framework for comparing the function of different types of respiratory organs and resulted in significant quantitative conclusions. For example, if ventilation, blood flow, diffusing capacity, blood-O_2 equilibrium curves, inspired and mixed venous PO_2 are equal, then oxygen uptake decreases in the following order: counter-current > cross-current > ventilated pool. The model analysis also allows interesting thought experiments such as "How much would ventilation have to increase in a dog to match the increased oxygen uptake that would result from transplanting a bird's cross-current lung into the mammal?"

As valuable as Piiper and Scheid's approach proved, it could not be used to answer at least two questions we had. Specifically, what are the average gas-blood partial pressure gradients driving diffusion in the different models and what are the effects of functional inhomogeneities in the different models? We attempted to answer these questions by elaborating on Piiper and Scheid's model analysis. Because we are focusing on oxygen transfer in lungs, only the cross-current and ventilated pool models are considered.

MODEL ANALYSIS

Background. Piiper and Scheid's (1975) models are based on the definition of a gas flux as the product of a conductance and partial pressure difference:

$$\dot{M} = G \; \Delta P \tag{1}$$

Convective conductances are the product of flows and the capacitance coefficient of the medium (cf. Piiper et al., 1971). For ventilation:

$$Gvent = \dot{V} \; \beta g \tag{2}$$

For blood flow:

$$Gperf = \dot{Q} \; \beta b \tag{3}$$

Diffusive conductance is the diffusing capacity:

$$Gdiff = DL \tag{4}$$

The model analysis assumes (1) steady state, (2) constant capacitance coefficients (i.e., linear blood-gas equilibrium curves), and (3) homogeneity. The validity of these assumptions and the consequences of violating them have been extensively discussed elsewhere (Burger et al., 1979; Hempleman and Powell, 1986; Piiper and Scheid, 1977; Powell and Scheid, 1987; Scheid, 1979).

Given these assumptions, the model analysis predicts relative partial pressure differences and limitations as functions of only Gvent, Gperf and Gdiff. The partial pressure differences defined are:

$$\Delta Pvent = (PI-PE)/(PI-Pv) \tag{5}$$

$$\Delta Pperf = (Pa-Pv)/(PI-Pv) \tag{6}$$

$$\Delta Ptr = (PE-Pa)/(PI-Pv) \tag{7}$$

Normalizing to PI-Pv results in partial pressure differences that are a fraction of the maximum possible and allows comparisons between cases with different PI and Pv. Limitations are defined as the relative difference between the maximum gas flux that could occur if the limitation was eliminated by raising the conductance in question to infinity ($\dot{M}max$) and the gas flux actually occurring with finite conductances ($\dot{M}act$):

$$L = (\dot{M}max - \dot{M}act)/\dot{M}max \tag{8}$$

Equations for Lvent, Lperf and Ldiff plus the ΔP's described above were presented by Piiper and Scheid (1975).

One of the most useful indices of gas transfer efficacy in the model predictions is ΔPtr. For mammalian lungs, this represents the alveolar-arterial gradient, (i.e., $AaDO_2$) which is a well known index of respiratory function. In an ideal ventilated pool model with infinite Gdiff, $\Delta Ptr = 0$. However, as Gdiff is decreased, say by gross thickening of blood-gas barrier, ΔPtr increases as would $AaDO_2$.

Fig. 1 shows that the degree of ΔPtr increase depends on Gvent/Gperf. This has important implications for comparing effects of different DLO_2 in, for instance, reptiles and mammals. One must also consider differences in $\dot{V}$, $\dot{Q}$, blood-O_2 equilibrium curves and body temperature, all of which determine Gvent/Gperf.

The same trends are observed in cross-current and in ventilated pool models, but there are quantitative differences (Fig. 1). In ideal cross-

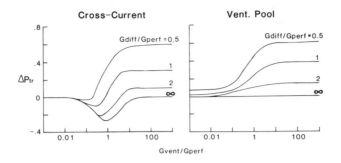

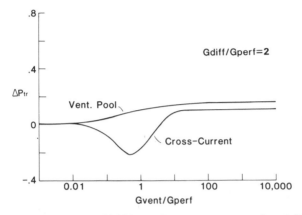

Fig. 1. Effect of Gdiff/Gperf on ΔPtr vs. Gvent/Gperf in cross-current and ventilated pool models. Models assumed to be homogeneous. Lower panel shows ΔPtr for both models at Gdiff/Gperf = 2. Lower ΔPtr in the cross-current model indicates greater efficacy.

current models ΔPtr can be negative at favorable values of Gvent and Gperf. As Gdiff is decreased in a cross-current model, ΔPtr becomes more positive but it may still be less than zero at relatively low Gdiff values depending upon Gvent/Gperf. For any given Gvent, Gperf and Gdiff, cross-current ΔPtr is ≤ ventilated pool ΔPtr (bottom of Fig. 1). However, to assess diffusion efficacy changes within a cross-current model one must be careful to reference to the ideal cross-current ΔPtr at the given Gvent/Gperf; ideal ΔPtr varies with Gvent/Gperf in cross-current exchange unlike ΔPtr in ideal ventilated pools, which is always zero.

Diffusive resistance. Generally, ΔP values have the meaning of relative resistances (cf. Piiper and Scheid, 1977). In ventilated pools, ΔPtr is the relative diffusion resistance. If Gdiff is infinite, ΔPtr = 0 and if Gdiff is 0, ΔPtr = 1. However, the fact the Δ Ptr can be negative in cross-current exchange makes its meaning more complex. Piiper and Scheid noted this and proposed that Δ Ptr be viewed as a composite of diffusion resistance and the inherent properties of the model:

$$\Delta Ptr = \Delta Pdiff + \Delta Po \qquad (9)$$

ΔPo is defined as Δ Ptr when Gdiff is infinite. Equations defining these ΔP (Piiper and Scheid, 1977) are presented in Table I.

A relative diffusion resistance is given by ΔPdiff but it is still conceptually difficult in cross-current models. For example, if Gvent/Gperf = 1 and Gdiff = 0, ΔPdiff = 1.26. This is "necessary" to force

Table I. Model analysis of cross-current and ventilated pool alveolar gas exchange ΔPtr, ΔPo and ΔPdiff from Piiper and Scheid (1977).

	Crosscurrent	Ventilated Pool
ΔPtr	$e^{-Z'} - X(1 - e^{-Z'})$	$\dfrac{X \cdot e^{-Y}}{X + 1 - e^{-Y}}$
ΔPo	$-X + (1 + X)e^{-1/X}$	0
ΔPdiff	$(X + 1) \cdot (e^{-Z'} - e^{-1/X})$	$\dfrac{X \cdot e^{-Y}}{X + 1 - e^{-Y}}$
ΔPD	$(X/Y) \cdot (1 - e^{-Z'})$	$\dfrac{X(1 - e^{-Y})}{Y(X + 1 - e^{-Y})}$

ΔPD is the average blood-gas partial pressure gradient during diffusion ($\dot{M} = DL \cdot \Delta PD$).
$X = \dot{V} \cdot \beta g / \dot{Q} \cdot \beta b = Gvent/Gperf$
$Y = DL / \dot{Q} \cdot \beta b = Gdiff/Gperf$
$Z' = 1/X(1 - e^{-Y})$

ΔPtr to one when there is a complete block at the diffusive transfer step. If Gdiff is infinite and Gvent/Gperf is still 1, ΔPtr = ΔPo = -0.26.

We thought that a meaningful index of diffusion resistance might be the average-blood-gas partial pressure difference, again normalized to the maximum difference:

$$\Delta PD = \overline{(Pg-Pb)} / (PI-Pv) \tag{10}$$

or:

$$\Delta PD = \dot{M}/Gdiff \tag{11}$$

Equations defining ΔPD in terms of Gvent, Gperf and Gdiff, analagous to Piiper and Scheid's P's, are given for cross-current and ventilated pool models in Table I.

Note that ΔPD is not truly a relative resistance even though it is normalized to PI-Pv and can vary from 0 to 1. This is because the average blood-gas partial pressure difference is greater than the difference between the maximum gradient and the ventilatory plus perfusive gradients:

$$\overline{(Pg-Pb)} > (PI-Pv) - [(PI-PE) + (Pa-Pv)]$$

However, ΔPD does help one understand how different models of gas exchange may require different Gdiff to meet metabolic demands.

Blood-gas diffusive PO_2 gradients. Fig. 2 (upper panels) shows the dependence of the average gradient driving diffusive gas flux across the blood-gas barrier (ΔPD) on Gvent/Gperf and Gdiff/Gperf for cross-current and ventilated pool models. As expected, lower Gdiff/Gperf, corresponding to a lower DL, results in a higher ΔPD for any given Gvent/Gperf in both models. A higher ΔPD is not necessarily better, however. Recall that $\dot{M} = Gdiff \cdot \Delta PD$; if ΔPD increases less than Gdiff decreases, then $\dot{M}$ decreases too.

The lower panel of Fig. 2 compares ΔPD for cross-current and ventilated pool models at the same Gdiff/Gperf. In this case, the higher ΔPD in the

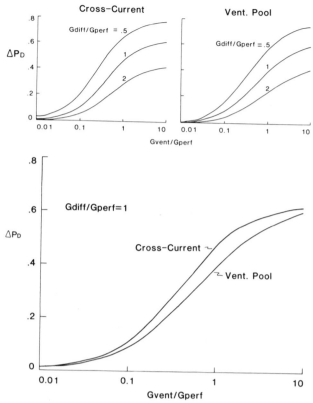

Fig. 2. Effect of Gdiff/Gperf on ΔPD vs. Gvent/Gperf in cross-current and ventilated pool models. Models assumed to be homogeneous. Lower panel shows ΔPD for both models at Gdiff/Gperf = 1. Higher ΔPD in the cross-current model indicates higher gas flux for given inputs.

cross-current model is advantageous because we are comparing different models at the <u>same</u> Gdiff. Thus, cross-current $\dot{M}$ is $\geq$ ventilated pool $\dot{M}$ for any Gvent/Gperf and Gdiff/Gperf. This is consistent with the general conclusions of Piiper and Scheid's original analysis that the efficacy of cross-current exchange is greater than ventilated pool exchange.

Although the lower panel of Fig. 2 indicates that diffusive gas flux is more favorable in cross-current lungs, the upper panels suggest that cross-current exchange is more susceptible to diffusion impairments. At a physiological Gvent/Gperf of 1, ΔPD decreases more in a cross-current than in a ventilated pool model as Gdiff/Gperf decreases from 2 to 1 and from 1 to 0.5. Diffusive gas flux is inherently better in the cross-current model but consequently "it has more to lose".

This aspect is demonstrated better in Fig. 3 which shows diffusion limitations, Ldiff, for both models. The upper panels show the dependence of Ldiff on conductances. For any Gvent/Gperf, Ldiff increases with decreasing Gdiff/Gperf as expected. What is not intuitively apparent is that for any Gdiff/Gperf, Ldiff increases as Gvent/Gperf increases. The lower panel compares Ldiff in the two models with Gvent/Gperf fixed at 1. As predicted from inspection of Fig. 2, Ldiff is greater for cross-current exchange at all but the highest Gdiff/Gperf values, where both models show no diffusion limitation. The larger Ldiff in cross-current exchange represents both a larger effect relative to ideal levels of exchange (cf. eq. 8)

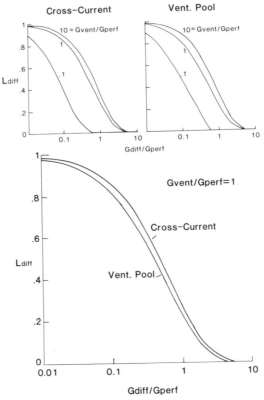

Fig. 3. Effect of Gvent/Gperf on Ldiff vs. Gdiff/Gperf in cross-current and ventilated pool models. Models assumed to be homogeneous. Lower panel compares Ldiff between the models at Gvent/Gperf = 1. Higher Ldiff in the cross-current model indicates greater susceptibility to diffusion limitation; i.e. larger decrease in gas flux from ideal conditions.

and in absolute terms (i.e., $\dot{M}max-\dot{M}act$). To our knowledge, this is the first time this prediction has been reported even though the equations necessary to make the prediction have been in print for over a decade. The physiological implications of this are considered in a later section.

Ventilation-Perfusion Inhomogeneity. The analysis to this point has considered homogeneous models and attributed all limitations to a blood-gas barrier diffusion resistance. In real lungs any of the three conductances may be mismatched between functional exchange units (e.g. mammalian acini or avian parabronchi). The most thoroughly studied functional inhomogeneity is ventilation-perfusion mismatch, or Gvent/Gperf inhomogeneity in terms of the analysis used here. For alveolar gas exchange, two to fifty compartment models of physiologic and inert gas exchange have been studied (for review see Wagner and West, 1980). The general finding is that Gvent/Gperf inhomogeneity decreases exchange efficacy and increases alveolar-arterial partial pressure difference. Studies on cross-current exchange are more limited (for review see by Powell and Scheid, 1987) but the conclusions are qualitatively similar.

To quantitatively compare the effects of Gvent/Gperf inhomogeneity in the two models we assumed $\dot{V}$ and $\dot{Q}$ were log-normally distributed against $\dot{V}/\dot{Q}$. This allows one to index increased inhomogeneity by the log standard deviation, σ, of either the $\dot{V}$ or $\dot{Q}$ distribution (cf. West, 1969). We used a 15

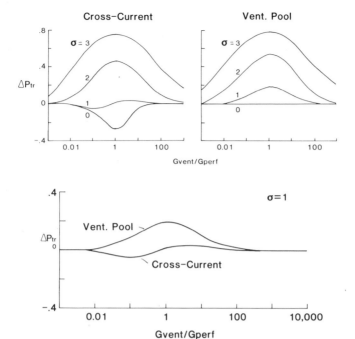

Fig. 4. Effect of $\dot{V}/\dot{Q}$ inhomogeneity (σ, log standard deviation of a log normal distribution of $\dot{Q}$ vs. $\dot{V}/\dot{Q}$) on ΔPtr vs. Gvent/Gperf. Gdiff assumed infinite. Lower panel compares ΔPtr between the models at $\sigma = 1$. Lower ΔPtr in cross-current model indicates greater efficacy for given conditions.

compartment model in which each compartment received $\dot{V}$ and $\dot{Q}$ predicted from a given mean $\dot{V}/\dot{Q}$ and σ, PI and Pv were the same for all compartments and it was assumed that no diffusion resistance existed (i.e., Gdiff = infinity). We assumed constant βb and solved for expired gas and end-capillary blood from each compartment, using equations for ΔPvent and ΔPperf. Flow-weighted averages of blood and gas leaving the lung were used to predict ΔPtr.

Fig. 4 shows the effect of increased Gvent/Gperf inhomogeneity in the two models. For the homogeneous case, ($\sigma = 0$), ΔPtr is identical to the ideal predictions of the original analysis with infinite Gdiff/Gperf; it is zero for all Gvent/Gperf in ventilated pools and reaches minimum value of -0.26 near Gvent/Gperf value of 1 in the cross-current model. Increasing inhomogeneity makes ΔPtr more positive in both models. This is the same qualitative effect that increasing diffusion resistance had in the original analysis. However, comparing Figs. 1 and 4 shows that the quantitative effects are different in the two models. Decreased Gdiff/Gperf in homogeneous exchangers causes large increases in ΔPtr at high Gvent/Gperf but increasing σ in nondiffusion-limited exchangers has much less effect on ΔPtr at high Gvent/Gperf. This is because of the effect of gas capacitance in blood (βb) on $\dot{V}/\dot{Q}$ inhomogeneity that is well known for alveolar gas exchange (e.g., West, 1969/70; Wagner, 1979).

The lower panel of Fig. 4 compares the effects of a similar degree of inhomogeneity on the two models. At this level of inhomogeneity, the cross-current model still shows a negative ΔPtr at some Gvent/Gperf and is always $\leq \Delta P$tr for the ventilated pool model. Thus, Gvent/Gperf inhomogeneity cannot make cross-current exchange efficacy less than ventilated pool efficacy. However, the magnitude of increase in ΔPtr for a given σ change

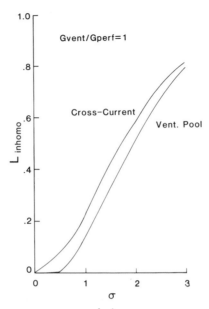

Fig. 5. Limitation due to $\dot{V}/\dot{Q}$ inhomogeneity vs. amount of
inhomogeneity (σ) for cross-current and ventilated pool models
at over all Gvent/(Gperf = 1). Gdiff assumed infinite. Higher
L inhomo for cross-current indicates larger effect of given
on gas flux.

is greater in cross-current than in ventilated pool models as evidenced from
inspection of the top panels. This is reminiscent of the situation reported
above for the blood-gas diffusion gradient and is also better seen with a
limitation index.

We calculated a Gvent/Gperf inhomogeneity limitation, Linhomo, by the
same formula described above (eq. 8). $\dot{M}$max is the gas flux in the ideal
case (infinite Gdiff, homogeneous) and $\dot{M}$act is the gas flux of the inhomoge-
neous, nondiffusion limited model. Fig. 5 shows the results for the two
models at a Gvent/Gperf of 1. The cross-current Linhomo is $\geq$ ventilated
pool Linhomo for any degree of inhomogeneity. Similar to the diffusion
limitation analysis the effect of inhomogeneity on cross-current $\dot{M}$ can be
both relatively (compared to ideal $\dot{M}$) and absolutely greater than the effect
on ventilated pool $\dot{M}$. Once again, it appears that the cross-current model
starts out with inherently higher efficacy and "has more to lose".

Diffusion and inhomogeneity interaction. Combining the effects of
diffusion limitations and $\dot{V}/\dot{Q}$ inhomogeneity can be very complicated because
there are so many ways to distribute the three relevant conductances to one
another and there is so little physiological data to guide us in our choice
of distributions (Piiper, 1961). Nevertheless, the problem is mathematically tractable and a single example reveals an important point. Fig. 6
shows ΔPtr for the two models in four conditions: (1) ideal with
Gdiff/Gperf = infinity and homogeneous, i.e., σ = 0; (2) Gdiff/Gperf = 1 and
σ = 0; (3) Gdiff/Gperf = infinity and σ = 1; (4) Gdiff/Gperf = 1 and σ = 1.
Diffusing capacity is distributed according to blood flow in the fourth
case. The first three cases have been shown earlier in Figs. 1 and 4. The
combined effects of inhomogeneity and diffusion limitation demonstrate two
important features. First, the combined effects are less than the sum of
the two individual effects. Second, cross-current exchange efficacy is
greater than in the ventilated pool even with the combined limitations.

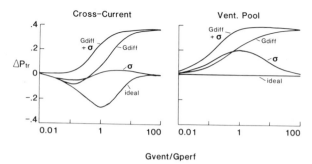

Fig. 6. Interaction of diffusion and inhomogeneity limitations effects on ΔP_{tr} in cross-current and ventilated pool models. Ideal -infinite Gdiff and homogeneous; Gdiff - homogeneous and Gdiff = 1; σ - infinite Gdiff and σ = 1; Gdiff + σ - Gdiff = 1 and σ = 1. Note that the combined effects are less than the sum of the individual effects and Ptr is less in cross-current than in ventilated pool exchange for all cases including combined limitations.

PHYSIOLOGICAL SIGNIFICANCE

The most obvious conclusions from the preceding analysis are, (1) for any fixed set of conditions efficacy in cross-current exchange is greater or equal to that in ventilated pool exchange, and (2) cross-current exchange is more susceptible than ventilated pool exchange to diffusion limitations. This latter conclusion especially sheds new light on a problem of comparative respiratory physiology. It is now well documented that the morphometric DLO_2 for birds is generally greater than for comparably sized mammals (for review see Powell and Scheid, 1987). It has been hypothesized that the high oxygen requirements of flight may have been a selective pressure for the evolution of a large DLO_2 in birds and this is supported by morphometric DLO_2 values in bats being higher than predicted for comparably sized nonflying mammals (Maina et al., 1982). However, an alternative hypothesis is that once birds evolved a parabronchial lung with cross-current exchange, then a higher DLO_2 was necessary as a safety factor against diffusion limitations. One need not envision this safety factor simply as protection against pathologic conditions. DLO_2 is a physiologic variable that changes with recruitment and/or distension of pulmonary capillaries, which apparently occurs in birds as it does in mammals (Hempleman and Powell, 1986).

Another conclusion that may be less obvious is that the relative efficacies of ventilated pool and cross-current exchangers are not fixed in nature because animals show a wide range of Gvent/Gperf ratios. Studying Fig. 3 one can imagine a ventilated pool having either a smaller or larger Ldiff compared to a cross-current exchanger depending on Gvent/Gperf. Remember that Gvent/Gperf may not vary only with $\dot{V}/\dot{Q}$ but also with blood-gas capacitance.

To explore the range of efficacy and of limitations in nature, we selected values from the literature to approximate a representative mammal, bird and reptile (Table II). Hypoxia data are used where possible to better satisfy the assumption of constant βb. Our representative bird is a composite of two studies with Gvent/Gperf of 0.7 and Gdiff/Gperf of 2.0. The hen was studied in hypercapnia resulting in relatively high Gvent/Gperf. However, the estimate of Gdiff/Gperf in the hen study is more accurate and better satisfies assumptions to calculate Gdiff than the duck study which provides a reasonable Gvent/Gperf. Reptilian data are often difficult to

Table II. Conductance ratios for a mammal, bird and reptile from the literature.

	DOG[1] (hypoxic)	HEN[1] (hypoxic-hypercapnic)	DUCK[2] (hypoxic)	NILE MONITOR[3] (25° C)
$\dfrac{Gvent}{Gperf}$	0.3	1.2	0.7	3.0
$\dfrac{Gdiff}{Gperf}$	3.0	1.9	0.8	1.0

References: 1 – Piiper and Scheid, 1975; 2 – Kiley et al., 1985; 3 – Hicks, Ishimatsu and Heisler, personal communication.

analyze because of irregular breathing patterns. However, the Nile monitor lizard (<u>Varanus niloticus</u>) is a slow but regular breather (ca. 3 breath/min at 25° C).

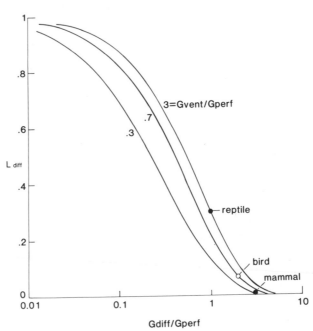

Fig. 7. Diffusion limitations for three representative lunged vertebrates. Models assumed homogeneous. Ldiff vs. Gdiff/Gperf for ventilated pool model at Gvent/Gperf = 0.3 (mammal) and 3.0 (reptile) and cross-current model at Gvent/Gperf = 0.7 (bird). This demonstrates that diffusion limitation in ventilated pools can be either larger or smaller than in cross-current exchange, depending upon Gvent/Gperf.

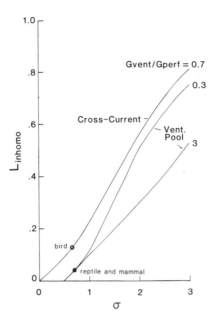

Fig. 8. V̇/Q̇ inhomogeneity limitations for three representative lunged vertebrates. Models assumed infinite Gdiff/Gperf. Linhomo vs. for ventilated pool models at Gvent/Gperf = 0.3 (mammal) and 3.0 (reptile) and cross-current model at Gvent/Gperf = 0.7 (bird). This demonstrates that inhomogeneity limitation is greater in the bird than in the mammal or reptile.

Fig. 7 shows the effects of diffusion limitations for the ventilated pool model at Gvent/Gperf of 0.3 and 3 and cross-current model at Gvent/Gperf = 0.7. The points indicate that indeed ventilated pool exchange can be either more or less diffusion-limited than cross-current exchange in nature as well as in theory. After inspecting the different lungs, one may expect the reptilian lung to be more diffusion-limited than the avian or mammalian lung, but the relation between diffusion-limitations and structure is not intuitively apparent for the mammalian and bird lung. A similar analysis can be made for the effects of Gvent/Gperf inhomogeneity. Estimates of inhomogeneity are poorer than for diffusion but most of the data indicates that V̇/Q̇ matching is similar in the different animals. Healthy dogs have very little inhomogeneity ($\sigma < 0.5$), Wagner, personal communication), birds have some more ($\sigma = 0.6$, Powell and Scheid, 1987) and the multicameral lungs of alligators have still a bit more ($\sigma = 0.7$, Powell and Gray, unpublished observations). The effects of hypoxia on V̇/Q̇ matching in mammals and birds are small (Sylvester et al., 1981; Powell and Hastings, 1983). Therefore, the difference in Gvent/Gperf inhomogeneity between animals is probably very small and one can compare the effects on the different models assuming a common degree of inhomogeneity.

Fig. 8 shows inhomogeneity limitations at the three representative Gvent/Gperf ratios; at the expected level of σ the two ventilated pool predictions overlap. The cross-current prediction is greater than both ventilated pools, so the effects of inhomogeneity are greater in birds than in both mammals and reptiles. This contrasts with the situation for diffusion limitations.

Finally, one can predict the combined effects of inhomogeneity and diffusion limitations on the three representative animals. Using the three

Gvent/Gperf ratios from Table II again, but assuming single values of Gdiff/Gperf = 2 and   = 1 we obtained the following: calculated $\dot{M}$ as a percent of ideal $\dot{M}$ was 80% for the reptile, 75% for the mammal and 68% for the bird. There are at least two ways to consider these results. One is to conclude the effects of combined limitations are similar in all animals. Considering the possible inaccuracies of inputs to our model analysis and assumption violations this is a valid conclusion.

Another view of the prediction that limitations are smallest in the reptile and largest in the bird, with the mammal somewhere in between, emphasizes the differences. Birds with the highest intrinsic efficacy operate farthest from ideal and the reptilian lung with a structure expected to transfer oxygen with less efficacy operates closest to ideal. Apparently the common requirement of pulmonary oxygen transfer has been solved by nature with a variety of solutions. Perhaps it is not different oxygen transfer requirements that provided the selective pressure for the evolution of different types of lungs because the different types seem to perform similarly at transferring oxygen. Differences in oviparity and viviparity have been proposed as an important determinant in lung evolution (Duncker, 1978). Natural selection may then simply result in lungs that arterialize blood at some minimal acceptable level and not in optimized lung function in the sense of reduced imperfections like diffusion limitations and $\dot{V}/\dot{Q}$ mismatch.

It is also possible that limitations have other physiological consequences that we have not considered and these may be subject to selection. $\dot{V}/\dot{Q}$ mismatching seems similar in most animals while there is wide variability in diffusion limitations for reasons that are not understood. Developing a better understanding of gas exchange limitation interactions in the lungs and with other organ systems provides a challenge to physiologists. As long as a comparative approach like the one presented here leads to significant questions and suggests ways to answer the questions, comparative respiratory physiology will remain a vital field.

REFERENCES

Burger, R.E., M. Meyer, W. Graf and P. Scheid (1979). Gas exchange in the parabronchial lung of birds: Experiments in unidirectionally ventilated ducks. Respir. Physiol. 36:19-37.
Duncker, H.R. (1978). V. Morphological and functional aspects of respiration and circulation, functional morphology of the respiratory system and coelemic subdivisions in reptiles, birds and mammals. Verh. Dtsch. Zool. 98:99-132.
Hemplemen, S.C. and F.L. Powell (1986). Influence of pulmonary blood flow and O_2 flux on DO_2 in avian lungs. Respir. Physiol. 59:285-292.
Kiley, J.P., F.M. Faraci and M.R. Fedde (1985). Gas exchange during exercise in hypoxic ducks. Respir. Physiol. 59:105-115.
Maina, J.N., A.S. King and D.Z. King (1982). A morphometric analysis of the lung of a species of bat. Respir. Physiol. 50:1-11.
Piiper, J. and P. Scheid (1977). Comparative physiology of respiration: Functional analysis of gas exchange organs in vertebrates. In: International Review of Physiology, edited by J.G. Widdicombe, Baltimore, University Park Press, pp. 219-253.
Piiper, J. and P. Scheid (1975). Gas transport efficacy of gills, lungs and skin: Theory and experimental data. Respir. Physiol. 23:209-221.
Piiper, J., P. Dejours, P. Haab and H. Rahn (1971). Concepts and basic quantities in gas exchange. Respir. Physiol. 13:292-304.
Piiper, J. (1961). Variations of ventilation and diffusing capacity to perfusion determining the alveolar-arterial O_2 difference: theory. J. Appl. Physiol. 16:507-516.

Powell, F.L. and P. Scheid (1987). Physiology of gas exchange in the avian respiratory system. In: Form and Function in Birds, Vol. 4, edited by A.S. King and J. McLelland, London, Academic Press.

Powell, F.L. and R.H. Hastings (1983). Effects of hypoxia on ventilation-perfusion matching in birds. Physiologist 26:A50.

Scheid, P. (1979). Mechanisms of gas exchange in birds. Rev. Physiol. Biochem. Pharmacol. 86:137-186.

Sylvester, J.T., A. Cymerman, G. Gurtner, O. Hottenstein, M. Cote and D. Wolfe (1981). Components of alveolar-arterial O_2 gradient during rest and exercise at sea level and high altitude. J. Appl. Physiol.: Respirat. Environ. Exercise Physiol. 50:1129-1139.

Wagner, P.D. and J.B. West (1980). Ventilation-perfusion relationships. In: Pulmonary Gas Exchange, Vol. 1, edited by J.B. West, New York, Academic Press, pp. 219-262.

Wagner, P.D. (1979). Susceptibility of different gases to ventilation-perfusion inequality. J. Appl. Physiol.: Respirat. Environ. Exercise Physiol. 46:372-386.

West, J.B. (1969/70). Effect of slope and shape of dissociation curve on pulmonary gas exchange. Respir. Physiol. 8:66-85.

West, J.B. (1969). Ventilation-perfusion inequality and overall gas exchange in computer models of the lung. Respir. Physiol. 7:88-110.

LIMITATIONS TO THE EFFICIENCY OF PULMONARY GAS EXCHANGE

DURING EXERCISE IN MAN

M. D. Hammond

Division of Pulmonary and Critical Medicine
Department of Internal Medicine
University of South Florida, College of Medicine
James A. Haley Veterans Hospital
Tampa, FL 33612

INTRODUCTION

The human lung must function over a wide range of metabolic demands and environmental conditions. It is not rare for oxygen consumption ($\dot{V}O_2$) to vary from 3-5 ml·kg·min^{-1} at rest to as much as 70 ml·kg·min^{-1} during exercise only moments later, or for inspired PO_2 (PIO_2) to range from 150 Torr (sea level) to 80 Torr (equivalent altitude 4500 meters) or less over a period of hours to days. The ability to function in these different circumstances comes with a small price: although the lung is remarkably efficient at rest at sea level, it becomes less so at higher $\dot{V}O_2$ (Asmussen and Nielsen, 1960), particularly at high altitude. For example, in healthy resting subjects the ideal alveolar-arterial PO_2 difference (A-aDO$_2$) is normally only 5-10 Torr, but it may increase to 25 Torr or more during heavy exercise (Dempsey et al., 1984). While this increased gradient has relatively little effect on arterial O_2 content at sea level, it can lead to substantial additional arterial desaturation at altitude, where subjects are operating on the steep descending slope of the oxyhemoglobin dissociation curve.

The reduced efficiency of pulmonary gas exchange observed during exercise or altitude exposure can be attributed to three potential mechanisms: 1) ventilation-perfusion inequality, 2) alveolar-end-capillary diffusion limitation for O_2 and 3) right to left shunts, both intrapulmonary and post-pulmonary. In the past, efforts to partition the A-aDO$_2$ have been limited by difficulties in measuring any of these potential components accurately during exercise. Recent studies employing the multiple inert gas elimination technique (MIGET) allow the relative roles of each of these potential mechanisms to be examined (Gale et al., 1985; Hammond et al., 1986a,b; Wagner et al., 1986a,b). Collectively, this work has cast interesting new light on our understanding of gas exchange during exercise both at sea level and simulated acute altitude exposure.

EXPERIMENTAL APPROACH

The general strategy used to partition the A-aDO$_2$ into its components rests upon use of the multiple inert gas elimination technique to estimate the ventilation-perfusion distribution (Wagner et al., 1974a,b). The measured $\dot{V}A/\dot{Q}$ distribution is then used to derive predicted values for arterial blood and alveolar gas tensions and the A-aDO$_2$ attributable to $\dot{V}A/\dot{Q}$ inequality alone. The component of the A-aDO$_2$ due to diffusion limitation can then be estimated as the residual difference between the actual observed A-aDO$_2$ and that predicted from the measured $\dot{V}A/\dot{Q}$ distribution. This derivation depends upon the assumption that post-pulmonary shunts are negligible, since these are not measurable by MIGET. Arguments which support this assumption have been presented in detail elsewhere (Torre-Bueno et al., 1985; Hammond et al., 1986a; Wagner et al., 1986a), and will not be discussed further here.

In the work reviewed here only normal subjects were studied, both at rest and during steady state exercise performed on a bicycle ergometer. Data were collected both at sea level and simulated altitude, the latter being achieved either in a chamber (hypobaric hypoxia - Gale et al., 1985; Wagner et al., 1986a,b) or by having subjects breathe hypoxic gas mixtures at 1 atmosphere pressure (normobaric hypoxia - Hammond et al., 1986b). In all circumstances baseline data were collected at rest and then further measurements were made at several levels of exercise up to the subject's predetermined maximal capacity. In summary, the raw data collected during each run included: expired ventilation ($\dot{V}E$), respiratory rate (RR), cardiac output (CO), heart rate (HR), mixed expired and arterial respiratory gas tensions, and mixed expired and arterial inert gas contents. In some experiments a pulmonary artery catheter was placed (Wagner et al., 1986a,b), permitting measurement of pulmonary vascular pressures, pulmonary capillary wedge pressure (PCWP) and mixed venous respiratory and inert gas tensions and contents. Details, which have varied slightly among these studies, have been described at length in the cited papers.

RESULTS

NORMOXIA: Inert gas data demonstrate increased $\dot{V}A/\dot{Q}$ mismatching during normoxic exercise at sea level (Hammond et al., 1986a; Wagner et al., 1986a). This is true both for indexes of $\dot{V}A/\dot{Q}$ inequality derived from the enforced smoothing algorithm used to estimate the ventilation-perfusion distribution [log standard deviation of perfusion (log SD$\dot{Q}$) and log standard deviation of ventilation (log SD$\dot{V}$)] and for dispersion indexes derived directly from the measured inert-gas retention and excretion curves (DISP$_R$, DISP$_E$ and DISP$_{R-E}$). For simplicity only log SD$\dot{Q}$ and log SD$\dot{V}$ will be shown in the figures and Table presented here; complete data sets are available in the original references cited above. It is important to put the degree of ventilation-perfusion inequality which develops into perspective. The recent work is consistent in this regard, showing a rise in logSD$\dot{Q}$ of 0.2 from rest to $\dot{V}O_2$ max in healthy normal subjects, as seen in Fig. 1. This amounts to a change in logSD$\dot{Q}$ of 0.05 per liter $\dot{V}O_2$. Care must be taken in presuming the relationship of logSD$\dot{Q}$ to $\dot{V}O_2$ to be linear, however. Fig. 1 does suggest a threshold beyond which $\dot{V}A/\dot{Q}$ mismatching increases more rapidly during normoxic exercise. It may be that $\dot{V}A/\dot{Q}$ inequality remains relatively constant at low levels of exercise, and then deteriorates more rapidly at higher $\dot{V}O_2$, CO, and, presumably, higher pulmonary vascular pressures. This could explain the discrepancy between recent studies and earlier work which had suggested either a minimal increase (Gledhill et al., 1977) or even a reduction (Derks, 1980) in $\dot{V}A/\dot{Q}$ mismatching in subjects exercising at moderate levels (i.e., up to a $\dot{V}O_2$ max of 2.2 l·min^{-1}).

The $\dot{V}A/\dot{Q}$ mismatching measured in normoxia would cause an increase in A-aDO$_2$ from 5 Torr at rest to 10 Torr at $\dot{V}O_2$ = 4 l·min^{-1}, as shown in the left hand panel of Fig. 2. This represents a minor impairment to gas exchange that can be demonstrated to have minimal effects on arterial O_2 content (Hammond et al., 1986b). Also note that below a threshold value ($\dot{V}O_2$ = 3 $l\cdot min^{-1}$) $\dot{V}A/\dot{Q}$ inequality is sufficient to explain the entire A-aDO$_2$. However, when $\dot{V}O_2$ rises above this threshold the observed A-aDO$_2$ rises beyond that attributable to $\dot{V}A/\dot{Q}$ inequality alone. This implies the development of diffusion limitation for oxygen. At $\dot{V}O_2$ = 4 $l\cdot min^{-1}$ diffusion limitation would add an additional 8-10 Torr to the A-aDO$_2$ by this analysis. Thus, at very high workloads in normoxia, the contribution of diffusion limitation to the increase in A-aDO$_2$ is essentially equal to that of $\dot{V}A/\dot{Q}$ mismatching.

HYPOXIA: Studies in acute hypoxia have shown similar trends to those in normoxia, although the results are quantitively different, and the relative contributions of $\dot{V}A/\dot{Q}$ inequality and diffusion limitation appear to change. As expected, $\dot{V}O_2$ max falls in hypoxia. Thus, although the slope of logSD$\dot{Q}$ vs $\dot{V}O_2$ is steeper in hypoxia, the absolute value reached is similar, as shown in Fig. 1. Note also that comparable values of logSD$\dot{Q}$ are reached in normoxia and hypoxia at similar levels of maximum cardiac output and minute ventilation. This is shown in Table I, which summarizes maximum performance data obtained during normoxia and hypoxia in two sets of experiments (Hammond et al., 1986a,b). Taken together, these data imply that at least some of the observed increase in $\dot{V}A/\dot{Q}$ mismatching may be explained by a close linkage between $\dot{V}A/\dot{Q}$ inequality and increased blood (cardiac output) and gas (minute ventilation) convection rates through the lung. However, such linkage is apparently not the entire explanation, because at high altitude some subjects demonstrate marked deterioration in $\dot{V}A/\dot{Q}$ relationships which cannot be explained by a rise in cardiac output or minute ventilation alone (Wagner et al., 1986a).

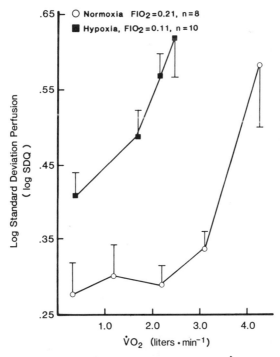

Fig. 1. Log standard deviation of perfusion (logSD$\dot{Q}$) versus $\dot{V}O_2$ for normal subjects during normoxic (open circles) and hypoxic (closed circles) exercise.

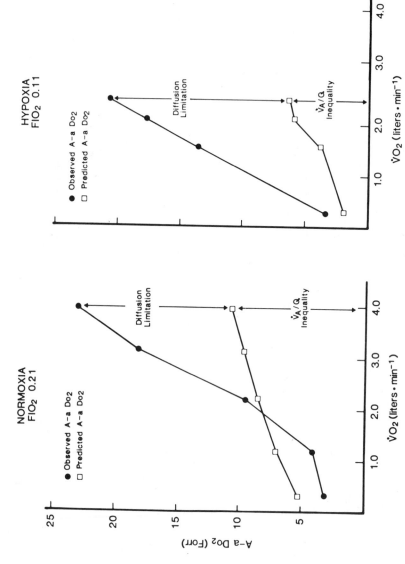

Fig. 2. Measured (closed circles) and predicted (open circles) A-aDO₂ versus V̇O₂ in normal subjects during normoxic (left panel) and hypoxic (right panel) exercise.

Table I

	Rest		Maximal Exercise		Rest		Maximal Exercise	
	$\bar{X}$	(SEM)	$\bar{X}$	(SEM)	$\bar{X}$	(SEM)	$\bar{X}$	(SEM)
HR	59	(1.0)	168	(2.1)	75	(3.3)	170	(2.2)
CO	6.2	(0.1)	24.9	(1.0)	8.3	(0.7)	23.3	(1.2)
$\dot{V}E$	8.5	(0.6)	117.0	(1.0)	8.3	(0.7)	126.5	(6.3)
$\dot{V}O_2$	0.28	(0.02)	3.97	(0.1)	0.29	(0.01)	2.31	(0.1)
PaO_2	99	(1.9)	90	(2.8)	37	(1.4)	36	(1.2)
$A\text{-}aDO_2$ (O)	2.4	(1.7)	23.0	(2.8)	3.4	(1.6)	20.5	(0.7)
$A\text{-}aDO_2$ (p)	5.2	(0.7)	10.7	(2.7)	1.8	(0.2)	6.1	(0.7)
$logSD\dot{Q}$	0.28	(.04)	0.58	(.10)	0.41	(.03)	0.62	(.09)
$logSD\dot{V}$	0.26	(.01)	0.36	(.03)	0.41	(.03)	0.47	(.04)

Diffusion limitation becomes much more pronounced in hypoxia, as shown in the right hand panel of Fig. 2. Note that diffusion limitation for oxygen is evident at all levels of exercise at PIO_2 of 80 Torr, and is likely present even at rest at extreme altitudes (i.e. above 6000 meters). Furthermore, because subjects exercising in hypoxia are functioning on the steep portion of the oxyhemoglobin dissociation curve, significant further arterial oxygen desaturation occurs as a result of diffusion limitation. The importance of this can be seen in Fig. 3 which shows the relationship between $\dot{V}O_2$ and the fall in O_2 content due to diffusion limitation. In hypoxia (solid circles) diffusion limitation contributes to a decrease in arterial O_2 content as large as 2-3 ml per 100 ml blood, or 16% of arterial O_2 capacity. This clearly would pose a significant limit to oxygen delivery. This is in marked contrast to the situation in normoxia, where subjects are operating on the plateau of the oxy-hemoglobin dissociation curve, and the effect of O_2 diffusion limitation on arterial O_2 content is minimal (Fig. 3, open diamond).

It is evident that in acute experiments hypobaric and normobaric hypoxia are qualitatively and quantatively similar at equivalent PIO_2 values, both in terms of $\dot{V}A/\dot{Q}$ inequality and diffusion limitation (Hammond et al., 1986b). This is evidence against important roles for either gas density changes or gas phase diffusion limitation in the development of $\dot{V}A/\dot{Q}$ inequality at the high minute ventilations required during exercise at altitude.

COMMENT

Although inert gas studies during human exercise allow the effects of $\dot{V}A/\dot{Q}$ inequality and diffusion limitation to be estimated, they do not define the primary physiologic events behind the observed disturbances in pulmonary gas exchange. Obviously, $\dot{V}A/\dot{Q}$ inequality can occur only when inhomogeneities of ventilation, perfusion or both develop in the lung. It is tempting to speculate that changes in airway or pulmonary vascular tone could explain our findings during exercise. Unfortunately, these

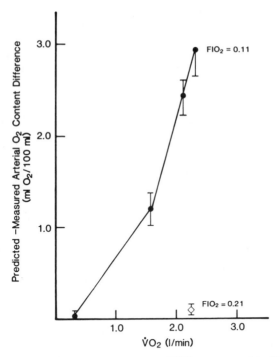

Fig. 3. Arterial O_2 content difference attributable to diffusion limitation in normal subjects during hypoxic exercise (closed circles) and normoxic exercise (open diamond).

mechanisms have not been fully explored in conjunction with the gas exchange experiments conducted to date, and therefore their potential importance cannot be quantitated. However, there is information available which points away from simple broncho- or vaso-active mechanisms and toward a third hypothesis: the development of minor interstitial pulmonary edema. In support of this concept, the slope of $logSD\dot{Q}$ is steeper in hypoxia than in normoxia (Fig. 1) but remains linked to CO, $\dot{V}E$, and also to pulmonary artery pressure (PAP) (Wagner et al., 1986a). Oxygen breathing improves the $\dot{V}A/\dot{Q}$ distribution at similar exercise levels in comparison to hypoxia (Hammond et al., 1986b), and also lowers PAP (Wagner et al., 1986a), but has no effect in normoxia (Hammond et al., 1986a). This strongly suggests that the effect of oxygen is related to the reduction in PAP. Together, these results imply that increased PAP and high CO may be the major factors behind the deterioration in $\dot{V}A/\dot{Q}$ matching observed during exercise. It is known that both elevated vascular pressures and high blood flow can increase fluid flux across the pulmonary vasculature, thereby producing a net increase in interstitial fluid in the lung. Such interstitial edema could cause subtle changes in lung micromechanics, and therefore might affect the $\dot{V}A/\dot{Q}$ distribution measured by MIGET without producing symptoms or other detectable physiologic abnormalities. Other findings which support this interstitial edema hypothesis include: 1) The persistence of $\dot{V}A/\dot{Q}$ abnormalities temporarily after exercise (Hammond et al., 1986a); 2) A reduction in carbon monoxide diffusing capacity following prolonged exercise (Miles et al., 1983); 3) The absence of changes in spirometric indexes following hypoxic exercise (Wagner et al., 1986b); and 4) The development of shunts in some subjects studied recently during chronic hypoxia (Wagner et al., 1986b).

As is the case for $\dot{V}A/\dot{Q}$ inequality, MIGET data cannot address the underlying cause for diffusion limitation, although it is possible to use the data to discount the significance of gas phase diffusion limitation. This is done by looking for systematic molecular-weight-dependent errors in the fit of gases of different molecular weights within the enforced smoothing algorithm used to transform raw retention data into a $\dot{V}A/\dot{Q}$ distribution. Gas phase diffusion limitation would cause systematic elevations in the retention of high molecular weight gases relative to the best fit retention curve. No such errors can be found (Wagner et al., 1986a).

Available data do not allow further partitioning of the site of diffusion limitation between the tissue (Dm) and blood (θVc) components. However, it is evident intuitively that during heavy exercise the high cardiac outputs and low mixed venous PO_2 values that subjects achieve will produce a profound change in the ratio of diffusive to perfusive conductance ($DL/\beta\dot{Q}$), and can push the lung beyond the theoretical limit where diffusion limitation can be predicted to occur (Piiper and Scheid, 1980). These problems are further accentuated in hypoxia, where the oxygen driving pressure is reduced, and the arterial and mixed venous gas tensions are both on the steep part of the oxyhemoglobin dissociation curve, which thus tends to maintain the O_2 tension gradient from alveolar gas to end-capillary blood because the blood capacitance for O_2 is so enhanced (Wagner, 1982). Lastly, although cardiac output increases over a 5 fold range, pulmonary capillary volume likely only doubles during exercise; as a result red blood cell transit times through the pulmonary capillary fall to less than 1/2 to 1/3 of normal resting values (to as low as 0.2 - 0.4 seconds), greatly reducing the time available for complete equilibration of oxygen tension between alveolar gas and end-capillary blood.

In conclusion, it is evident that both $\dot{V}A/\dot{Q}$ inequality and diffusion limitation impair the efficiency of pulmonary gas exchange in humans during heavy exercise. These effects are accentuated by hypoxia and, thus, are greater in subjects at altitude. It is apparent that these mechanisms may pose a limit to O_2 delivery during extreme exercise in normoxia, and clearly pose an increasingly severe limit to O_2 delivery as man ascends to altitude.

ACKNOWLEDGEMENT

Material in Figures 1, 2 and 3 adapted from: J. Appl. Physiol. 60(5):1590-1598, 1986 and 61(5):1749-1757, 1986. Copyright - American Physiological Society, 1986. Used with permission.

REFERENCES

Asmussen, E. and M. Nielsen (1960). Alveolar-arterial gas exchange at rest and during work at different O_2 tensions. Acta. Physiol. Scand. 50:153-166.

Dempsey, J.A., P.G. Hanson and K.S. Henderson (1984). Exercise induced arterial hypoxemia in healthy human subjects at sea level. J. Physiol. Lond. 355:161-175.

Derks, C.M. (1980). Ventilation-perfusion distribution in young and old volunteers during mild exercise. Bull. Eur. Physiopathol. Respir. 16:145-154.

Gale, G.E., J. Torre-Bueno, R. Moon, H.A. Saltzman and P.D. Wagner (1985). $\dot{V}A/\dot{Q}$ inequality in normal man during exercise at sea level and simulated altitude. J. Appl. Physiol. 58:978-988.

Gledhill, N., A.B. Froese and J.A. Dempsey (1977). Ventilation to perfusion distribution during exercise in health. In: Muscular Exercise and the Lung, edited by J. Dempsey and C.E. Reed. Madison WI: Univ. of Wisconsin Press, p. 325-342.

Hammond, M.D., A.E. Gale, K.S. Kapitan, A.R. Ries and P.D. Wagner (1986a) Pulmonary gas exchange in humans during exercise at sea level. J. Appl. Physiol. 60:1590–1598.

Hammond, M.D., G.E. Gale, K.S. Kapitan, A.R. Ries and P.D. Wagner (1986b). Pulmonary gas exchange in humans during normobaric hypoxic exercise. J. Appl. Physiol. 61:1749–57.

Miles, D.S., C.E. Doerr, S.A. Schonfeld, D.E. Sinks and R.W. Gotshall (1983). Changes in pulmonary diffusing capacity and closing volume after running a marathon. Respir. Physiol. 52:349–359.

Piiper, J. and P. Scheid (1980). Blood-gas equilibration in lungs. In: Pulmonary Gas Exchange. Vol. 1 Ventilation, Blood Flow and Diffusion, edited by J.B. West, New York, Academic Press. pp. 131–171.

Torre-Bueno, J., P.D. Wagner, H.A. Saltzman, G.E. Gale and R.E. Moon (1985). Diffusion limitation in normal man during exercise at sea level and simulated altitude. J. Appl. Physiol. 58:989–995.

Wagner, P.D., R.B. Laravuso, R.R. Uhl and J.B. West (1974a). Continuous distributions of ventilation-perfusion ratios in normal subjects breathing air and 100% O_2. J. Clin. Invest. 54:54–68.

Wagner, P.D., H.A. Saltzman and J.B. West (1974b). Measurement of continuous distributions of ventilation-perfusion ratios: theory. J. Appl. Physiol. 36:588–599.

Wagner, P.D. (1982). Influence of mixed venous PO_2 on diffusion of O_2 across the blood:gas barrier. Clin. Physiol. 2:205–215.

Wagner, P.D., G.E. Gale, R.E. Moon, J.R. Torre-Bueno, B.W. Stolp and H.A. Saltzman (1986a). Pulmonary gas exchange in humans exercising at sea level and simulated altitude. J. Appl. Physiol. 61:260–270.

Wagner, P.D., J.R. Sutton, J.T. Reeves, A. Cymerman, B.M. Groves and M.K. Malconian (1986b). $\dot{V}A/\dot{Q}$ inequality at rest and during exercise throughout a simulated ascent of Mt. Everest. Fed. Pro. 45:4232.

TWO-COMPARTMENT ANALYSIS OF $\dot{V}_A/\dot{Q}$ INEQUALITY

AND TRANSIENT BLOOD-GAS DIFFERENCES

J.A. Loeppky, S.A. Altobelli, A. Caprihan and T.W. Chick

Research Division
Lovelace Medical Foundation and Veterans Administration Hosp.
Albuquerque, New Mexico 87108

INTRODUCTION

Numerous $\dot{V}_A/\dot{Q}$ models have been proposed and utilized to explain steady state gas exchange in mammalian lungs. These models vary in the number of conceptualized lung compartments, the experimental sophistication required to obtain the necessary input data, and the complexity of numerical solutions. We have incorporated classical gas exchange concepts and mathematical techniques from other recent models to obtain a simplified two-compartment model which requires only standard gas exchange measurements as input variables. The $\dot{V}_A/\dot{Q}$ ratios and other relevant blood and gas values are then obtained for each compartment which are compatible with the input data (Loeppky et al., 1987).

There were three objectives of these investigations: (a) to test the ability of this two-compartment model to describe steady state changes in $\dot{V}_A/\dot{Q}$ inequality induced in dogs by aerosolized carbachol, known to cause constriction in the small airways; (b) to determine whether this increase in $\dot{V}_A/\dot{Q}$ inequality will potentiate the transient negative blood-gas PCO_2 differences $[(a-A)PCO_2]$ and positive gas-blood PO_2 differences $[(ET-a)PO_2]$ during non-steady states following step changes of inspired PO_2 and PCO_2 by rebreathing as previously reported (Loeppky et al., 1985); and (c) to simulate the experimentally measured time courses of these PCO_2 and PO_2 differences with gas exchange equations assuming that total $\dot{V}_A$, $\dot{Q}$, and the $\dot{V}_A/\dot{Q}$ distribution pattern obtained during steady state remain constant during rebreathing.

METHODS

Gas exchange measurements. Eighteen control experiments (C) were performed on seven anesthetized (Thiopental Na) and paralyzed (Pavulon) dogs. Eight rebreathing experiments were repeated on four of these dogs following the administration of carbachol (CARB) by nebulizer until inspiratory pressures increased by 30%. During baseline steady state in both sets of experiments, duplicate blood samples were taken 15 sec apart just preceding rebreathing and analyzed for pHa, PaO_2, and $PaCO_2$ (Radiometer-BMS 3) at $37.0^\circ C$ and corrected to measured blood temperature (Loeppky et al., 1985). End-tidal PCO_2 ($PACO_2$), PO_2, ($PETO_2$), and mixed expired PO_2 and PCO_2

75

were simultaneously measured by Medspect-II (Chemetron) mass spectrometer. Expired gases were collected in Douglas bags and volumes measured by spirometer. Total cardiac output ($\dot{Q}T$) was calculated from measured $\dot{V}O_2$, assuming a constant arteriovenous O_2 difference. Subsequent studies in our laboratory have shown that the mean ($\pm SE$) arteriovenous O_2 difference and true shunt fraction in six similar dogs were 4.6(0.3) vol% and 0.038(0.007), respectively. The mean PB was 631.1(± 0.5) Torr and FIO_2 was 0.2094.

Numerical techniques for two-compartment $\dot{V}A/\dot{Q}$ analysis. The $\dot{V}A/\dot{Q}$ model requires the simultaneous solution of three gas exchange equations for each compartment at particular $\dot{V}A/\dot{Q}$ ratios. These equations are recent variations (Olszowka and Wagner, 1980) of classical equations (Rahn and Fenn, 1955), taking into account N_2 inequality between alveolar gas and blood in high and low $\dot{V}A/\dot{Q}$ compartments.

$$PN_2 = \frac{P_IN_2[(\dot{V}_A/\dot{Q})PO_2 + 8.63 \ (cO_2 - c_vO_2) + 0.0017 \cdot 8.63 \cdot P_IO_2]}{P_IO_2[(\dot{V}_A/\dot{Q}) + 0.0017 \cdot 8.63]} \tag{1}$$

$$(\dot{V}_A/\dot{Q})PCO_2 = 8.63 \ (c_vCO_2 - cCO_2) \tag{2}$$

$$P_B - 47.1 - PO_2 - PCO_2 - PN_2 = 0 \tag{3}$$

where PO_2, PCO_2, and PN_2 are values in the alveolar gas phase (PO_2 and PCO_2 assumed equal to end-capillary blood gas pressures) with corresponding contents (cO_2 and cCO_2) in vol%, PIN_2 (assumed equal to PvN_2) and PIO_2 are inspired values and cvO_2 and $cvCO_2$ are mixed venous contents. The CO_2 and O_2 dissociation curves are also required. For CO_2, the relationship described by Loeppky et al. (1983) was utiized, where cCO_2 is a function of PCO_2, pH, O_2 saturation (SO_2), and Hb, incorporating a Haldane factor of 0.28. For dog blood, a value of 0.3780 was substituted for 0.4542 in equation 6 of that report as established from other in vitro experiments (Rodkey et al., 1971). For O_2, the dissociation curve for dog blood as described by Reeves et al. (1982) was used with a Bohr factor of 0.498.

The $V_A/\dot{Q}$ ratios for each compartment are obtained as follows:

$$\dot{Q}_T = f_s(\dot{Q}_T) + f_Q \cdot \dot{Q} + (1-f_Q)\dot{Q} \tag{4}$$

where fs is the true shunt fraction, $\dot{Q}$ is total pulmonary flow and fQ is the fractional flow to one compartment. Similarly, $\dot{V}A$ is divided into two compartments

$$\dot{V}_A = f_v \cdot \dot{V}_A + (1-f_v)\dot{V}_A \tag{5}$$

These divisions then establish the $\dot{V}_A/\dot{Q}$ ratio in each compartment, whereby:

$$(\dot{V}_A/\dot{Q})_1 = f_v \cdot \dot{V}_A/f_Q \cdot \dot{Q} \tag{6}$$

and

$$(\dot{V}_A/\dot{Q})_2 = (1-f_v)\dot{V}_A/(1-f_Q)\dot{Q} \tag{7}$$

Table I. Mean (±SE) steady state measurements in control and carbachol-treated dogs prior to rebreathing.

	Control (n=18)	CARB (n=8)
Hb (g%)	15.4 (0.4)	15.4 (0.7)
pH_a	7.490 (0.028)	7.402 (0.047)
P_aO_2 (Torr)	95.3 (3.8)	58.0 (6.4)
P_aCO_2 (Torr)	25.8 (3.0)	28.8 (1.7)
$\dot{V}CO_2$ (ml/min)	71 (5)	59 (8)
$\dot{V}O_2$ (ml/min)	83 (6)	79 (9)
P_ACO_2 (Torr)	22.0 (2.9)	21.8 (1.5)
$\dot{Q}_T$ (L/min)	1.80 (0.12)	1.72 (0.20)
Wt (Kg)	15.1 (1.1)	15.9 (1.0)
temp (°C)	36.2 (0.3)	36.1 (0.3)
(a–A) PCO_2 (Torr)	3.8 (0.6)	7.0 (0.9)
*(a–A) PCO_2 (Torr)	3.1 (0.4)	5.7 (0.6)
$P_{ET}O_2$ (Torr)	97.0 (3.6)	90.4 (3.7)
(ET–a) PO_2 (Torr)	1.7 (2.2)	32.4 (2.0)
*(ET–a) PO_2 (Torr)	7.6 (1.8)	36.0 (5.5)

Note: *Corrected to blood temperature; all other values at 37.0°C

Solutions are obtained at two $\dot{V}A/\dot{Q}$ ratios such that the sum of the two sets of pulmonary capillary blood O_2 and CO_2 contents multiplied by their respective flow fractions, taking into account fs, will equal the contents in the mixed arterial blood. Similarly, the sum of PO_2 and PCO_2 values in the alveolar gas phase of each compartment, multiplied by their respective flow fractions, will equal values measured in the mixed alveolar gas. The two ratios are obtained by a systematic trial-and-error search until the difference (error) between calculated and measured sums of $\dot{V}O_2$ and $\dot{V}CO_2$ is minimal (unique solution). The simultaneous solution of equations at each ratio was performed by an IBM PC with computer routine CO5NBF in the NAG PC50 Fortran library (Loeppky et al., 1987). The model assumes no errors in the measured input variables, no diffusion limitation, and that $PETCO_2$ = $PACO_2$ (Luft et al., 1979). $PETO_2$ is not required.

Rebreathing procedures. During baseline steady state measurements, dogs were ventilated by a constant volume ventilator. Rebreathing was initiated by switching the airway to a bag-in-box initially containing 7% CO_2, 37% O_2 in N_2 in a 3.0-L volume with constant ventilation before and during rebreathing. Rebreathing was continued for 3 min with blood PCO_2 and PO_2 obtained serially from a carotid artery and corrected to blood temperature (Loeppky et al., 1985).

RESULTS AND DISCUSSION

   Steady State. Table I shows the values during steady state in control
and CARB, which served as input to the model. CARB caused a pronounced
hypoxemia and significantly increased (a-A)PCO$_2$ and (ET-a)PO$_2$ by 2.6 and
28.4 Torr, respectively, and also reduced V̇CO$_2$ by 17%. The V̇A/Q̇ analysis
(Table II) from the mean data indicates that in control the low V̇A/Q̇
compartment (L) predominated, accounting for 94% of gas transfer, indicating a nearly "ideal" lung (defined as a single compartment or two compartments with fv and fQ̇ values of 0.5 in each). With CARB, L accounted for 54%
of gas transfer while the underperfused high V̇A/Q̇ compartment (H), now
receives most of the ventilation and causes the net hypoventilation
relative to C. The skewed distribution of fv and fQ̇ between H and L can be
numerically described by weighting H and L by their fractional contribution
to gas exchange (W) to estimate the mean Ln(V̇A/Q̇)

$$M = \text{mean Ln}(\dot{V}_A/\dot{Q}) = W_1[\text{Ln}(\dot{V}_A/\dot{Q})_L] + W_2[\text{Ln}(\dot{V}_A/\dot{Q})_H] \tag{8}$$

The dispersion of the H and L ratios from M is estimated from the second
moment

$$\text{SD Ln}(\dot{V}_A/\dot{Q}) = \sqrt{W_1[\text{Ln}(\dot{V}_A/\dot{Q})_L - M]^2 + W_2[\text{Ln}(\dot{V}_A/\dot{Q})_H - M]^2} \tag{9}$$

These indices were also calculated for V̇A and Q̇ separately by substituting
fv and (1-fv), and fQ̇ and (1-fQ̇) for W_1 and W_2, respectively. These results
(Table II) indicate that the V̇A/Q̇ distribution [SD Ln(V̇A/Q̇)] was doubled by
CARB because of an increase in the distribution of Q̇. This is better
depicted in the graphical representation of the distributions in Fig. 1,
obtained by assuming a Ln normal distribution for V̇A and Q̇. The excessive
perfusion of low V̇A/Q̇ regions is now obvious, a finding previously
documented by Wagner and West (1980) in asthmatic patients and experimental
asthma in dogs (Rubinfeld et al., 1978) with the inert gas elimination
method. They reported bimodal V̇A/Q̇ distributions, which our model can not
discern.

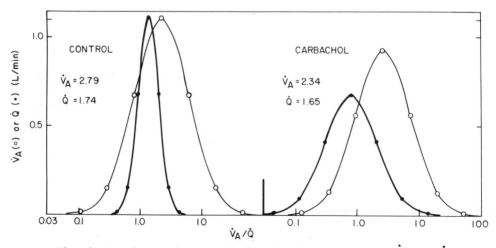

Fig. 1 Continuous Ln normal distribution curves for V̇A and Q̇
        from data in tables.

78

Table II. Results of two-compartment $\dot{V}_A/\dot{Q}$ analyses using first 8 values in Table I as input.

	Control		CARB	
	Low	High	Low	High
$\dot{V}_A/\dot{Q}$	1.27	13.85	0.47	4.72
PO_2 (Torr)	92.3	117.7	56.0	108.1
PCO_2 (Torr)	26.0	9.7	31.9	18.4
pH	7.488	7.785	7.375	7.517
S_{O_2}	0.97	1.00	0.85	0.98
$\dot{V}O_2$ (ml/min)	77	2	53	26
$\dot{V}CO_2$ (ml/min)	64	7	22	37
f_V	0.77	0.23	0.26	0.74
$f_{\dot{Q}}$	0.97	0.03	0.78	0.22
	Mean	SD	Mean	SD
$Ln(\dot{V}_A/\dot{Q})$	0.39	0.59	0.30	1.15
$Ln(\dot{V}_A)$	0.79	1.01	0.95	1.01
$Ln(\dot{Q})$	0.30	0.39	-0.24	0.96

The relative sensitivity of SD $Ln(\dot{V}A/\dot{Q})$ to errors in the input values was tested by assuming a systematic 5% overestimation in each of the 8 input values in Table I and fs while holding all others constant. The resulting average absolute difference in SD $Ln(\dot{V}A/\dot{Q})$ for C and CARB was greatest for $PACO_2$ (17%), followed by $PaCO_2$ (9%), pH (7%) and PaO_2 (5%), with all others below 4%. The error in calculated and measured sums of $\dot{V}O_2$ and $\dot{V}CO_2$ for CARB was zero and 4 ml/min for C.

Rebreathing. In order to achieve a representative display in Figs. 2 and 3, the Pa values are plotted so that their vertical distances from the PET curves correspond to the measured (a-A)PCO_2 and (ET-a)PO_2. These arterial points have been fitted with exponential equations. The (a-A)PCO_2 becomes negative in C by about 3 Torr after 20 to 30 sec of rebreathing, but approaches zero after one min. With CARB this difference is approximately doubled, and equality is not achieved within 3 min. The (ET-a)PO_2 increases in C, but becomes even wider in CARB during rebreathing due to the time lag of PaO_2 relative to $PETO_2$. The time courses of these differences are plotted in Fig. 4, corrected for a time lag of 5 sec (Loeppky et al., 1985). The differences for both PO_2 and PCO_2 are still approximately doubled at 25-30 sec by CARB, similar to the increase in $\dot{V}A/\dot{Q}$ inequality noted during steady state before rebreathing.

Simulation of non-steady state time courses. By assuming that the $\dot{V}A/\dot{Q}$ distributions measured during steady state (Table II) remain constant during rebreathing, gas exchange equations were derived and simultaneously solved at finite time steps independently for CO_2 and O_2.

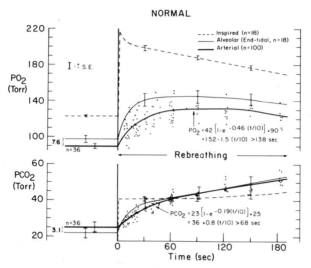

Fig. 2 Mean inspired and end-tidal curves and arterial values
of PO_2 and PCO_2 obtained before and during rebreathing
in 18 experiments.

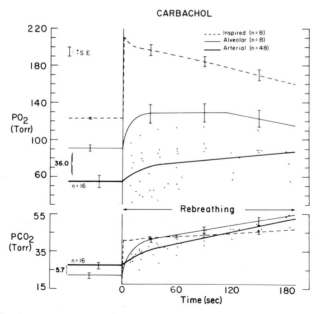

Fig. 3 Mean curves and arterial values obtained just after
aerosolized carbachol in 8 experiments.

$$dP_I/dt = [\dot{V}_H \cdot P_H + \dot{V}_L \cdot P_L - (\dot{V}_H + \dot{V}_L)P_I]/V_{bag} \qquad (10)$$

$$dP_H/dt = [\dot{Q}_H(P_v-P_H)\beta_b/\beta_g - \dot{V}_H \cdot P_H + \dot{V}_H \cdot P_I]/V_H \qquad (11)$$

$$dP_L/dt = [\dot{Q}_L(P_v-P_L)\beta_b/\beta_g - \dot{V}_L \cdot P_L + \dot{V}_L \cdot P_I]/V_L \qquad (12)$$

$$dP_v/dt = [\dot{V}CO_2/\beta_b - P_v(\dot{Q}_H + \dot{Q}_L) + \dot{Q}_H \cdot P_H + \dot{Q}_L \cdot P_L]/V_{bl} \qquad (13)$$

where P is partial pressure (Torr), $\dot{V}$ is ventilation (L/min,BTPS), V is volume (L, BTPS), and $\dot{Q}$ is flow (L/min). VH and VL are alveolar volumes with the total equal to wt x 0.050 and assuming that $VH/\dot{Q}H = VL/\dot{Q}L$ (Briscoe et al., 1960). Vbl is venous blood volume (0.064 x wt). The subscripts I, H, L, and v refer to the inspired (bag), high $\dot{V}A/\dot{Q}$ compartment, low $\dot{V}A/\dot{Q}$ compartment and mixed venous blood. βb is the capacitance coefficient in blood (dissociation curve slope) in ml/L/Torr (Piiper et al., 1971) and βg is the equivalent gas phase capacitance coefficient (1.16 ml/L/Torr). These equations express the conservation of mass and were solved by a 4th order Runge-Kutta method. At each time step PET and Pa are obtained by weighting PH and PL by relative ventilation and perfusion fractions, respectively, which are assumed to remain constant from steady state. βb was obtained at each time step from the derivative of the dissociation curves.

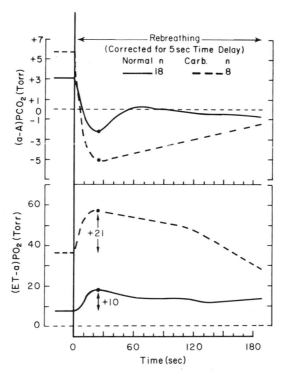

Fig. 4 Mean PCO_2 and PO_2 differences between arterial blood and end-tidal gas observed in Figs. 2 and 3. Arterial values are corrected for an assumed 5-sec lung-to-carotid artery sampling time delay.

The resulting curves are shown in Figs. 5 and 6 for C and CARB. There is good agreement between the simulated and experimental curves in terms of temporal patterns and absolute values. The negative (a-A)PCO_2 values are predicted early during rebreathing and the 3-min values are within 10% of the measured values. A more comprehensive non-steady state model would include Bohr and Haldane factors, a CO_2 storage compartment, and a mixing time delay in Vbl. These would improve the match with the experimental curves (Figs. 2 and 3). This non-steady state model pointed out the importance of the alveolar volume distribution (Piiper and Scheid, 1980) and indicated that a volume distribution in proportion to perfusion gives a reasonable approximation of the CO_2 curves for CARB (Fig. 3) but VH and VL should be more similar in C. The O_2 curves are relatively insensitive to volume distribution.

Conclusions. The two-compartment lung model serves to describe the increase in $\dot{V}_A/\dot{Q}$ inequality during steady state resulting from aerosolized carbachol. The changes are generally consistent with the results obtained from the higher resolution inert gas washout technique. The model serves to predict negative (a-A)PCO_2 differences and increased positive (ET-a)PO_2 differences from $\dot{V}A/\dot{Q}$ inequality following step increases in these gases. Enhanced $\dot{V}A/\dot{Q}$ inequality will potentiate these differences. The prediction of the non-steady state results during rebreathing indicates that this two-compartment description of pulmonary gas exchange is a reasonable alternative to more sophisticated techniques in view of its simplicity.

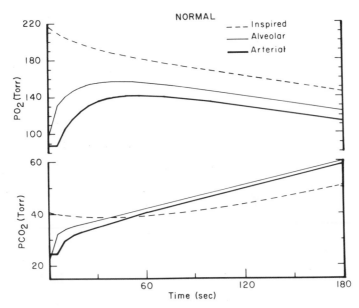

Fig. 5 Rebreathing curves corresponding to Fig. 2, simulated by computer using equations 10-13 and $\dot{V}A/\dot{Q}$ distributions in Table II. A 5-sec time delay is included for arterial values and alveolar volumes were assumed to be proportional to perfusion.

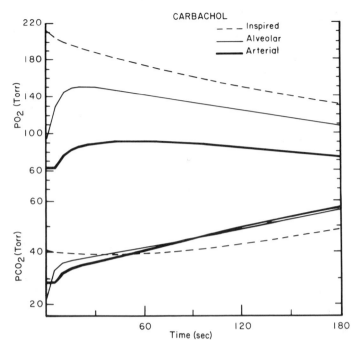

Fig. 6   Simulated curves corresponding to Fig. 3.   See legend of Fig. 5.

REFERENCES

Briscoe, W.A., E.M. Cree, J. Filler, H.E. Houssay and A. Cournand (1960). Lung volume, alveolar ventilation and perfusion interrelationships in chronic pulmonary emphysema. J. Appl. Physiol. 15:785-795.

Loeppky, J.A., A. Caprihan and U.C. Luft (1987). $\dot{V}A/\dot{Q}$ inequality during clinical hypoxemia and its alterations. In: Man in Stressful Environments, edited by K. Shiraki and M.K. Yousef. Springfield, IL, C.C. Thomas, in press.

Loeppky, J.A., U.C. Luft and E.R. Fletcher (1983). Quantitative description of whole blood CO_2 dissociation curve and Haldane effect. Respir. Physiol. 51:167-181.

Loeppky, J.A., P. Scotto and J. Piiper (1985). Transient PO_2 and PCO_2 differences between end-tidal gas and arterial blood during rebreathing in awake dogs. Respir. Physiol. 60:135-144.

Luft, U.C., J.A. Loeppky and E.M. Mostyn (1979). Mean alveolar gases and alveolar-arterial gradients in pulmonary patients. J. Appl. Physiol. 46:534-540.

Olszowka, A.J. and P.D. Wagner (1980). Numerical analysis of gas exchange. In: Pulmonary Gas Exchange, Vol. I, Ventilation, Blood Flow, and Diffusion, edited by J.B. West. New York, Academic Press, pp. 263-306.

Piiper, J., P. Dejours, P. Haab and H. Rahn (1971). Concepts and basic quantitaties in gas exchange physiology. Respir. Physiol. 13:292-304.

Piiper, J. and P. Scheid (1980). Blood-gas equilibration in lungs. In: Pulmonary Gas Exchange, Vol. I, Ventilation, Blood Flow, and Diffusion, edited by J.B. West. New York, Academic Press, pp. 131-171.

Rahn, H. and W.O. Fenn (1955). A Graphical Analysis of the Respiratory Gas Exchange: The O_2-CO_2 Diagram. Washington, D.C., Amer. Physiol. Society.

Reeves, R.B., J.S. Park, G.N. Lapennas and A.J. Olszowka (1982). Oxygen affinity and Bohr coefficients of dog blood. J. Appl. Physiol. 53:87-95.

Rodkey, F.L., H.A. Collison, J.D. O'Neal and J. Sendroy, Jr. (1971). Carbon dioxide absorption curves of dog blood and plasma. J. Appl. Physiol. 30:178-185.

Rubinfeld, A.R., P.D. Wagner and J.B. West (1978). Gas exchange during acute experimental canine asthma. Am. Rev. Resp. Dis. 118:525-536.

Wagner, P.D. and J.B. West (1980). Ventilation-perfusion relationships. In: Pulmonary Gas Exchange. Vol. I, Ventilation, Blood Flow, and Diffusion, edited by J.B. West. New York, Academic Press, pp. 219-262.

OXYGEN TRANSFER IN THE CIRCULATION

DISTRIBUTION OF SKELETAL MUSCLE BLOOD FLOW DURING LOCOMOTORY EXERCISE

M. Harold Laughlin

Department of Biomedical Sciences
College of Veterinary Medicine and Dalton Research Center
University of Missouri
Columbia, MO 65211

INTRODUCTION

If submaximal, rhythmic exercise is to be maintained for an extended period of time (60 min for example), the ATP used to fuel muscular contractions must be provided primarily by oxidative metabolism. As a result, oxygen transfer from air to active skeletal muscle tissue must be matched to the metabolic rate of the muscles. In this regard, up to this point in this symposium, we have considered the determinants of blood oxygenation. The transfer of the oxygenated blood to the skeletal muscle capillaries is the next step in the oxygen transport process. After this convective process, oxygen transfer into the myocytes is determined by the perfused capillarity of the tissue and the determinants of diffusion (i.e., O_2 concentrations in blood and tissue, diffusion distance, etc.).

The purpose of this presentation is to consider one key component of the process of oxygen transport to skeletal muscle during dynamic exercise: the distribution of skeletal muscle blood flow. There is a dramatic increase in blood flow to skeletal muscle at the initiation of dynamic exercise that is accomplished by two primary mechanisms: changes in the distribution of blood flow throughout the body and increases in cardiac output (Rowell, 1974; Clausen, 1976). During intense exercise, blood flow to relatively inactive tissues is decreased. Indeed, it appears that there is an almost linear relationship between exercise intensity and increased resistance to blood flow in the kidneys and splanchnic vascular beds (Rowell, 1986). Cardiac output has also been shown to increase linearly with increasing exercise intensity up to $\dot{V}O_2$ max (Rowell, 1974). This increment in cardiac output combined with the redistribution of blood flow away from metabolically less active tissue can result in 10 to 50 fold increases in blood flow to skeletal muscle (Laughlin and Armstrong, 1985). It appears that the increased oxygen extraction from arterial blood, evidenced by the decreased mixed venous O_2 content, is the result of an increase in the relative amount of cardiac output distributed to active skeletal muscle fibers and to an increase in the number of skeletal muscle capillaries that are perfused with red blood cells. The assumption that skeletal muscle is homogeneous and/or that skeletal muscle perfusion is homogeneous is not uncommon. For example, some of the classical mathematical models describing transport in skeletal muscle vasculature have assumed homogeneous blood flow distribution throughout a homogeneous capillary bed

(Renkin, 1959; Crone, 1963). However, it is clear that skeletal muscle blood flow is not homogeneous, either temporally or spatially within muscle tissue. Since a thorough understanding of skeletal muscle blood flow and its distribution is necessary for a complete understanding of oxygen transfer during sustained locomotory exercise, a second purpose of this paper is to consider the magnitude, importance and causes of regional variations in blood flow within skeletal muscle of mammals during locomotory exercise. It will be helpful to first consider the distribution of blood flow in resting skeletal muscle.

BLOOD FLOW IN RESTING SKELETAL MUSCLE

Blood flow in resting or active muscle is heterogeneous on at least 2 levels of consideration: 1) gross or macrocirculatory and 2) microcirculatory. The first type (gross heterogeneity) refers to the differences in measured blood flow seen among muscles and in 0.1 - 2.0 g samples of muscle with various whole organ or tissue sampling techniques (i.e., microsphere technique). Thus, some muscles have blood flows that are much higher than others under some conditions. Microcirculatory flow heterogeneity refers to differences in flow among individual blood vessels observed in vivo. For example, variations in red blood cell (RBC) velocity among capillaries in a given muscle.

Blood flow has been shown to vary within and among resting skeletal muscles (Hudlicka, 1973; Laughlin and Armstrong, 1985; Mackie and Terjung, 1983; Piiper et al., 1985). Indicator dilution experiments also indicate that blood flow is not uniformly distributed in resting skeletal muscle (Paradise et al., 1971; Sparks and Mohrman, 1977). Some investigators have attempted to deal with the potential influence of blood flow heterogeneity upon solute transport in model analyses as if this is a random phenomenon (Paradise et al., 1971; Kjellmer et al., 1967; Sparks and Mohrman, 1977). However, the causes of this heterogeneity have not been established and, as will be discussed below, not all muscle blood flow heterogeneity is due to "random causes".

Microcirculatory observations indicate that there is considerable heterogeneity of capillary perfusion in resting skeletal muscle. Many of us have observed vasomotion in some microcirculatory experiment or teaching laboratory exercise. Similar vasomotion can be observed in resting skeletal muscles (Tyml and Groom, 1980; Sarelius, 1986). Although there is evidence that the control of capillary perfusion is located in the terminal arterioles (Damon and Duling, 1984; Duling and Klitzman, 1980; Gorczynski et al., 1978; Honig et al., 1982; Klitzman et al., 1982), the mechanisms responsible for the control of capillary perfusion heterogeneity have yet to be established. Thus, while the mechanisms remain to be revealed, blood flow heterogeneity exists in resting skeletal muscle and this heterogeneity can generally be described as a random phenomenon (Damon and Duling, 1984; Duling and Klitzman, 1980; Honig et al., 1982; Kjellmer et al., 1967; Paradise et al., 1971; Sparks and Mohrman, 1977). When skeletal muscles start to contract, due to electrical stimulation or in normal locomotory exercise, the description of muscle blood flow heterogeneity and the causes of heterogeneity become even more complex.

MUSCLE BLOOD FLOW DURING LOCOMOTORY EXERCISE

At the initiation of locomotory exercise there is a rapid increase in muscle blood flow. The magnitude of this increase is related to the intensity of exercise and the relative level of activity of the skeletal muscle tissue being studied. Blood flow during the initial period of the hyperemic response often exceeds the value seen later during sustained exercise (Laughlin and Armstrong, 1983, 1985). Thus, muscle blood flow changes as

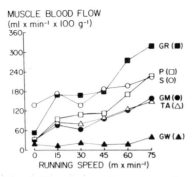

MUSCLE BLOOD FLOW
(ml x min⁻¹ x 100 g⁻¹)

RUNNING SPEED (m x min⁻¹)

Fig. 1. Muscle blood flows as a function of treadmill running
speed for male rats. GR, GM, and GW are red, middle and white
gastrocnemius muscle samples, respectively. P, S, and TA are
plantaris, soleus, and tibialis anterior muscles, respectively. Data are from Laughlin and Armstrong (1982) reprinted
with permission of the American Physiological Society.

a function of time during exercise of constant intensity. The fact that
blood flow is not equally distributed to all skeletal muscle tissue in
conscious subjects during locomotory exercise has been

appreciated for several years (for review, see Rowell, 1986). For example,
blood flow to the muscles of the arm has been shown to decrease in humans
during leg exercise (Blair et al., 1961; Bevegard and Shepherd, 1966).
However, blood flow variation within and among skeletal muscles during
dynamic exercise is more complex than one would predict based simply upon
which muscles are active in a specific type of exercise. Let us consider
the example of a rat walking or running on a motor driven treadmill.

Fig. 1 presents blood flow (BF) data from the leg muscles of rats
performing treadmill exercise (0 degree incline) at different speeds. It
can be seen that blood flow varies among muscles before exercise (i.e.,
soleus BF > all others) and at all intensities of exercise (i.e., at 75
m/min red gastrocnemius muscle BF > all others). It is also clear that the
relative amount of total muscle blood flow going to each muscle varies with
exercise intensity and that flow varies within muscles (i.e., at 15 m/min BF
was increased in the red portion of the gastrocnemius but decreased in the
white portion of this muscle). Similar results can be shown for thigh
muscles of rats and in the extensor muscles of other species of mammal.
Thus, blood flow to skeletal muscle during exercise is distributed within
muscle tissue in a very heterogeneous manner.

What causes skeletal muscle blood flow heterogeneity? As will become
clear from the discussion below, a satisfactory answer to this question can
not be provided with the available data. However, some of the determinants
have been demonstrated: 1) fiber type composition of the muscle; 2) muscle
fiber recruitment patterns within and among muscles; 3) central and local
blood flow control mechanisms; 4) the method of stimulation and the resulting type of contraction of the muscle; and 5) the location of the muscle in
relation to other muscles (i.e., deep versus superficial).

1.   Muscle Fiber Type: Mammalian skeletal muscles are composed of fibers
with different physiologic, morphologic, and biochemical characteristics.
Since several excellent reviews of the properties of the different fiber
types have been written, only a brief overview will be given here (Burke,
1981; Burke and Edgerton, 1975; Saltin and Gollnick, 1983). Although it is
clear that there is a continuum among fibers for each given functional
property, it is common to divide mammalian skeletal muscle fibers into 3

general categories based upon contractile (often myofibrillar ATPase activity) and biochemical characteristics. One commonly used classification system is that proposed by Peter et al. (1972). According to this system of nomenclature, the fibers are identified as slow-twitch oxidative (SO) (also known at type I), fast-twitch oxidative glycolytic (FOG) (also known as type IIa) and fast-twitch glycolytic (FG) (also known as type IIb). Although not all mammals have all three fiber types (i.e., dogs do not have FG fibers and FOG fiber type is less obvious in human muscles (Saltin and Gollnick, 1983), in general, the muscle fiber types are distributed within and among synergistic groups of extensor muscles in similar patterns in most mammalian species (Armstrong, 1980; Armstrong and Laughlin, 1983; Collatos et al., 1977). For example, the deepest muscle in an extensor muscle group is usually composed of a high proportion of SO fibers. The muscles adjacent to the deep SO muscles are usually composed of high proportions of SO and FOG fibers. In the rest of the muscles of the group, the relative percentage of FOG fibers is seen to decrease as one goes from deep to superficial while the percentage of FG fibers increases. The degree of stratification of fiber types among the muscles varies greatly among species. The rat is an example of a mammal with a high degree of muscle fiber type stratification that makes it possible to obtain muscle samples composed primarily of a given fiber type (Armstrong, 1980). This has made it convenient to study relationships between blood flow and fiber type in the skeletal muscles of rats (Armstrong and Laughlin, 1983; Laughlin and Armstrong, 1982; Mackie and Terjung, 1983).

In studies of blood flow to resting and contracting skeletal muscle in situ, it is common to observe a relationship between muscle fiber type and blood flow with the red, high-oxidative muscles having higher blood flows (Mackie and Terjung, 1983; Folkow and Halicka, 1968; Hudkicka, 1973). In addition, morphological evidence indicates that high oxidative skeletal muscles (SO and FOG) have higher capillary densities (Saltin and Gollnick, 1983). Thus, within a given species, the capacity of skeletal muscle for blood flow would be expected to be related to the fiber type composition of the muscle. This is true in the rat. As shown in Fig. 2, we have found that blood flow to resting rat skeletal muscles during maximal papaverine vasodilation is different in muscles composed of various fiber types (Laughlin and Ripperger, 1987). Similarly, the highest blood flows attainable in different rat skeletal muscles with treadmill exercise are linearly related to the succinate dehydrogenase activity (oxidative capacity) of the muscles (Armstrong et al., 1987; Laughlin and Armstrong, 1985). These data clearly indicate that one cause of muscle blood flow heterogeneity within a muscle group, even with uniform vasodilator stimuli, is the relationship between muscle fiber type composition and vascular blood flow capacity. Thus, in the presence of maximal vasodilation, one would not expect blood flow to be distributed homogeneously if the different fiber types are not distributed homogeneously. This applies to a skeletal muscle or muscle group that is stimulated such that all fibers are active. Even if maximal exercise hyperemia is produced, blood flow will not be homogeneously distributed if the fiber type composition of the muscle tissue is heterogeneous. In locomotory exercise the skeletal muscles are not uniformly activated to contract.

2. <u>Muscle Fiber Recruitment Patterns</u>: It is intuitively obvious that the most efficient method for O_2 transport to active muscle during exercise will result from increasing blood flow to the active skeletal muscles rather than increasing flow to all muscle tissue. For example, the data in Fig. 1 indicate that prior to exercise, the soleus muscle, which is most active in maintaining posture (Laughlin and Armstrong, 1982), received much more blood flow than any other muscle. Although it may be less obvious, the same rationale applies within a muscle that is not uniformly active during a bout of exercise. As shown in Fig. 1, blood flow seems to follow muscle fiber

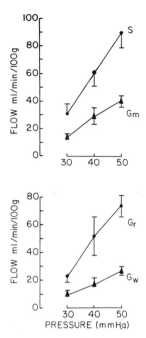

Fig. 2. Blood flow as a function of perfusion pressure in isolated perfused rat hindquarters. Flows were measured during maximal papaverine vasodilation. S = soleus muscle. Gr, Gm, and Gw are red, middle, and white portions of the gastrocnemius muscle. Data are taken from Laughlin and Ripperger (1987).

recruitment patterns both within and among muscles (Laughlin and Armstrong, 1985). Consider the changes in blood flow seen in the different portions of the gastrocnemius muscle when the rats went from rest to walking at 15 m/min (Fig. 1). The blood flow increased to the red portion of the gastrocnemius muscle (GR) and decreased to the superficial white (GW) portion. It should also be noted that blood flow remained high in the soleus muscle when the rats started to walk (Fig. 1). We know from the work of Armstrong et al. (1977) and Sullivan and Armstrong (1978) that the muscle fibers in the GR region of the rat are recruited at this intensity of exercise while those in the GW are not. Thus, just as there appears to be sympathetic vasoconstriction of the vascular beds in inactive skeletal muscles and muscle groups (i.e., the arms during leg exercise, [Blair et al., 1961; Bevegard and Shepherd, 1966]), there also appears to be increased vascular resistance and decreased blood flow to inactive fibers (GW) within muscles of rats during exercise. The result is that the increased blood flow is directed specifically to the active muscle fibers. When rats run at 75 m/min all of the muscle fibers of the ankle extensor muscles should be active (Sullivan and Armstrong, 1978). As expected, blood flow is increased to all muscle tissue (Fig. 1) with the blood flows being greatest in the more oxidative muscle tissue. The data in Fig. 1 illustrate the interactions of the effects of muscle fiber type composition and muscle fiber recruitment patterns upon muscle blood flow distribution patterns. We have observed that skeletal muscle blood flow distribution patterns also follow apparent fiber recruitment patterns in rats during swimming (Laughlin et al., 1984).

Muscle blood flow distribution also changes with time during sustained treadmill exercise (Laughlin and Armstrong, 1983). This may be due to local metabolic and/or temperature effects or to changes in central hemodynamics.

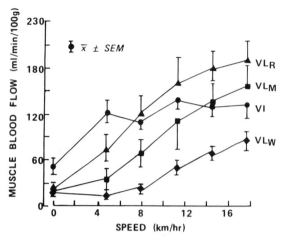

Fig. 3. Muscle blood flows as a function of treadmill running speed in miniature swine. VI is the vastus intermedius muscle and VLr, VLm, and VLw are red, middle, and white portions of vastus lateralis muscle, respectively. Data are from Armstrong et al. (1987) with permission of the American Physiological Society.

However, these other influences seem to be superimposed upon the more dominant effects of fiber type and fiber activity levels.

Finally, as mentioned above, the rat shows an extreme degree of fiber type stratification within and among its skeletal muscles. It would therefore not be surprising to observe that muscle fiber type and recruitment patterns do not have the same effects upon blood flow distribution in other mammals. However, we have observed relationships among blood flow, fiber type, expected fiber recruitment patterns and time during exercise in the skeletal muscles of miniature swine that are qualitatively similar to those of rats (Armstrong et al., 1987). As shown in Fig. 3, the blood flow distribution pattern within and among skeletal muscles and the way this distribution changes with increasing exercise intensity in miniature swine is similar to that shown in Fig. 1 for rats. The major difference between the response of rats (Fig. 1) as compared to pigs (Fig. 3) is that several of the deep high-oxidative muscles of the pig showed plateaus in the relationship between blood flow and exercise intensity. At high speeds, only the peripheral less oxidative muscle tissue continued to show increases in blood flow (Fig. 3). In rats, blood flow continued to increase with running speed in all muscles (Armstrong and Laughlin, 1986). Horses and cattle have blood flow distribution patterns within their muscles that are similar to those of the miniature swine during treadmill exercise (unpublished observations). Thus, heterogeneity of skeletal muscle blood flow during locomotory exercise appears to be a general mammalian phenomenon.

3.   Blood Flow Control: There is evidence that the control of blood flow varies with skeletal muscle fiber type. While it is unlikely that the control mechanisms are totally different in different types of muscle, the relative importance of each control factor may differ among muscles composed of different fiber types. For example, considering central control of blood flow, the vasculature of SO muscle is much less influenced by sympathetic alpha adrenergic activity than is the vasculature of fast-twitch muscle (Folkow and Halicka, 1968; Gray, 1971; Hilton et al., 1970; Laughlin and Armstrong, 1987), whereas fast muscles show greater vasodilation in response to epinephrine infusions (Hilton et al., 1978). In reference to local metabolic control of blood flow, Mellander (1981) has

suggested that blood flow to muscles composed primarily of the FG fiber type may be influenced more by tissue osmolarity than is blood flow to high-oxidative muscle and Hilton et al. (1978) proposed that K^+ may be an important factor in the control of blood flow in FG muscle but not in SO muscle. Finally, we have observed that dipyridamole produces increased blood flow to SO skeletal muscle and respiratory muscle of pigs during exercise but has no effects upon fast-twitch skeletal muscle tissue. These results indicate that the SO and respiratory muscle release more adenosine during exercise and are therefore similar to myocardial muscle (Laughlin and Armstrong, 1986). These examples indicate that caution should be applied in using the assumption that a vasoactive drug will have a homogeneous effect throughout the vasculature of skeletal muscle.

4. **Type of Contraction:** If a skeletal muscle is stimulated so that a maximal sustained tetanic contraction is produced, blood flow will be observed to stop. The hyperemia will only be seen upon relaxation (post-exercise hyperemia [Hudlicka, 1972]). Increases in blood flow are observed with rhythmical stimulation that produces either twitch type contractions or trains of tetanic contractions (Mackie and Terjung, 1983). Under these conditions the interactions among blood flow and the muscle's tension time index and the time between contractions (diastole) are similar to those seen in the coronary circulation.

The magnitude of the exercise hyperemia observed also varies with the fiber type of the skeletal muscle in a manner that would be predicted from the blood flow capacity measured in resting muscle (Mackie and Terjung, 1983). In addition, within a given muscle, the magnitude of the hyperemia varies with the type of contraction produced by the stimulation parameters. For example, Folkow and Halicka (1968) compared the hyperemic responses to muscle contractions at various frequencies in the soleus (SO) and gastro-cnemius (FOG and FG) muscles of cats. They found that soleus blood flow increased linearly with frequency of stimulation up to a frequency of 6-8 Hz. At higher frequencies of stimulation, soleus flow was higher immediately after contraction stopped than during contraction. The highest flow observed in the soleus was 120 ml/min/100g. The gastrocnemius muscle showed a more rapid increase in blood flow with increasing stimulation frequency and a peak flow of about 50 ml/min/100g. The peak flow was the same for twitch or following tetanic contractions in the gastrocnemius muscle. These results demonstrated 3 differences between these 2 types of skeletal muscle: 1) the SO muscle (soleus) had a higher resting blood flow; 2) the SO muscle had a higher peak blood flow; and 3) maximal blood flow was seen in the SO muscle between tetanic contractions while the gastrocnemius muscle had similar peak flows following tetanic contractions and during twitch type contractions.

The aforementioned data raise another important question: What is maximal blood flow? The data of Folkow and Halicka (1968) indicate that maximal muscle blood flows are in the range of 50 to 120 ml/min/100 g. We have seen that muscle flows during treadmill exercise in rats are in the range of 60 to 350 and in pigs 70 to 250 ml/min/100 g. Also, Rowell et al. (1987) have shown that blood flow to the human quadriceps muscle is in the range of 250 - 300 ml/min/100 g during maximal one-leg work in hypoxia. Thus, the skeletal muscle blood flows recorded in models of muscle in exercise (i.e., muscles stimulated to contract under various conditions) are often considerably less than those seen in locomotory exercise. This difference may be partially due to the type of contractions involved in the different forms of muscle activity.

Fig. 4 illustrates the fact that the magnitude of the exercise hyperemia in a given muscle is different when the muscles perform twitch type contractions, trains of tetanic contractions or locomotory exercise.

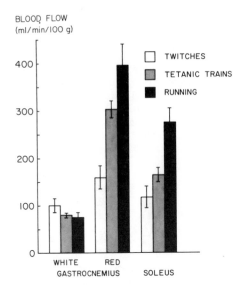

Fig. 4. Peak blood flow to 3 types of rat skeletal muscle during three different types of contraction. The data for the muscles during twitch and tetanic stimulation conditions are from Mackie and Terjung (1983) and those for running rats are from the 75 m/min running speeds of Laughlin and Armstrong (1982). The blood flow values for twitch type stimulation were taken after 10 min of contraction and those for tetanic stimulation were taken after 1 min of contraction (Mackie and Terjung, 1983).

The data in Fig. 4 were all obtained from the muscles of rats. Mackie and Terjung (1983) determined the stimulation parameters that would produce the highest blood flow with each type of contraction. The stimulation was provided via the sciatic nerve. In the white gastrocnemius tissue (85% FG), twitch type contractions produced greater blood flows than tetanic trains or treadmill exercise. In the red portion of the gastrocnemius (30% SO, 60% FOG), blood flow was higher during trains of tetanic contractions (Fig. 4). However, locomotory exercise produced even higher blood flows. Locomotory exercise also produced the highest blood flows in the soleus muscle (Fig. 4). Thus, depending upon the fiber type composition of the muscle, the type of muscle contraction performed may determine the magnitude of the exercise hyperemia observed.

5. Location of the muscle: What causes these differences? It is possible that some of these differences result from mechanical factors due to the location of the muscle within the group in relation to other muscles and to the pennation of the muscles. These mechanical factors could affect blood flow in at least two important ways: increased resistance to blood flow due to mechanical interference during contraction or differences in the efficiency of the muscle pump in assisting perfusion of the muscle.

It has been shown that sustained tetanic contraction can stop blood flow due to mechanical interference (Shepherd, 1983). This effect is greatest in the deep areas of large muscle groups. It also appears that the effect is different in fusiform muscles such as the soleus as compared to pennate muscles such as the gastrocnemius. Since intramuscular pressures vary throughout muscles (with pressures being higher deep in muscle groups and lower in superficial areas; Kirkebo and Wisnes, 1982), these variations will produce different degrees of interference with blood flow throughout the muscle and, therefore, could contribute to blood flow heterogeneity.

On the other hand, the mechanical effect of muscle contraction upon the veins combined with the valves within the veins can also counteract the mechanical interference with blood flow via the "muscle pump" effect. Contraction of skeletal muscle raises intramuscular pressures (Kirkebo and Wisnes, 1982; Petrofsky and Hendershot, 1984). This results in compression of the veins causing blood to flow out of the compressed segments. The venous valves and their orientation toward the heart allow blood to flow out of the compressed venous segments in only one direction. If the muscle maintains a tetanic contraction, the net effect is to increase the resistance to blood flow through the muscle as described above. Since capillaries and precapillary resistance vessels do not appear to be compressed during muscle contraction the increased resistance to blood flow is believed to be due to compression of large arteries and all sizes of veins passing through the muscle (Gray et al., 1967). Rhythmical contractions cause a pumping action on the veins in that the kinetic energy imparted to the venous blood during muscle contraction causes blood to flow out of the veins during contraction (venous outflow) and the pump is refilled from the capillaries and arteries during muscle relaxation (Pollack and Wood, 1949). Thus, as in the coronary vascular bed, arterial inflow occurs during relaxation and venous outflow occurs during contraction (Folkow et al., 1970).

If we return to Fig. 4, it seems possible that the reason that muscle blood flows in high-oxidative muscle tissue are higher during tetanic trains of contraction as compared to twitch type contraction is that the muscle pumping action during twitching is less efficient. It is also possible that the muscle pump is less effective in the white muscle tissue due to its superficial location or to the fact that the vascularity per gram of muscle is less. This could be the reason that twitch contractions produce as much blood flow in this muscle as do the other forms of contraction. Kirkebo and Wisnes (1982) have shown that muscle tissue pressures decrease from deep to superficial within the calf muscles of rats during contraction. Therefore, the vessels in the white superficial muscle may be exposed to a small increment in tissue pressure during contraction and perhaps less muscle pumping action. These data suggest that any type of rhythmic muscle contraction can produce a similar amount of muscle pump contribution to perfusion in FG (white) muscle. However, the data in Fig. 4 also indicate that the situation is different for the high-oxidative muscle tissue. The SO (soleus) and FOG (red gastrocnemius) muscle both appear to have the highest blood flows during treadmill exercise. It is important to note that the arterial perfusion pressures were about 130 mmHg in both studies that produced the data presented in Fig. 4.

Folkow et al. (1970) have shown that the muscle pump only works when venous transmural pressure and/or venous distending pressures were greater than 0. If venous pressures were 0 or negative, muscle contaction only produced an increase in vascular resistance. The relationship between venous pressure and muscle pump efficiency has not been described in further detail. However, it seems reasonable that there would be an optimum venous distending pressure and that this would vary with the hydrostatic column height from the muscle to the animal's heart. The rat clearly does not have much of a hydrostatic column. However, venous pressures are usually above 0 during exercise in rats, so the pump could still work. Since data in this regard are not available, speculation does not seem productive at this time. Suffice it to say that it appears that the muscle pump mechanism may play a role in the perfusion of skeletal muscle during exercise and that this effect appears to be different in various types of muscle tissue.

Blood flow heterogeneity within skeletal muscle can not be completely explained by muscle fiber type, muscle fiber recruitment patterns, types of muscle contraction, and other mechanical effects. Piiper et al. (1985) have shown that considerable blood flow heterogeneity exists within resting and uniformly stimulated dog gastrocnemius muscle. Pendergast et al. (1985) have reported similar amounts of heterogeneity within several skeletal muscles of dogs during treadmill exercise. The blood flow heterogeneity described by these studies does not appear to be related to muscle fiber type, recruitment order or other known functional parameters. These data suggest that this is a random form of blood flow heterogeneity in addition to or superimposed upon the factors discussed above.

Microvascular Flow Heterogeneity: To this point the discussion of blood flow heterogeneity in skeletal muscle has been at the level of whole muscles or in 0.2 -1.0 g samples within muscles. In relation to O_2 transport, much important heterogeneity exists at the microcirculatory level within tissue samples of this size. The microsphere technique may not have the resolution to detect true capillary flow heterogeneity.

The capillary beds of skeletal muscles are heterogeneous in that the capillary lengths, interconnections, tortuosities, permeabilities and extravascular environment vary within and among muscles. Flow within the capillaries is also non-homogeneous. At the microvascular level, it is even difficult to define a perfused capillary. For example, is a capillary that is perfused with plasma but no RBC's a perfused capillary? In reference to O_2 transport, it may be best to consider only capillaries that have RBC's moving through them as "perfused capillaries". However, even with this definition of a perfused capillary, it is necessary to determine how much movement of the red cells constitutes flow.

If capillary flow is defined as RBC movement through the capillary, flow through the capillaries has been observed to be very heterogeneous (Renkin, 1984). As mentioned above, in resting muscle, vasomotion can be seen with groups of capillaries being perfused for a period of time while others are not and vice versa. Damon and Duling (1984) have shown that the number of RBC's per length of capillary varies both in capillaries with RBC movement (flow) and in capillaries with no RBC movement. RBC velocity varies among capillaries and the distribution of RBC velocities changes as a function of time following muscle contraction (Tyml, 1986). Sarelius (1986) has shown that the path taken by RBC's as they pass through a branching capillary network in muscle is longer than would be predicted from anatomical measurements of capillary lengths and capillary RBC velocities. These experiments also demonstrate that RBC mean transit times are longer than have been estimated from more indirect methods and that the distribution of RBC mean transit times is shifted by vasodilation. Also, as shown in Fig. 5, while there is clearly a distribution of RBC functional path lengths in cremaster muscles, neither the path lengths nor their distributions changed during adenosine induced vasodilation (Sarelius, 1986).

The cause and potential controlling factors for microvascular flow heterogeneity are currently of intense interest in the microcirculatory field. For example, Honig et al. (1982) proposed that vasomotor control of capillary density (i.e., the density of perfused capillaries) in skeletal muscle is switched off in an all or none fashion at the start of muscle contraction. They proposed that when a terminal arteriole dilates, all the capillaries down stream are perfused. This conclusion was based upon measurements of the number of capillaries containing RBC's in dog gracilis muscles quick-frozen under various conditions in-situ (Honig et al., 1982). However, Gorczynski et al. (1978), Klitzman et al. (1982) and Damon and Duling (1984) report that perfused capillaries and nonperfused capillaries are often supplied by the same terminal arteriole. Sarelius (1986) proposed that successive segments along branching capillary networks have larger diameters. Thus, the distribution of flow in the network is determined by the interactions of path length and vessel diameters. Sarelius (1986) also indicated that hyperemia may be associated with recruitment of new

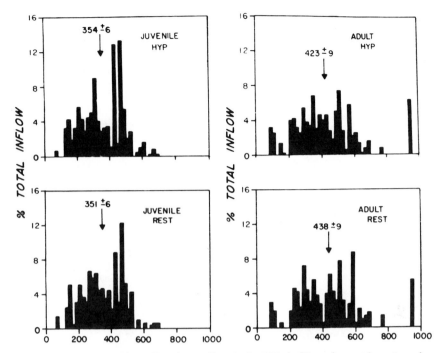

Fig. 5. The distribution of total RBC inflow in each network among different functional flow path lengths in juvenile and adult cremaster muscles of golden hamsters. Resting data for each age are presented on the bottom illustrations and data obtained during hyperemia (HYP) in response to local application of adenosine are presented on top. Mean ± SEM values for functional RBC path length are indicated on each histogram. This figure is reprinted from Sarelius (1986) with permission.

capillary networks. While the amount of information available is astounding, it is clear that much is left to be learned about the control of capillary perfusion and perfusion heterogeneity in active skeletal muscle.

In relation to oxygen transport, the most important parameters probably are the red cell capillary transit time through the portion of the vascular bed that allows exchange, the number of RBC's per length of capillary, the relationship between RBC transit time and surface area available for gas exchange, microvascular oxygen content, and capillary density. If the exchange surface area is heterogeneously distributed, the most efficient blood flow distribution may also need to be heterogeneous. That is, long capillaries should receive more flow than short capillaries (Renkin, 1984). The RBC transit times should be matched to gas exchange area. The transit time for the RBC's varies with the route through the bed (path length). Indeed, measured RBC transit times (2-3 sec) appear to be much longer than whole organ techniques would predict and may be an order of magnitude greater than the minimal time required for unloading O_2 from the RBC (Sarelius, 1986). If a red cell goes through branches or cross-connections between capillaries its transit time and the capillary surface area to which the cell is exposed will be increased as compared to transit directly through the shortest path. It is also likely that the location within the microcirculation where gas exchange occurs will change with flow (i.e., some gas may exchange in the precapillary vessels under resting conditions whereas very little will take this path under high flow conditions).

Many studies are currently being conducted that are making the measurements necessary to allow a more thorough description of capillary blood flow heterogeneity in active muscle. Understanding these phenomena in locomotory muscles will be even more difficult because of obvious technical limitations for applying available microcirculatory techniques to these muscles during exercise. We can not be sure that the muscles best suited for microvascular study are representative of all skeletal muscles. Thus, we do not have adequate data to allow a complete model analysis for locomotory exercise at this time.

In conclusion, for the sake of argument, I have perhaps over-emphasized perfusion heterogeneity. However, blood flow is distributed within and among skeletal muscles in a heterogeneous manner during exercise and, it appears, under most conditions. Some blood flow heterogeneity is due to the fact that blood flow capacity varies with muscle fiber type. Thus, blood flow distribution is often related to muscle fiber type distribution. In active muscles, blood flow distribution is related to both the patterns of fiber type distribution and muscle fiber recruitment. In locomotory exercise some sections of the musculature receive several times as much blood flow as do others. It is clear that many factors can influence muscle blood flow distribution within skeletal muscles. During exercise random factors that produce heterogeneity appear to be dominated by the spatial distribution of the muscle fiber types and recruitment patterns of the fibers within the muscles. In addition, it is likely that considerable heterogeneity exists in capillary perfusion within the muscles during exercise. However, this has not been established because the observation of capillary perfusion within active skeletal muscles during locomotory exercise is still beyond available technical abilities. Depending upon the temporal characteristics of the heterogeneity and the techniques being applied to study blood flow, random heterogeneity can be corrected for with appropriate statistics. The non-random sources of heterogeneity must be described and understood if oxygen transport in muscle is to be completely understood. This is of particular importance in interpreting oxygen transport data dependent upon techniques that measure blood flow at the macrocirculatory level since these observations can not directly assess RBC flux, velocity or spacing within the capillaries.

ACKNOWLEDGMENTS

I thank my good friends and colleagues, R.B. Armstrong and R.J. Korthuis for the many discussions we have had concerning the perfusion of skeletal muscle during exercise. I also thank Don Connor for preparing the illustrations. The author's work is supported by NIH grants HL 36088 and HL 36531 and Research Career Development Award HL-01774.

REFERENCES

Armstrong, R.B. (1980). Properties and distributions of the fiber types in the locomotory muscles of mammals. In Comparative Physiology: Primitive Mammals. K. Schmidt-Neilsen and C.R. Taylor, eds. Cambridge:Cambridge University Press. pp. 243-254.

Armstrong, R.B. and M.H. Laughlin (1983). Blood flows within and among rat muscles as a function of time during high speed treadmill exercise. J. Physiol. 344:189-208.

Armstrong, R.B., M.D. Delp, E.F. Goljan and M.H. Laughlin (1987). Distribution of blood flow in muscles of miniature swine during exercise. J. Appl. Physiol. In Press.

Armstrong, R.B., P. Marum, C.W. Saubert IV, H.W. Seeherman and C.R. Taylor (1977). Muscle fiber activity as a function of speed and gait. J. Appl. Physiol. 43:672-677.

Armstrong, R.B., C.W. Saubert, W.L. Sembrowich, R.E. Shepherd and P.D. Gollnick (1974). Glycogen depletion in rat skeletal muscle fibers at different intensities and durations of exercise. Pflugers Arch. 352:243-256.

Bevegard, B.J. and J.T. Shepherd (1966). Reaction in man of resistance and capacity vessels in forearm and hand to leg exercise. J. Appl. Physiol. 21:123-132.

Blair, D.A., W.E. Gloves and I.C. Roddie (1961). Vasomotor response in the human arm during leg exercise. Cir. Res. 9:264-274.

Burke, R.E. (1981). Motor units: Anatomy, physiology and functional organization. In Handbook of Physiology, The Nervous System, Sect. 1, Bethesda, MD:American Physiological Society, pp. 345-422.

Burke, R.E. and V.R. Edgerton (1975). Motor unit properties and selecive involvement in movement. Exercise Sport Sci. Rev. 3:31-81.

Clausen, J.P. (1976). Circulatory adjustments to dynamic exercise and effect of physical training in normal subjects and in patients with coronary artery disease. Prog. Cardiovas. Diseases 18:359-395.

Collatos, T.C., V.R. Edgerton, J.L. Smith, and B.R. Botterman (1977). Contractile properties and fiber type compositions of flexors and extensors of elbow joint in cat: Implications for motor control. J. Neurophysiol. 40:1292-1300.

Crone, C. (1963). The permeability of capillaries in various organs as determined by use of the "indicator diffusion" method. Acta Physiol. Scand. 58:292-305.

Damon, D.H. and B.R. Duling (1984). Distribution of capillary blood flow in the microcirculation of the hamster: An in vivo study using epifluorescent microscopy. Microvas. Res. 27:81-95.

Duling, B.R. and B. Klitzman (1980). Local control of microvascular function: Role in tissue oxygen supply. Ann. Rev. Physiol. 42:373-382.

Folkow, B., P. Gaskell and B.A. Waaler (1970). Blood flow through limb muscles during heavy rhythmic exercise. Acta Physiol. Scand. 80:61-72.

Folkow, B. and H.D. Halicka (1968). A comparison between red and white muscle with respect to blood supply, capillary surface area and oxygen uptake during rest and exercise. Microvas. Res. 1:1-14.

Gorczynski, R.J., B. Klitzman and B.R. Duling (1978). Interrelations between contracting striated muscle and precapillary microvessels. Am. J. Physiol. 235:H494-H504.

Granger, H.J., A.H. Goodman and D.N. Granger (1976). Role of resistance and exchange vessels in local microvascular control of skeletal muscle oxygenation in the dog. Cir. Res. 38:379-385.

Gray, S.D. (1971). Responsiveness of the terminal vascular bed in fast and slow skeletal muscle to adrenergic stimulation. Angiologica 8:285-296.

Gray, S.D., E. Carlsson and N.C. Staub (1967). Site of increased vascular resistance during isometric muscle contraction. Am. J. Physiol. 213:683-689.

Gruner, J.A. and J. Altman (1980). Swimming in the rat: Analysis of locomotor performance in comparison to stepping. Exp. Brain Res. 40:374-382.

Henneman, E. and L.M. Mendell (1981). Functional organization of moto neuron pool and its inputs. In Handbook of Physiology. Sec. 1, Vol. II. Bethesda, MD, American Physiological Society, pp. 423-507.

Hilton, S.M., M.G. Jefferies and G. Vrbova (1970). Functional specializations of the vascular bed of soleus. J. Physiol. 206:545-562.

Hilton, S.M., O. Hudlicka and J.M. Marshall (1978). Possible mediators of functional hyperemia in skeletal muscle. J. Physiol. 282:131-147.

Honig, C.R., C.L. Odoroff and J.L. Frierson (1982). Active and passive capillary control in red muscle at rest and in exercise. Am. J. Physiol. 243:H196-H206.

Hudlicka, O. (1973). Muscle Blood Flow: Its Relation to Muscle Metabolism and Function. Amsterdam: Swets and Zeitlinger.

Kirkebo, A. and A. Wisnes (1982). Regional tissue fluid pressure in rat calf muscle during sustained contraction or stretch. Acta Physiol. Scand. 114:551-556.

Kjellmer, I., I. Lindbjerg, I. Prerovsky and H. Tonnesen (1967). The relation between blood flow in an isolated muscle measured with the 135 Xe clearance and a direct recording technique. Acta Physiol. Scand. 69:69-78.

Klitzman, B., D.N. Damon, R.J. Gorczynski and B.R. Duling (1982). Augmented tissue oxygen supply during striated muscle contraction in the hamster: Relative contributions of capillary recruitment, functional dilation and reduced tissue PO_2. Cir. Res. 51:711-721.

Laughlin, M.H. and R.B. Armstrong (1982). Muscular blood flow distribution patterns as a function of running speed in rats. Am. J. Physiol. 243:H296-H306.

Laughlin, M.H. and R.B. Armstrong (1983). Rat muscle blood flows as a function of time during prolonged slow treadmill exercise. Am. J. Physiol. 244:H814-H824.

Laughlin, M.H. and R.B. Armstrong (1985). Muscle blood flow during locomotory exercise. Exercise Sport Sci. Revs. 13:95-136.

Laughlin, M.H. and R.B. Armstrong (1986). The effects of dipyridamole on the distribution of muscle blood flow during treadmill exercise in miniature swine. Fed. Proc. 45:1152.

Laughlin, M.H. and R.B. Armstrong (1987). Adrenoreceptor effects on rat muscle blood flow during treadmill exercise. J. Appl. Physiol. 62:in press.

Laughlin, M.H., S.J. Mohrman and R.B. Armstrong (1984). Muscular blood flow distribution patterns in the hindlimb of swimming rats. Am. J. Physiol. 246:H398-H403.

Mackie, B.G. and R.L. Terjung (1983). Blood flow to different skeletal muscle fiber types during contraction. Am. J. Physiol. 245:H265-H275.

Mellander, S. (1981). Differentiation on fiber composition, circulation and metabolism in limb muscles of dog, cat and man. In Vasodilation. P.M. Vanhoutte and I. Lenses, eds., New York:Raven Press, pp. 243-254.

Mia, J.V., V.R. Edgerton and R.J. Barnard (1970). Capillarity of red, white and intermediate muscle fibers in trained and untrained guinea pigs. Experientia 26:1222-1223.

Paradise, N.F., C.R. Swayze, D.H. Shin and I.J. Fox (1971). Perfusion heterogeneity in skeletal muscle using tritiated water. Am. J. Physiol. 220:1107-1115.

Petrofsky, J.S. and D.M. Hendershot (1984). The interrelationship between blood pressure, intramuscular pressure, and isometric endurance in fast and slow twitch skeletal muscle in the cat. Eur. J. Appl. Physiol. 53:106-111.

Pendergast, D.R., J.A. Krasney, A. Ellis, B. McDonald, C. Marconi and P. Cerretelli (1985). Cardiac output and muscle blood flow in exercising dogs. Respiration Physiol. 61:317-326.

Peter, J.V., R.J. Barnard, V.R. Edgerton, C.A. Gillespie and K.E. Stemel (1972). Metabolic profiles of three types of skeletal muscle in guinea pigs and rabbits. Biochemistry 11:2627-2633.

Piiper, J., D.R. Pendergast, C. Marconi, M. Meyer, N. Heisler and P. Cerretelli (1985). Blood flow distribution in dog gastrocnemius muscle at rest and during stimulation. J. Appl. Physiol. 58:2068-2074.

Pollack, A.A. and E.H. Wood (1949). Venous pressure in the saphenous vein at the ankle in man during exercise and changes in posture. J. Apl. Physiol. 1:649-662.

Renkin, E.M. (1959). Transport of potassium-42 from blood to tissue in isolated mammalian skeletal muscles. Am. J. Physiol. 197:1205-1210.

Renkin, E.M. (1984). Control of microcirculation and blood-tissue exchange. Chapter 14, In Handbook of Physiology, Sec. 2, Cardiovascular System. Vol. IV, Microcirculation, Part 2, pp. 627-687.

Rowell, L.B. (1974). Human cardiovascular adjustments to exercise and thermal stress. Physiol. Rev. 54:75-159.

Rowell, L.B. (1986). Human Circulation: Regulation During Physical Stress. Oxford Univ. Press., New York, NY, pp. 1-327.

Rowell, L.B., B. Saltin, B. Kiens and N.J. Christensen (1987). Is peak quadriceps blood flow in humans even higher during exercise with hypoxemia? Am. J. Physiol. 251:H1038-H1044.

Saltin, B. and P.D. Gollnick (1983). Skeletal muscle adaptability: significance for metabolism and performance. In Handbook of Physiology, Skeletal Muscle. Bethesda, MD:Am. Physiological Society, pp. 555-631.

Sarelius, I.H. (1986). Cell flow path influences transit time through striated muscle capillaries. Am. J. Physiol. 250:H899-H907.

Shepherd, J.T. (1983). Circulation to skeletal muscle. In Handbook of Physiology, The Cardiovascular System, Sec. 2, Vol. III: Peripheral Circulation. Bethesda, MD:Am. Physiological Society.

Sparks, H.V. and D.E. Mohrman (1977). Heterogeneity of flow as an explanation for the multi-exponential washout of inert gas from skeletal muscle. Microvas. Res. 13:181-184.

Sullivan, T.E. and R.B. Armstrong (1978). Rat locomotory muscle fiber activity during trotting and galloping. J. Appl. Physiol. 44:358-363.

Tyml, K. (1986). Capillary recruitment and heterogeneity of microvascular flow in skeletal muscle before and after contraction. Microvas. Res. 32:84-98.

Tyml, K. and A.C. Groom (1980). Regulation of blood flow in individual capillaries of resting skeletal muscle in frogs. Microvas. Res. 20:346-357.

Walmsley, B., J.A Hodgson and R.E. Burke (1978). Forces produced by medial gastrocnemius and soleus muscles during locomotion in freely moving cats. J. Neurophysiol. 41:1103-1216.

CIRCULATORY ADJUSTMENTS TO ANEMIC HYPOXIA

Stephen M. Cain and Christopher K. Chapler

Department of Physiology and Biophysics
University of Alabama at Birmingham
Birmingham, Alabama 35294
                    and
Department of Physiology
Queen's University
Kingston, Ontario K7L 3N6

Oxygen transfer in the circulation can be reduced in a number of ways, as Barcroft described many years ago (Barcroft, 1920). Following his nomenclature, we will describe some of the circulatory responses to hypoxic hypoxia (Barcroft's anoxic type), which is a condition of lowered arterial PO_2, and to anemic hypoxia, which is a condition of lowered arterial O_2 concentration. Most attention will be directed toward the latter type because, in the case of experimental hemodilution particularly, there are some fascinating differences from hypoxic hypoxia that occur for reasons that are not immediately obvious.

We start by showing the situation for hypoxic hypoxia in an anesthetized, paralyzed, pump-ventilated dog. By ventilating with 9% O_2 we caused both whole body and limb skeletal muscle O_2 uptake to be supply-limited (Cain and Chapler, 1980). In Fig. 1 are shown the vascular resistances for whole body and hindlimb skeletal muscle in two groups of animals, one of which was given an α-adrenergic blocker with some volume expansion to maintain cardiac output. In the nonblocked group, whole body peripheral resistance fell significantly in the first 20 min of severe hypoxia in which arterial PO_2 was lowered to 25 Torr. This was mostly the result of an increase in cardiac output. At the same time, limb resistance was unchanged. By the end of the 40 min hypoxic period, total resistance fell even further but limb resistance also showed a significant decrease by that time. In the case of the α-blocked group, both limb and total resistance fell together with onset of hypoxia. This illustrates a protective circulatory response to severe hypoxic hypoxia. An increase in sympathetic vasoconstrictor tone redistributes blood flow to essential areas such as heart and brain where that increased tone is less effective than in regions such as gut, kidney, and resting skeletal muscle.

Redistribution of blood flow by differential resistance responses in parallel circuits is a primary circulatory defense against hypoxic hypoxia that may be related to a vestigial diving reflex (Hochachka, 1986). With prolonged time in hypoxia, the increased vasoconstrictor tone is counterbalanced by local vasodilatory factors that are presumably related to the time and intensity of the local imbalance in O_2 supply and demand. The

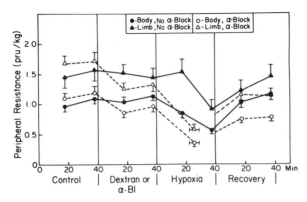

Fig. 1. Total (circles) and limb peripheral resistance (triangles) in nonblocked (filled symbols) and α–block groups (unfilled symbols). Means ± SE. From Cain and Chapler, 1980. (With permission from J. Appl. Physiol.)

centrally mediated increase in sympathetic tone is primarily attributable to peripheral chemoreceptor activity which is increased by any lowering of arterial PO_2 (Daly, 1983).

Contrast the results from hypoxic hypoxia with those from anemic hypoxia (Cain and Chapler, 1978). In these experiments, anesthetized, paralyzed, pump-ventilated dogs were exchange-transfused with 6% dextran to hematocrits of 16% and then to 10%. Whole body O_2 uptake was unaffected at the first level but was significantly decreased at the more severe level in which total O_2 transport (the product of cardiac output and arterial O_2 concentration) was comparable to that measured in severe hypoxic hypoxia. Limb O_2 uptake was unaffected at either level because blood flow to the limb increased in proportion to the increase in cardiac output. The vascular resistance shown in Fig. 2 decreased with hemodilution both in the whole body and in the limb but the percent change was not different between the two measurements. In other words, there was no preferential redistribution of blood flow away from skeletal muscle in severe anemic hypoxia. The reasons for that have occupied us for several years and we don't have all the answers yet (Chapler and Cain, 1986).

One reason that blood flow was apparently not redistributed preferentially in anemic hypoxia may have been related to the fact that hemodilution lowered blood viscosity. The relationship of viscosity to hematocrit was recalculated from data of Fan et al. (1980) and is shown in Fig. 3. In going from an hematocrit of 45% to 10%, viscosity is approximately halved. Vascular resistance is the product of hindrance and viscosity so resistance would have been halved by hemodilution as well. This may so lessen the effect of relatively small changes in resistance vessel diameters that this regulatory response to anemic hypoxia may have become ineffectual.

We decided to test whether increased sympathetic tone could alter vascular resistance and blood flow to hind limb skeletal muscle in anemic hypoxia (Chapler and Cain, 1981). To do so we infused pharmacologic doses (2 μg/kg·min) of norepinephrine (NE) intravenously into anesthetized, paralyzed, pump-ventilated dogs. After a normocythemic control period, NE was infused for 20 min followed by another control period without NE. Hematocrit was then lowered to 9% by exchange of blood for dextran. After 20 min of anemic hypoxia, NE infusion was begun and continued for 40 min. After 20 min, autologous red cells were exchanged until hematocrit was raised to about 25%. A final period of 20 min without NE infusion ended the experiment. A second group was treated exactly the same after being given 1

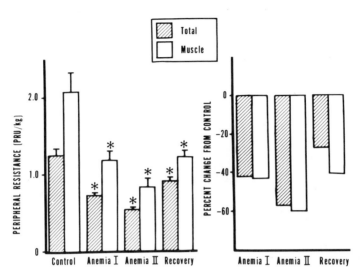

Fig. 2. Total and limb peripheral resistances per unit weight and as percent change from control. Means ± SE. Asterisk denotes a significant difference (P < 0.05) from control. From Cain and Chapler, 1978. (With permission from J. Appl. Physiol.)

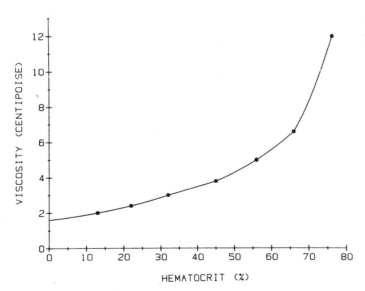

Fig. 3. Relationship between hematocrit and blood viscosity (recalculated from data of Fan et al., 1980).

mg/kg of propranolol to block β-adrenergic receptors (β-block). The reason for this was two-fold. First, β-receptor activity is increased by baroreceptor buffering of any increase in systemic arterial pressure with NE so that peripheral blood flow actually increases and peripheral resistance changes but little. This was not the effect we wished to achieve with NE. Second, Chapler et al. (1979) had shown that anemia actually increases O_2 demand in resting skeletal muscle, an effect that was eliminated by β-block. This would have created an extra metabolic stimulus for local vasodilation during anemic hypoxia in the unblocked group. The use of propranolol solved both of these problems.

Once again skeletal muscle fared better with respect to its O_2 uptake ($\dot{V}O_2$) than did the body as a whole. Whereas whole body $\dot{V}O_2$ decreased significantly with anemia in both groups, limb $\dot{V}O_2$ did not decrease in either group. There was evidence that muscle O_2 demand was limited by supply in the unblocked group because there was a significant rise in limb $\dot{V}O_2$ after hematocrit was raised to 25%, but limb $\dot{V}O_2$ remained unchanged throughout the experiment in the β-blocked group. During anemic hypoxia, NE failed to increase limb vascular resistance in either group. There was, however, a large difference in limb resistance between the groups during recovery at hematocrit of 25% with that in the β-block group being greater.

In the β-block group, there was no evidence of any imbalance between O_2 supply and demand during the anemic period because there was no overshoot in $\dot{V}O_2$ with recovery. Also, the vasodilation seen with NE in the unblocked group was reversed to a strong vasoconstriction during normocythemia but not during anemia. We concluded that even with a strong increase in sympathetic vasoconstrictor tone there was little or no tendency for preferential redistribution of blood flow in anemic hypoxia to favor heart and brain at the expense, in particular, of skeletal muscle.

If that were the case, would any difference be seen if adrenergic vasoconstrictor tone was removed entirely by use of α-adrenergic blockers such as phenoxybenzamine during anemic hypoxia? In another series of experiments, (Chapler and Cain, 1982) $\dot{V}O_2$ and hemodynamic parameters in whole body and limb skeletal muscle were measured sequentially under the following conditions. We first made measurements at normal hematocrit and then exchanged blood for dextran until hematocrit was 15%. Following that set of measurements we gave 3 mg/kg phenoxybenzamine and made the third set of measurements. We measured again after volume expansion with blood-dextran (12 ml/kg). We next exchanged autologous packed cells until hematocrit was 25%. Cardiac output and limb blood flow increased with anemia, decreased with α-block, were restored by volume expansion, and decreased again when hematocrit was raised. After an initial increase in heart rate with hemodilution, the changes in cardiac output were entirely attributable to the changes in stroke volume of the heart. The effects of changes in cardiac output on $\dot{V}O_2$ are shown in Fig. 4. Whole body $\dot{V}O_2$ increased slightly but significantly with anemia, an effect we have seen in other experiments (Cain, 1977; Cain and Chapler, 1985), then decreased with α-block, and returned to the initial value with volume expansion. $\dot{V}O_2$ in limb skeletal muscle did not change at all. Furthermore, after a decrease with hemodilution, neither total nor limb peripheral resistance changed thereafter during anemia.

There was a notable lack of effect of α-block on resistance vessels when blood viscosity was lowered by hemodilution in contrast to the marked effect seen in normocythemic animals in hypoxic hypoxia (Cain and Chapler, 1980). Conversely, there was a very noticeable effect of α-block on capacitance vessels in that stroke volume and cardiac output fell so much that whole body $\dot{V}O_2$ could not be maintained because of the consequent fall in total O_2 delivery. When we compensated for the peripheral pooling by

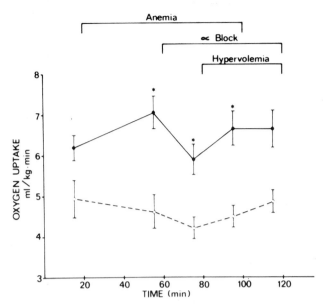

Fig. 4. Whole body (filled symbols, solid line) and limb (unfilled symbols, dashed line) O_2 uptake. Means ± SE. Significant difference from preceding sampling period denoted by the asterisk. From Chapler and Cain, 1982. (With permission from J. Appl. Physiol.)

expanding blood volume, venous return was increased and cardiac output once again became adequate to meet the O_2 demand. Since an increase in cardiac output is the primary compensatory response to anemic hypoxia, the important role of increased sympathetic vasoconstrictor tone is not on the resistance vessels, which would be counterproductive, but rather it is to increase tone in capacitance vessels and to increase venous return thereby. Chapler et al. (1981) provided direct evidence that this can occur even in peripheral circulations when they showed a translocation of fluid volume from the hindlimb to the central circulation during anemic hypoxia.

The story to this point has been consistent in that decreased viscosity associated with anemic hypoxia tended to diminish any regulatory scheme based on differential resistance responses to distribute O_2 transport in a preferential manner. Skeletal muscle, at least, certainly appeared to get its share of a lowered O_2 supply. The NE infusion should have supplied sufficient vasoconstrictor tone but it was non-discriminatory and may not have followed any natural selection to produce a differential effect in different organ systems. In other words, those experiments did not eliminate the possibility that skeletal muscle simply did not receive sufficient vasoconstrictor stimulation during anemic hypoxia to decrease its share of cardiac output. We tried yet another way to augment sympathetic tone during experimental hemodilution using bilateral carotid clamping to activate baroreceptors (Cain and Chapler, 1985). Once again we used β-block to prevent the buffering action of β-vasodilator receptors. Peripheral vascular resistance was raised significantly in the blocked animals by carotid clamping even after hemodilution to hematocrit of 15%. The notable event was that the combination of carotid clamping and anemic hypoxia limited blood flow to limb skeletal muscle sufficiently to decrease its VO_2 significantly even though whole body VO_2 remained unchanged. This result indicated to us that the potential for redistribution of blood flow away from skeletal muscle during anemic hypoxia was present but that there was insufficient stimulus to accomplish it.

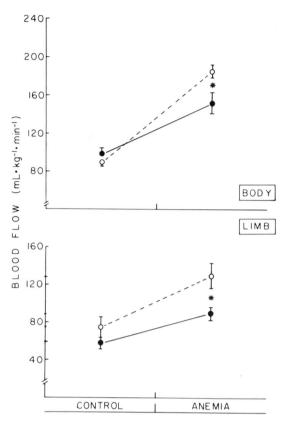

Fig. 5. Cardiac output and limb blood flow before and after
isovolemic hemodilution. Means ± SE. The unfilled symbols
are the sham-operated group and the filled symbols are the
aortic denervated group. Asterisk denotes a significant
difference between groups. From Szlyk et al, 1984. (With
permission from Can. J. Physiol. Pharmacol.)

The chemoreflex mediated by the aortic chemoreceptors does increase
sympathetic activity, however, in response to a decrease in blood O_2 concentration as well as in PO_2 (Daly et al., 1963; Hatcher et al., 1978; Lahiri
et al., 1981). It appeared to be less effective on resistance vessels than
when the carotid body was also stimulated, as in hypoxic hypoxia. Because
the aortic chemoreflex appeared to be mostly involved with the cardiac
output response to anemic hypoxia, we next looked to see how much that
response would be modified by denervating the receptor (Szlyk et al., 1984).
Much of the cardiac output response to anemic hypoxia derives from the
immediate mechanical unloading of resistance vessels when blood is diluted
with cell-free solutions and its viscosity is lowered (Murray and Escobar,
1968; Murray et al., 1969). However, reflex vasoconstriction was deemed a
valuable component by virtue of the shift of volume from capacitance vessels
into the central circulation, as we have described. The aortic body was
deafferented by identifying the vagus nerve in the vagosympathetic bundle
and severing everything else in the manner described by Krasney (1971).
Denervation was tested by brief injection of sodium cyanide into the left
ventricle. The result of aortic body denervation before and after hemodilution to hematocrit of 13% is shown in Fig. 5. The increase in cardiac
output was approximately halved in comparison with a sham operated group.
This had the physiological consequence that VO_2 was decreased significantly
by hemodilution in the denervated group but was not affected in the sham-

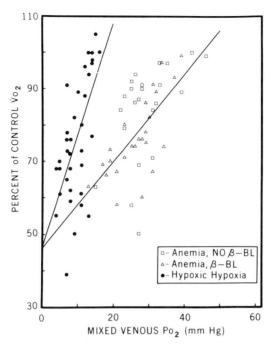

Fig. 6. The O_2 uptake ($\dot{V}O_2$) as a percent of normoxic control value is shown in relation to mixed venous PO_2 in anemic (unfilled symbols) and hypoxic (filled symbols) hypoxia. From Cain, 1977. (With permission from J. Appl. Physiol.)

operated group. Aortic chemoreceptor activity, therefore, was judged to play an essential role in the circulatory responses to experimental hemodilution.

In terms of regional responses to lowering of hematocrit, we have discussed skeletal muscle because it is the area that we have studied most. It also happens to be a major organ system that accounts for 25% of resting blood flow (Wade and Bishop, 1962). Nevertheless, other organ systems may behave differently in their responses to experimental hemodilution. Fan et al. (1980) used microsphere deposition to follow the responses of heart, brain, intestine, liver, kidney and spleen as hematocrit was manipulated over the range of 78% to 12%. When compared to the control hematocrit of 45%, blood flow increased significantly in heart as well as in brain with each lowering of hematocrit below that level. The other organ systems did not share in the general increase of cardiac output and only maintained their flow except in the spleen where flow actually decreased. Small increases in vascular hindrance were observed in intestine and kidney with a marked increase in the spleen. The changes in flow as hematocrit was lowered were never sufficient to compensate for the loss of O_2-carrying capacity in every organ system except the heart which succeeded in maintaining its total O_2 transport down to the lowest measured hematocrit. Indeed, the heart itself will begin to fail as hematocrit is lowered below 10% (Von Restorff et al., 1975) and this sets the critical hematocrit level for the whole body (Cain, 1978).

This brings us to a comparison between critical limits for anemic vs. hypoxic hypoxia. For the first look at this problem, the obvious answer was to seek a common comparator on the venous side of the circulation, the mixed venous PO_2 (Cain, 1977). This assumed that mixed venous PO_2 would approximate the integrated mean tissue PO_2, an idea that was reinforced by Tenney's

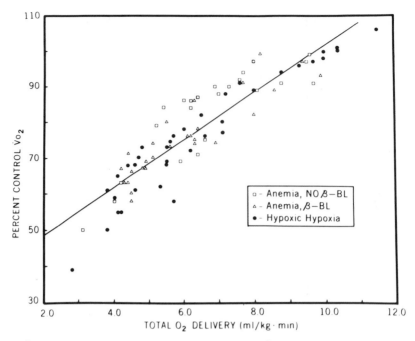

Fig. 7. Percent of control O_2 uptake ($\dot{V}O_2$) in relation to the total O_2 transport in anemic (unfilled symbols) and hypoxic (filled symbols) hypoxia. From Cain, 1977. (With permission from J. Appl. Physiol.)

(1974) theoretical analysis. The experiments were carried out on anesthetized, paralyzed, pump-ventilated dogs. One group was made hypoxic in stages by ventilation with low O_2 gas mixtures. Two other groups were progressively hemodiluted and one of those was pretreated with β-block to produce a further range of cardiac outputs. The results are shown in Fig. 6. A roughly linear relationship between mixed venous PO_2 and the percent of control $\dot{V}O_2$ was found, once $\dot{V}O_2$ was decreased below the normoxic level. The intersection of the line fitted to that relationship with 100% $\dot{V}O_2$ identified a critical PO_2 value; that is, the value of PO_2 below which O_2 uptake was limited by O_2 availability. The fitted lines were distinctly different for the two forms of hypoxia and yielded very different critical values of PO_2, 44.8 torr for anemic and 17.3 torr for hypoxic hypoxia. Conversely, when the relationship between total O_2 transport and percent of control $\dot{V}O_2$ was examined (Fig. 7), both anemic and hypoxic hypoxia data were fitted equally well by a single line. The intersection of that line with 100% $\dot{V}O_2$ occurred at a total O_2 transport of 9.8 ml/kg·min. The reason that there was a common critical transport value but different critical PO_2 values is still not crystal clear. Some contributing factors can be discussed and there may even be a glimmer of a reasonable conclusion.

The implication of the results that were just presented is that O_2 delivery in either anemic or hypoxic hypoxia is not diffusion-limited. If one takes the fact that the critical PO_2 within mitochondria is very small, less than 1 torr, then the venous PO_2 comprises nearly the total diffusion gradient at the venous end of the tissue capillary. This is where O_2 uptake would first become limited by O_2 availability in the classic Krogh analysis. If the system is diffusion-limited below the critical PO_2, then the ratio of $\dot{V}O_2$ and venous PO_2 is the O_2 diffusing capacity or transfer capacity. If they were as different as the two linear relationships shown before for the two kinds of hypoxia, then it seems unlikely that the same critical level of total O_2 transport would have been found. The ratio of $\dot{V}O_2$ and total O_2

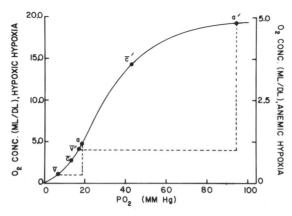

Fig. 8. Oxyhemoglobin dissociation curve (ODC) with normal (left ordinate) and one-fourth normal (right ordinate) O_2 carrying capacity. Arterial (a), mixed venous (v), and mean capillary (c) positions are shown for hypoxic and anemic (primed symbols) hypoxia. From Cain, 1983. (With permission from Clinics in Chest Medicine.)

transport reduces down to the arteriovenous difference in O_2 concentration divided by the arterial O_2 concentration. This defines the O_2 extraction ratio. Since the $\dot{V}O_2$ at the supply-independent plateau was about the same for the two kinds of hypoxia, that meant that there was a common critical O_2 extraction ratio as well. In fact, the arterial and venous O_2 concentrations at the critical levels were similar in the two kinds of hypoxia. The venous O_2 concentration, therefore, represents about the same unused O_2 reserve but the PO_2 at that point has to be higher for anemic than for hypoxic hypoxia because of differences in the oxyhemoglobin dissociation curves (ODC).

The ordinates for the ODC shown in Figure 8 are in units of O_2 concentration, that on the left for hypoxic hypoxia and that on the right for anemic hypoxia. The arterial and mixed venous points on the curve approximate values that were measured in the two types of hypoxia, the primed symbols representing anemia. For the same values of arterial and mixed venous O_2 concentrations, the mixed venous and estimated mean capillary PO_2 are much greater for anemic hypoxia simply because more of the ODC is utilized. With sufficient driving pressure for diffusion, why is not more O_2 unloaded in anemic hypoxia?

This question brings us to a consideration of other factors that may intervene between the tissue and mixed venous PO_2 to cause a shunt-like effect in the periphery. One possibility is that there is such a fast transit of red cells through capillaries in anemic hypoxia that an equilibrium between the red cell and plasma PO_2 is not achieved while the red cell is in the capillary. This would cause tissue PO_2 to be lower than the mixed venous PO_2, once equilibrium has been reestablished, because so little O_2 needs to be dissociated from hemoglobin to raise plasma PO_2 during passage of the blood back to the heart. The quantitative importance of such a process would be very hard to assess except at the microscopic level. There is some evidence against such kinetic disequilibrium creating a large difference between tissue and mixed venous PO_2. Messmer et al. (1973) made direct measurements with multiple microelectrodes and found that tissue PO_2 was maintained or even raised in most organ systems at the lowest level of hematocrit that they used, 20%. Blood viscosity is not very different between hematocrits of 20% and 10%, the critical level. Presumably, red cell transit times would not differ very much either.

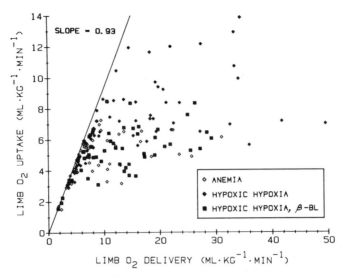

Fig. 9. Limb skeletal muscle O_2 uptake in relation to total O_2 delivery during anemic (unfilled symbols) and hypoxic (filled symbols) hypoxia.

Yet another possibility that might contribute to functional shunting within a regional circulation is diffusional shunting of O_2 between arterioles and venules that may be in proximity to one another. This would be more apt to occur in anemic than hypoxic hypoxia because of the higher gradient for PO_2 between the two sites. The information presented in Fig. 9 would rule against any significant shunting, at least in skeletal muscle. In dogs that were hemodiluted below the critical level of O_2 transport, the limb skeletal muscle was able to extract up to 93% of the available O_2 just as did the dogs that were made hypoxic by ventilation with low O_2 mixtures. A result like this would not have been possible if there was any significant O_2 wastage at the muscle tissue level. The fact that mixed venous PO_2 was always higher than the muscle venous PO_2 (Cain and Chapler, 1978) would indicate that blood with higher O_2 concentration was being added from other circulations. The most likely is the kidney which we know does not decrease its blood flow with hemodilution and which normally receives a much greater blood flow than it needs for its own O_2 demand to subserve its filtering functions.

There is other supportive evidence for the idea that the critical value of mixed venous PO_2 is higher in anemic than in hypoxic hypoxia because of interorgan distribution of O_2 transport. When vasoconstrictor tone was prevented from increasing by α-adrenergic receptor blockade in hypoxic hypoxia, O_2 extraction was never able to improve to the same point that was found in unblocked animals with time in hypoxia (Cain, 1978). This can be seen in Fig. 10 by the much more shallow slope of the solid line in comparison with that of the unblocked animals represented by the dashed line. We propose that an analogous situation existed for anemic hypoxia because much less arterial vasoconstriction was evoked in the absence of carotid body activation. This may have been why blood from highly perfused areas with relatively low O_2 demand, such as the kidneys, contributed more heavily to the mixed venous pool in anemic hypoxia than was the case in hypoxic hypoxia even as O_2 supply became limiting to O_2 uptake in both cases.

To summarize, O_2 transfer from the atmosphere to the tissues is lowered by anemic hypoxia but oxygenation of arterial blood at the lung is normal. A ventilatory response to this form of hypoxia, therefore, would serve

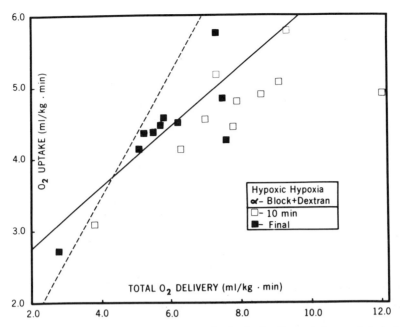

Fig. 10. Relationship between whole body O_2 uptake and total O_2 delivery during severe hypoxic hypoxia. The solid line and all the experimental points were obtained after the animals had been given phenoxybenzamine to block α-adrenergic receptors. The dashed line was obtained from experiments with nonblocked animals at the end of the hypoxic period. From Cain, 1978. (With permission from J. Appl. Physiol.)

little purpose and the primary response is circulatory in nature. Our studies of experimental hemodilution in anesthetized, paralyzed, pump-ventilated dogs may not faithfully characterize chronic forms of anemia found in humans but they have provided us with some fascinating clues to the integrative behavior of regulatory systems when faced with this form of hypoxia. The reflex activity of the aortic chemoreceptor, the mechanical unloading of resistance by the change in viscosity, and possible local metabolic effects are the principal mechanisms to accomplish compensatory actions. Adrenergic controls seem to be of greater importance to increase cardiac output rather than to alter its distribution in the periphery. Mixed venous PO_2 stays higher in anemic hypoxia than in hypoxic hypoxia and this is partly the result of functional shunting at the organ system level. We as yet have no direct information pertaining to the microcirculation but it obviously responds to any threat of tissue hypoxia by arranging itself for efficient utilization of any available O_2. The extent to which capillary hematocrits are altered by hemodilution in the ranges that we have studied still needs to be ascertained and will be necessary to complete our analysis of O_2 transport to tissue in anemic hypoxia.

ACKNOWLEDGMENT

The original research reported here was supported in part by Grant HL 14693 from the National Heart Lung and Blood Institute and by grants from the Medical Research Council of Canada.

REFERENCES

Barcroft, J. (1920). Presidential address on anoxaemia. Lancet 199 (II):485–489.

Cain, S.M. (1977). Oxygen delivery and uptake in dogs during anemic and hypoxic hypoxia. J. Appl. Physiol.: Respirat. Environ. Exercise Physiol. 42:228–234.

Cain, S.M. (1978). Effects of time and vasoconstrictor tone on O_2 extraction during hypoxic hypoxia. J. Appl. Physiol.: Respirat. Environ. Exercise Physiol. 45:219–224.

Cain, S.M. (1983). Peripheral oxygen uptake and delivery in health and disease. Clin. Chest. Med. 4:139–148.

Cain, S.M. and C.K. Chapler (1978). O_2 extraction by hind limb versus whole dog during anemic hypoxia. J. Appl. Physiol.: Respirat. Environ. Exercise Physiol. 45:966–970.

Cain, S.M. and C.K. Chapler (1980). O_2 extraction by canine hindlimb during α-adrenergic blockade and hypoxic hypoxia. J. Appl. Physiol.: Respirat. Environ. Exercise Physiol. 48:630–635.

Cain, S.M. and C.K. Chapler (1985). Hindlimb vascular responses to sympathetic augmentation during acute anemia. Can. J. Physiol. Pharmacol. 63:782–786.

Chapler, C.K. and S.M. Cain (1981). Blood flow and O_2 uptake in dog hindlimb with anemia, norepinephrine, and propranolol. J. Appl. Physiol.: Respirat. Environ. Exercise Physiol. 51:565–570.

Chapler, C.K. and S.M. Cain (1982). Effects of α-adrenergic blockade during acute anemia. J. Appl. Physiol.: Respirat. Environ. Exercise Physiol. 52:16–20.

Chapler, C.K. and S.M. Cain (1986). The physiologic reserve in oxygen carrying capacity: studies in experimental hemodilution. Can. J. Physiol. Pharmacol. 64:7–12.

Chapler, C.K., S.M. Cain and W.N. Stainsby (1979). Blood flow and oxygen uptake in isolated canine skeletal muscle during acute anemia. J. Appl. Physiol.: Respirat. Environ. Exercise Physiol. 46:1035–1038.

Chapler, C.K., W.N. Stainsby and M.A. Lillie (1981). Peripheral vascular response during acute anemia. Can. J. Physiol. Pharmacol. 59:102–107.

Daly, M. de B. (1983). Peripheral arterial chemoreceptors and the cardiovascular system. In: Physiology of the Peripheral Arterial Chemoreceptors. H. Acker and R.G. O'Regan (eds.). Elsevier Science Publishers, B.V. Amsterdam, pp. 329–393.

Daly, M. de B., J.L. Hazeldine and A. Howe (1963). Stimulation of the isolated perfused aortic bodies in the dog: reflex peripheral vascular responses. J. Physiol. (London). 169:89P–90P.

Fan, F.-C., R.Y.Z. Chen, G.B. Schuessler and S. Chien (1980). Effect of hematocrit variations on regional hemodynamics and oxygen transport in the dog. Am. J. Physiol. 238:H545–H552.

Hatcher, J.D., L.K. Chiu and D.B. Jennings (1978). Anemia as a stimulus to aortic and carotid chemoreceptors in the cat. J. Appl. Physiol.: Respirat. Environ. Exercise Physiol. 44:696–702.

Hochachka, P.W. (1986). Balancing conflicting metabolic demands of exercise and diving. Fed. Proc. 45:2948–2952.

Krasney, J. (1971). Cardiovascular responses to cyanide in awake sinoaortic denervated dogs. Am. J. Physiol. 220:1361–1366.

Lahiri, S., E. Mulligan, T. Nishino, A. Mokashi and R.O. Davies (1981). Relative responses of aortic and carotid body chemoreceptors to carboxyhemoglobinemia. J. Appl. Physiol.: Respirat. Environ. Exercise Physiol. 50:580–586.

Messmer, K., L. Sunder-Plassmann, F. Jesch, L. Goernandt, E. Sinagowitz and M. Kessler (1973). Oxygen supply to the tissues during limited normovolemic hemodilution. Res. Exp. Med. 159:152–166.

Murray, J.F. and E. Escobar (1968). Circulatory effects of blood viscosity: comparison of methemoglobinemia and anemia. J. Appl. Physiol. 24:594–599.

Murray, J.F., E. Escobar and E. Rapaport (1969). Effects of blood viscosity on hemodynamic responses in acute normovolemic anemia. Am. J. Physiol. 216:638-642.

Szlyk, P.C., C. King, D.B. Jennings, S.M. Cain and C.K. Chapler (1984). The role of aortic chemoreceptors during acute anemia. Can. J. Physiol. Pharmacol. 62:519-523.

Tenney, S.M. (1974). A theoretical analysis of the relationship between venous blood and mean tissue oxygen pressures. Respir. Physiol. 20:283-296.

Von Restorff, W., B. Hoefling, J. Holtz and E. Bassenge (1975). Effect of increased blood fluidity through hemodilution on coronary circulation at rest and during exercise in dogs. Pfleugers Arch. 357:15-24.

Wade, O.L. and J.M. Bishop (1962). Cardiac Output and Regional Blood Flow. Blackwell Scientific Publications Ltd., Oxford.

RED CELL PROPERTIES AND OPTIMAL OXYGEN TRANSPORT

Robert M. Winslow

Blood Research Division
Letterman Army Institute of Research
Presidio of San Francisco
California 94129-6800

INTRODUCTION

It is clear that the oxygen-binding behavior of hemoglobin can affect overall oxygen transport, because certain genetic alterations in the molecule may lead to polycythemia (Winslow and Anderson, 1982). Furthermore, an extreme left-shift in the oxygen equilibrium curve (OEC) caused by severe respiratory alkalosis was an essential feature in oxygen transport on the summit of Mt. Everest (Winslow et al., 1984), confirming the importance of blood oxygen affinity.

However, demonstration that subtle changes in the position or shape of the OEC may affect oxygen transport have been more theoretical than real (Neville, 1976; Bencowitz et al., 1982), and often based on inadequate models of hemoglobin oxygenation. Other known red cell properties that affect oxygen transport include the rates of binding of physiological ligands (O_2, CO_2, H^+, 2,3-DPG), buffering capacity, the barrier to diffusion presented by the red cell membrane, the layer of unstirred plasma immediately surrounding the red cells, and hematocrit-dependent viscosity.

In this paper, the interaction of some of these red cell properties with other physiological determinants of oxygen transport will be considered. After a description of a mathematical model for these interactions, the model will be evaluated using human data.

Human gas exchange data were collected over several years of studies in Chronic Mountain Sickness, or Monge's disease. This interesting clinical entity appears to be an overreaction of the body to produce red cells in response to a hypoxic stimulus (Winslow and Monge, 1986). If untreated, patients develop cardiac failure resulting in part from an extremely large volume of viscous blood. Treatment consists of descent to low altitude or phlebotomy. To date, no etiology has been identified and we currently believe that this "disease" is simply an overexuberant normal mechanism. The interaction of red cell, circulatory, and pulmonary variables in this disorder makes it an ideal test of any comprehensive model of O_2 transport.

COMPUTATIONS

The general approach used in these calculations is to compute gas exchange profiles along the pulmonary capillary. These calculations are based on the work of Wagner (1977) as recently described (Winslow, 1986). While it may be more useful to compute gas exchange along tissue capillaries, this is a more complex procedure because of the additional countercurrent exchange that occurs between arterioles and venules. In the present calculations, we first estimate the mixed venous acid-base conditions, the cardiac output, and the oxygen equilibrium curves under arterial and venous conditions. Then, over discreet finite time elements, new conditions are estimated as the red cell traverses the pulmonary capillary.

Acid-base Calculations

Equations from Thomas (1972) are used for acid-base calculations, including base excess (BE) and blood buffer base (BBB). In all cases, temperature is assumed to be $37^{\circ}C$. Significant differences from the Thomas procedures, however, are the inclusion of the SO_2 calculated from the OEC and PO_2, as described below, and the inclusion of the equations of Kelman (1966, 1967) for the calculation of total CO_2 concentration, $[CO_2]$, and its distribution across the red cell membrane.

Conversions between pH and PCO_2 are carried out using Kelman's equations, in which $[CO_2]$ and pH are known, and the distribution $[HCO_3^-]_i/[HCO_3^-]_o$ is calculated from the hemoglobin saturation.

Oxygen Equilibrium Curves

The model we use for the binding of O_2 to hemoglobin was first described by Adair (1925). The functional form of this model is given by the equation:

$$Y = \frac{a_1 p + 2a_2 p^2 + 3a_3 p^3 + 4a_4 p^4}{4(1 + a_1 p + a_2 p^2 + a_3 p^3 + a_4 p^4)} \tag{1}$$

where p is PO_2 and the a_i's are related to the equilibrium constants for the successive binding of O_2 molecules. We previously described computer methods for fitting this model to experimental data (Winslow et al., 1977).

Also, in a previous publication, we described algorithms for the computation of a set of Adair a's for any combination of 2,3-DPG/Hb molar ratio, PCO_2, and pH (Winslow et al., 1983). Thus, using the a's for the chosen conditions, we can calculate SO_2 from PO_2 using the Adair equation, or PO_2 from SO_2 using Newton-Raphson approximation. $[CO_2]$ is calculated from pH, PCO_2 and hematocrit, using Kelman's equations.

Cardiac Output

The data of Guyton et al., (1973) obtained in dogs suggests a nearly linear relationship between resting cardiac output and hematocrit, between hematocrit values of 0 and about 60%. We use this relation to derive an empirical equation for prediction of the cardiac output.

Resting cardiac output $(\dot{Q}_0)$, taken from Guyton et al., (1973) can be defined by the third order polynomial

$$\dot{Q}_0 = A + B(Hct) + C(Hct)^2 + D(Hct)^3 \tag{2}$$

where the empirically-derived parameters are:

$$A = 176.1971$$
$$B = 195.3331$$
$$C = 194.1772$$
$$D = 211.858$$

and $\dot{Q}_0$ is ml/min·kg. Jones et al., (1975) showed a linear relationship between $\dot{Q}$ and $\dot{V}O_2$ with a slope

$$d\dot{Q}/d\dot{V}O_2 = 7.6 \tag{3}$$

If the data obtained by Jones et al., (1975) are derived from subjects with hematocrits of approximately 45%, and if $d\dot{Q}$ follows the same Hct relationship as resting $\dot{Q}$, then

$$d\dot{Q}/d\dot{V}O_2 = 7.6 \times (\dot{Q}_0/\dot{Q}_{45}) \tag{4}$$

$\dot{Q}_{45}$, the cardiac output at hematocrit 45%, is a constant. From eq. 2, its value is 68.3 ml/min·kg. Also, the relation between $\dot{Q}$ and $\dot{V}O_2$ has a $\dot{Q}$ intercept of about 40 ml/min·kg. Thus, the final expression for $\dot{Q}$ at Hct, $\dot{V}O_2$, and body weight (wt) becomes

$$\dot{Q} = [0.1113 \times \dot{Q}_0 \times \dot{V}O_2 + 40] \times wt \tag{5}$$

where $\dot{Q}_0$ is determined by eq. 2. Both $\dot{Q}_0$ and $\dot{V}O_2$ are ml/min·kg.

Mixed Venous Blood

Oxygen concentration of mixed venous blood can be obtained from the Fick Equation:

$$\dot{V}O_2 = \dot{Q} \times (CaO_2 - C\bar{v}O_2) \tag{6}$$

where CaO_2 is

$$CaO_2 = 1.34 \times [Hb] \times SaO_2 \tag{7}$$

and SaO_2 is a function of PaO_2, pH, [2,3-DPG]/[Hb], and $PaCO_2$. $C\bar{v}O_2$ can be calculated using the equation

$$C\bar{v}O_2 = CaO_2 - [\dot{V}O_2/\dot{Q}] \tag{8}$$

Mixed venous O_2 saturation is

$$S\bar{v}O_2 = C\bar{v}O_2/O_2CAP \tag{9}$$

119

where $O_2CAP = [Hb] \times 1.34$)

Using the gas exchange ratio, R, $C\bar{v}CO_2$ is also defined by the Fick Equation:

$$\dot{V}CO_2 = R \times \dot{V}O_2 = \dot{Q} \times (CaCO_2 - C\bar{v}CO_2) \qquad (10)$$

The change in oxygen saturation alters the BE. If we assume mixed venous blood is at thermochemical equilibrium, then pH and PCO_2 can be estimated by an iterative procedure from the equations of Thomas, knowing $C\bar{v}CO_2$ and BE.

Bohr Integration

The O_2 uptake progress curve is calculated in the classical manner, at time steps of 0.02 sec. Our procedure differs from previous ones because we are not restricted by the Hill equation to describe the OEC; this procedure is only accurate over a narrow saturation range. We also consider the cardiac output dependence on hematocrit and $\dot{V}O_2$. A new OEC is calculated at each step, so that the new saturation can be used in the acid-base calculation.

The pulmonary transit time (TCAP) is calculated from the cardiac output and the pulmonary capillary volume:

$$TCAP = Vc \; / \; \dot{Q} \qquad (11)$$

pH

We assume a rather slow rate of pH equilibration in the post-capillary blood (Hill et al., 1977; Bidani and Crandall, 1978; Bidani et al., 1978). For the present purpose, we do not consider the "downstream" equilibration problem (Jones et al., 1975). Both Bidani et al., (1978) and Hill et al., (1977) found that the lack of carbonic anhydrase results in a slow pH equilibration in plasma. We have used the experimentally derived value of 7s for a half time of pH change, based on the data of Hill et al., (1977).

Oxygen

The flux of O_2 into the pulmonary capillary blood in the ith element of the Bohr integration is proporational to the diffusing capacity of the lung and the difference in alveolar and capillary O_2 concentrations.

$$\frac{d(O_2)_i}{dt} = \frac{100}{Vc} \times \frac{DLO_2}{60} \times (PAO_2 - P_cO_2) \qquad (12)$$

Roughton and Forster (1957) described the components of DLO_2

$$\frac{1}{DLO_2} = \frac{1}{DMO_2} + \frac{1}{\phi O_2 \times V_c} \qquad (13)$$

where ϕO_2 is the reaction rate with hemoglobin, and Vc is the volume of capillary blood. DLO_2 and DMO_2 are the lung and membrane diffusion capacities, respectively.

120

These authors also provided the data available for the rate of reaction of O_2 with blood (ϕO_2). ϕO_2 can be calculated from O_2CAP (Staub et al., 1962). We use linear interpolation to find k'c for a given saturation, then calculate ϕO_2

$$\phi O_2 = k'c \times (1-SO_2) \times O_2CAP \times (0.779 \times 10^{-3}) \qquad (14)$$

where ϕO_2 has the dimension ml O_2/ml blood/min/torr.

Carbon dioxide

Changes in total CO_2 concentration follow an equation analogous to that for O_2.

$$\frac{d[CO_2]_i}{dt} = \frac{100 \times}{Vc} \times \frac{DLCO_2}{60} \times (P_cCO_2 - PACO_2) \qquad (15)$$

Where the definition of $DLCO_2$ is analogous to that of DLO_2:

$$DLCO_2 = \frac{1}{DMCO_2} + \frac{1}{\phi CO_2 \times Vc} \qquad (16)$$

We assume $DMCO_2 = 20 \times DMO_2$ (Wagner, 1977). ϕCO_2, the overall reaction rate for CO_2 in blood, is taken as a simple exponential with half time of 150 ms (Wagner 1977).

Estimation of Alveolar Gas Composition

From Asmussen and Nielsen (1956)

$$PAO_2 = PIO_2 - \frac{PACO_2 \times (PIO_2 - PEO_2)}{PECO_2} \qquad (17)$$

and, using the Bohr equation

$$PaCO_2 = \frac{PECO_2}{(1-VD/VT)} \qquad (18)$$

PAO_2 becomes a function of PIO_2, PEO_2, and the VD/VT ratio

$$PAO_2 = PIO_2 - \frac{(PIO_2 - PEO_2)}{1-(VD/VT)} \qquad (19)$$

VD/VT was measured in one of our high-altitude subjects and found to vary between 0.1 and 0.075 during exercise. In the subsequent calculations of PAO_2 it was assumed to be 0.1 and constant during exercise.

Computer Program

Using input arterial blood data, the program uses the Kelman routines, the OEC simulation, and cardiac output calculation to compute arterial and venous blood gas contents, tensions, and pH. Mixed venous blood gas tensions and pH are computed by an iterative procedure. First O_2 concentration is calculated from $\dot{Q}$ and $\dot{V}O_2$, then CO_2 concentration is calculated from O_2 concentration and the gas exchange ratio, R. Second, pH is corrected for oxyhemoglobin saturation, and third, PCO_2 is systematically increased until a pH-PCO_2 pair is found that satisfies the BE. After mixed venous conditions are set, the diffusion equations are used to calculate new O_2 and CO_2 contents. The new acid-base parameters and a new OEC are calculated and new equilibrium values are determined as above. The process is repeated until the capillary transit time is reached.

Oxygen uptake progress curves were computed at hematocrits from 5 to 85% over the 2,3-DPG/Hb ratio range 0.3-1.5 mol/mol. At each 2,3-DPG/Hb value, curves were calculated at each hematocrit in steps of 5%. The constraints used were that 1) $C\bar{v}O_2$ and $P\bar{v}O_2$ must be positive, 2) end capillary PO_2 must reach the measured PaO_2, and 3) acid-base conditions must be found such the BEBv = BEBa. In the case that condition 2) was not satisfied, $\dot{V}O_2$ was decreased by 10% of its initial value and the computations were retried. If all conditions could not be met, a value of 0 was assigned to $P\bar{v}O_2$ and the process continued.

To study the interaction of red cell oxygen affinity with hematocrit and viscosity (cardiac output), triaxial (hematocrit, 2,3-DPG/Hb, $P\bar{v}O_2$) plots were made. The gas transport measurements in high altitude natives are summarized in Tables I-III and the input values to the optimization program are shown in Table IV.

GENERAL DESCRIPTION OF MODEL OUTPUT

In order to describe the utility of the model, we will briefly describe the effects of variation in input parameters on the model results. A preliminary version of this model was described in a previous publication (Winslow, 1986).

Red Cell Variables

2,3-DPG/Hb

Study of variations in the 2,3-DPG/Hb molar ratio is a much more instructive procedure than simply shifting the OEC without altering its shape (Bencowitz et al., 1982) since under physiologic conditions 2,3-DPG/Hb is the main modulator of the position and shape of the OEC. Increasing 2,3-DPG/Hb has the effect of increasing $P\bar{v}O_2$, but also of increasing the time required for oxygenation, since the oxygen affinity is decreased. During exercise these two effects are illustrated more clearly. As 2,3-DPG/Hb increases, the time needed to oxygenate approaches the pulmonary capillary transit time, and at a 2,3-DPG/Hb ratio of 1.4 mol/mol Hb, the program could not satisfy the input conditions without decreasing $\dot{V}O_2$.

In hypoxia, the effect of 2,3-DPG/Hb is even more apparent. At rest, decreasing 2,3-DPG/Hb allows the blood to become oxygenated within the allowed transit time, but increasing even above 0.4 mol/mol leads to arterial desaturation. In exercise, this can become limiting: $P\bar{v}O_2$ must decrease to dangerously low levels at all 2,3-DPG/Hb values, but when 2,3-DPG/Hb is less than 0.4 mol/mol, arterial desaturation also occurs.

Hematocrit

With increasing hematocrit, cardiac output decreases, producing lower $P\bar{v}O_2$. However, at the same time, the time allowed for capillary transit increases, which could be of some benefit to overall gas transport.

$\emptyset O_2$

Reduction of $\emptyset O_2$ lowers $P\bar{v}O_2$. The reduction is, however, not regular, since the effect of reducing $\emptyset O_2$ is complex: the value of $\emptyset O_2$ depends on SO_2, so the result of its reduction will not be immediately apparent. PaO_2 will fall precipitously when $\emptyset O_2$ falls to below 10% of its normal value. In exercise, the effect will be even more pronounced. It is well known that arterial desaturation occurs during exercise when O_2 uptake is diffusion-limited (Dempsey et a., 1984).

RED CELL FUNCTION AT HIGH-ALTITUDE: HUMAN STUDIES

Subjects

The data used to test the model were obtained in previous studies on high-altitude natives in Cerro de Pasco, Peru (altitude 4300 m). In addition to histories and physical examinations, measurements of pulmonary function, hematologic parameters, P50, 2,3-DPG/Hb and arterial blood gases were made.

All of the subjects were born at high altitude, and lived there all their lives. They had significantly increased vital capacities, and typical "barrel" chest configuration. Complete exercise test data are available on 9 subjects, 5 of whom we consider "normal" (Table I), and 4 whom we consider to have excessive polycythemia (Table II). In making these determinations, we use a hematocrit of 55% as the upper limit of expected normal hematocrit in Cerro de Pasco (Winslow et al., 1981).

Eight of the polycythemic subjects were chosen for hemodilution studies (Table III). The choice was based on willingness to participate, and ability to perform a cycle ergometer exercise test satisfactorily.

The input data for calculation of the gas exchange profiles are shown in Table IV. The table is a combination of our measurements shown in Tables I-III and values taken from the literature where data on our subjects are not available.

Maximal O_2 Uptake

The 1-minute ramp exercise test protocol used does not provide a true measure of $\dot{V}O_2$max because 1 minute is insufficient to achieve a steady state. Nevertheless, for the purpose of the calculations, the maximal achieved $\dot{V}O_2$ is used. In a study of mountaineers working at extreme altitude, Pugh (1964) found a $\dot{V}O_2$max of approximately 40 ml/min· kg, and West et al., (1983) later found a value of approximately 45 ml/min·kg. Our normal high-altitude subjects achieved a mean of 43.1 ml/min·kg during their maximal work. However, this is probably an overestimation, since it represents the averaged value over the last minute of the 1-minute incremental exercise test. It is likely that our subjects could not have sustained this level in the steady state.

Blood pH

Blood pH (pH_B) values have been corrected (Winslow et al., 1981) to a hematocrit of 45% using the equation

$$pH_p = pH_B + 0.0105(Hct) + 0.0889(Hct)^2 - .023 \qquad (20)$$

where Hct is a fraction.

Table I. Normal Subjects (n=5)

no.	wt kg	work kpm	PAO$_2$ torr	PaO$_2$ torr	PaCO$_2$ torr	pH	VO$_2$ ml/min kg	Hct %	2,3-DPG/Hb mol/mol
1979	62	0	54	45	32	7.429	6.5	53	0.95
		1000	55	36	34	7.405	45.0		
4279	68	0	54	39	24	7.443	11.8	53	1.00
		1350	55	32	26	7.337	50.2		
3779	63	0	61	36	23	7.428	6.9	55	0.96
		800	66	33	25	7.368	23.2		
5079	62	0	63	42	24	7.449	6.4	49	1.28
		1050	69	42	19	7.348	44.0		
6179	65	0	61	48	25	7.418	11.3	46	0.96
		1200	64	48	24	7.346	53.3		
Means	64	0	57	42	27	7.419	8.9	51.2	1.03
		1080	60	38	28	7.357	43.1		

Table II. Polycythemic Subjects (n=4)

no.	wt kg	work kpm	PAO2 torr	PaO2 torr	PaCO2 torr	pH	VO2 ml/min kg	Hct %	2,3-DPG/Hb mol/mol
2279	70	0	64	52	26	7.407	7.4	68	
		800	62	41	29	7.330	29.5		
3979	64	0	56	35	30	7.351	9.5	66	1.38
		700	58	30	37	7.304	34		
4579	60	0					14	75	1.14
		300			20.1				
3980	61	0	42	40	34	7.428	6.4	66	
		500	48	42	33	7.415	31		
Means	64	0	54	42	30	7.395	11.2	69	1.26
		575	56	38	33	7.350	28.6		

Table III. Postphlebotomy Subjects (n=4)

no.	wt kg	work kpm	PAO$_2$ torr	PaO$_2$ torr	PaCO$_2$ torr	pH	V̇O$_2$ ml/min kg	Hct %	2,3-DPG/Hb mol/mol
3979	64	0 900	76 61	39 36	28 36	7.396 7.287	5.0 33.4	45	1.13
2279	70	0 800	61 64	43 37	23 26	7.437 7.362	10.5 26.6	55	0.73
4579		0 600	56 60	36 35	24 27	7.397 7.357	10.9 26.9	58	0.90
3980	61	0 500	54 52	43 46	31 32	7.416 7.401	5.6 23.1	47	
Means	64	0 700	62 59	40 38	26 30	7.412 7.352	8 27.5	51	0.92

Table IV. Input Values

	normal		polycythemia		postbleed	
	Rest	Exercise	Rest	Exercise	Rest	Exercise
wt kg	64	64	64	64	64	64
PAO_2 torr	57	60	54	56	62	59
pHa	7.419	7.357	7.395	7.350	7.412	7.352
PaO_2 torr	42	38	42	38	40	38
$PaCO_2$ torr	27	28	30	33	26	30
LPH^1 sec	7	7	7	7	7	7
DPG/Hb	1.03	1.03	1.26	1.26	.92	.92
PP^2 g/l	70	70	70	70	70	70
Hct	.512	.512	.69	.69	.51	.51
R	.8	.8	.8	.8	.8	.8
$\dot{V}O_2\ \dfrac{mL}{min\cdot kg}$	8.9	43.1	11.2	28.6	8	27.5
$DMO_2\ \dfrac{mL}{min\cdot kg}$	83	104	83	104	83	104
V_c ml	106	136	106	136	106	136
Output						
$\dot{V}O_2\ \dfrac{mL}{min\cdot kg}$	8.9	38.8*	11.2	11.1*	8.0	27.5
$\dot{Q}$ L/min	5.97	17.4	4.04	4.02	5.65	13.2
TCAP sec	1.066	0.469	1.574	2.027	1.126	0.619

*Simulation program was unable to find conditions for input $\dot{V}O_2$.

[1] LPH is the half-time of pH change in postcapillary blood.

[2] PP is plasma protein concentration.

127

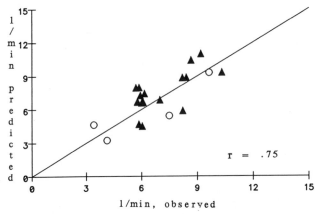

Fig. 1. Experimental verification of cardiac output predictions. A high altitude native with catheters in the pulmonary
and radial arteries was studied at rest and exercise. Fick
cardiac output (circles) were calculated by the measured $\dot{V}O_2$
and $(a-\bar{v})O_2$ difference before and after reduction of hematocrit from 62 to 42%. Also shown are data taken from the
literature (VanderElst and Kreukniet, 1982) for patients (8-45
patients per point, a total of 112) with chronic lung disease
(triangles) studied at different hematocrits and levels of
exercise.

$\underline{DMO_2}$

At sea level, the commonly accepted value for DMO_2 is 40 ml/min·torr
(Wagner, 1977). DeGraff et al., (1970) measured 2 subjects at sea level and
found that during maximal exercise DMO_2 averaged 84 ml/min·torr. They also
reported DMO_2 for natives of 3100 m to be about 83 ml/min·torr at rest,
increasing to about 104 ml/min·torr during exercise.

Pulmonary Capillary Volume

The volume of the pulmonary capillary blood was found by DeGraff et
al., (1970) to be 177 ml for a sea level population during exercise. For
natives of 3100m, they found 106 ml at rest and 136 ml during maximal
exercise.

Cardiac Output Estimation

In one subject, exercise studies were done at 2 different hematocrits
and 2 steady state work levels with arterial and pulmonary artery catheters
in place (Brown et al., 1985; Winslow et al., 1985). These catheters were
equipped with fiberoptic and electrode devices so that arterial and mixed
venous PO_2 and SO_2 could be simultaneously monitored. This allowed the
direct calculation of the cardiac output by the Fick equation. Thus, the
data obtained could be used as a check of the method to estimate the
prediction of cardiac output from equation (2). Fig. 1 shows that the
agreement between predicted and observed values is good. Also shown in Fig.
1 are data from patients with chronic lung disease (VanderElst and
Kreukniet, 1982). These data also are in reasonable agreement with our
predictive formulas, but the $\dot{V}O_2$ range covered is not as broad as desired.
The overall correlation between the predicted and observed $\dot{Q}$ is 0.75.

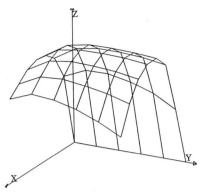

Fig. 2.   Hematocrit, 2,3-DPG/Hb, $P\bar{v}O_2$ surface, normoxia, rest.
X axis is hematocrit, 5-85%, Y axis is DPG/Hb, 0.3 - 1.5
mol/mol, Z axis is $P\bar{v}O_2$.   0 -50 torr.

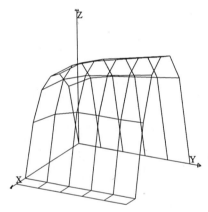

Fig. 3. O_2 transport optimization, normoxia, exercise.   Axes
are as in Fig.   2.

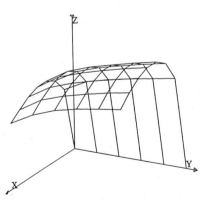

Fig. 4. O_2 transport optimization, hypoxia, rest.   Axes are
as in Fig.   2.

Prediction of Optimal $P\bar{v}O_2$

Fig. 2 shows the results for normoxia at rest. The hematocrit reaches an optimum at approximately 45%. In this example, increasing 2,3-DPG/Hb increases $P\bar{v}O_2$ steadily over the entire range studied. During exercise, Fig. 3, the hematocrit optimum becomes sharper: the reduction of $P\bar{v}O_2$ above hematocrit 45% is very pronounced, but there is still steady improvement with increasing 2,3-DPG/Hb.

Under hypoxic conditions, Fig. 4, the hematocrit optimum is slightly higher, now 45-50%, but the interaction with 2,3-DPG is very complex. For example, at a hematocrit close to 45%, no improvement in $P\bar{v}O_2$ is achieved by increasing the 2,3-DPG/Hb above 0.8 mol/mol, the sea level normal value. On either side of the hematocrit optimum, $P\bar{v}O_2$ decreases as the 2,3-DPG/Hb increases above this value. The sharpest optima are seen during hypoxic exercise, Fig. 5. Here, the hematocrit optimum decreases to about 40%, while the 2,3-DPG/Hb optimum is about 0.6 mol/mol. Deviations from these values will severely curtail maximal $\dot{V}O_2$.

High Altitude Natives

Fig. 6 shows the blood PO_2 as a function of time in the pulmonary capillaries in high-altitude natives, at rest and during exercise. It will be noted that the differences include a lower $P\bar{v}O_2$, a shorter capillary transit time, and lower end-capillary PO_2 during exercise.

That the transit time has increased at high altitude can be appreciated by the decreased cardiac output caused by increased hematocrit (Table IV). Since the oxygenation is already limited by the capillary transit time, exercise can only be performed at the expense of $P\bar{v}O_2$, which is forced to very low levels.

Phlebotomy Experiments

Tables II and IV show the measurements made in a subgroup of our high-altitude subjects that underwent isovolumic hemodilution. The O_2 uptake curves for this group are shown in Figs. 7 and 8. When the hematocrit is 69%, the capillary transit time is much longer because of reduced cardiac output. The model predicts that this should result in a smaller $(A-a)O_2$ gradient. In fact the data in Table IV suggest that, while this is the case at rest, during exercise the gradient is about the same in both cases. Thus, the actual situation might be more complex than the model predicts.

Figs. 7 and 8 also show that the $P\bar{v}O_2$ is lower in the high hematocrit group and increases markedly after phlebotomy, despite the drop in O_2-carrying capacity. These findings are generally confirmed by measurements in a single individual (Fig. 9) in whom a pulmonary artery catheter recorded arterial and mixed venous O_2 contents during rest and exercise.

DISCUSSION

The relative role of red cell properties in optimal oxygen transport is a complex matter and has been the subject of many studies in the past. Therefore, only a few selected comments are possible in a limited space.

The traditional view is that shifting the OEC to the right facilitates oxygen unloading in tissue sites (Aste-Salazar and Hurtado, 1944). However, in certain situations the opposite could be true. For example, Barcroft et al., (1923) believed that increased O_2 affinity was important in adaptation to high altitude. Their reasoning was based on analogy with the

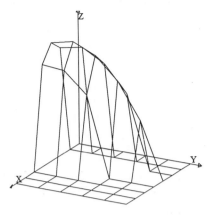

Fig. 5. O$_2$ transport optimization, hypoxia, exercise. Axes are as in Fig. 2.

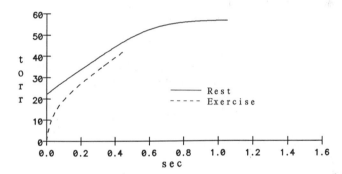

Fig. 6. O$_2$ uptake progress curves, high-altitude natives at rest (solid) and exercise (dashed).

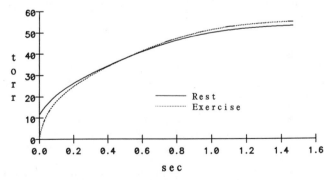

Fig. 7. O$_2$ uptake progress curves, polycythemic subgroup at rest (solid) and exercise (dashed).

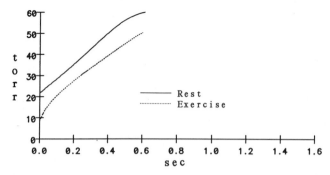

Fig. 8. O_2 uptake progress curves, postphlebotomy high-altitude natives, rest (solid) and exercise (dashed).

placental circulation, where fetal blood has a higher affinity than that of the mother.   Modification of hemoglobin to increase its affinity conveys superior survival on rats (Eaton et al., 1974), and mutant hemoglobins with increased affinity may precondition subjects to hypoxia (Hebbel et al., 1978).   We recently reported that on the summit of Mt. Everest, the P50 of whole blood is about 19 torr, protecting arterial saturation (Winslow et al., 1984).

An increase in 2,3-DPG is usually thought to augment O_2 delivery to the tissues by a right-shift of the OEC.   Our calculations support this view, but only at sea level.   At altitude, because of the diffusion limitation of O_2 uptake in the lung, decreased O_2 affinity seems to limit oxygen uptake. Whether or not this is offset by enhanced tissue unloading remains to be seen, because too little data are available to indicate how critical is venous PO_2 for the uptake of O_2 in the pulmonary capillaries under hypoxic conditions.   At this point it seems that the 2,3-DPG effect is a sea-level mechanism, not suited to O_2 delivery at altitude.

The results of the calculations presented here are in agreement with experimental findings in which arterial desaturation is seen during extreme hypoxic exercise (Dempsey et al., 1984).   The calculations predict that this desaturation can be prevented, to some extent, by increasing the oxygen affinity of the blood by decreasing 2,3-DPG/Hb.   They also point out that if arterial oxygenation is normal, particularly at rest, the level of 2,3-DPG may be relatively unimportant.   It is slightly more important during exercise, but becomes critical in hypoxia.

It is of interest that in the high-altitude normal and polycythemic simulations (Table IV and Figs. 5 and 7), the model could only be satisfied if the $\dot{V}O_2$ was less than that actually observed.   One explanation for this is that our subjects did not achieve steady state gas exchange at the highest exercise levels.   We discussed this possibility in a previous publication (Winslow et al., 1985), suggesting that O_2 reserves, including myoglobin and, possibly, the expanded red cell mass itself, could be accessed for short periods of intense exercise.   Furthermore, Balke (1964) noted that high-altitude natives are capable of short bursts of very high work output, which cannot be sustained for prolonged periods.   We believe that attention should be directed to whether or not a true steady state can be achieved in exercise studies involving high-altitude natives.

Increased hematocrit under conditions of hypoxia is widely believed to be a mainstay of high altitude adaptation (Erslev, 1981).   Our theoretical considerations suggest that the optimal hematocrit is about 45%, regardless

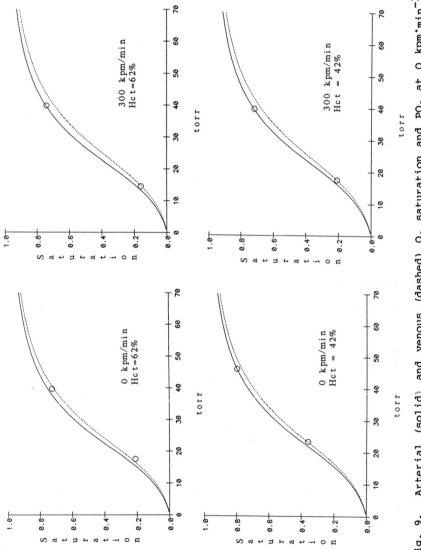

Fig. 9. Arterial (solid) and venous (dashed) O_2 saturation and PO_2 at 0 kpm·min^{-1} (left) and 300 kpm·min^{-1} (right) before (top) and after (bottom) hemodilution. Continuous OEC's were simulated from known pH, 2,3-DPG/Hb, and PCO_2.

of the altitude. This is because the effect of increased viscosity resulting from polycythemia on cardiac output seems to outweigh the effect of
increased oxygen carrying capacity. However, it must be kept in mind that
our cardiac output calculations are based on data from dogs, taken at rest,
and the exercise extrapolations are somewhat derived. Direct measurements
of cardiac output under hypoxic conditions are necessary to answer this
question.

Chronic mountain sickness is a condition of advanced age. Our view is
that symptoms in this condition result from long-term polycythemia and a
consequent burden on the circulatory system. It is quite possible that mild
polycythemia in younger persons is not a serious problem. In fact, a mild
elevation of hematocrit may improve the maximal VO_2, as was recently
suggested (Buick et al., 1980). The effect of increased red cell mass on
blood buffering has not received adequate attention in the literature.

The rate of red cell uptake of O_2 has not been adequately studied,
because of the difficulties in its measurement. One of the problems is that
there is a layer of unstirred plasma surrounding the cell membrane: the _in_
vivo importance of this layer is still not appreciated (Vandegriff and
Olson, 1985). It appears, however, that the reaction of O_2 with hemoglobin
itself is not rate-limiting.

Physiological CO_2 reactions are also not usually believed to play
important roles in gas exchange. However, CO_2 has many important effects.
First, its differential affinity between oxy- and deoxyhemoglobin serves as
a modulator for O_2 transport. The greater the difference in this affinity,
the greater the ability of hemoglobin to take up O_2 in the lung and deliver
it in the tissues. CO_2 also affects O_2 binding by hemoglobin by its effect
on pH.

An important feature of these simulations, hitherto not appreciated,
is the interesting effect of phlebotomy on PCO_2. Under all conditions, as
hematocrit increased, PCO_2 also increased. The explanation for this is that
at a given VO_2, the CO_2 concentration is independent of hematocrit. The
distribution expressions in the Kelman subroutines predict that to satisfy
equilibrium, PCO_2 must increase as plasma volume decreases. This could have
important regulatory consequences for central CO_2 receptors.

In general, exercise sharpens the optima for hematocrit and 2,3-DPG.
Extremes in either of these variables will have stronger effects in exercise
than at rest, and the optima are still sharper at altitude. In this regard,
it is of note that the surfaces for optimization are broad, particularly at
sea level, allowing for considerable variation from normal.

REFERENCES

Adair, G.S. (1925). The hemoglobin system VI. The oxygen dissociation
     curve of hemoglobin. J. Biol. Chem. 63:529-545.
Asmussen, E. and M. Nielsen (1956). Physiological dead space and alveolar
     gas pressures at rest and during muscular exercise. Acta Physiol.
     Scand. 38:1-21.
Aste-Salazar, H. and A. Hurtado (1944). The affinity of hemoglobin for
     oxygen at sea level and at high altitudes. Am. J. Cardiol. 142:733-
     743.

Balke, B. (1964). Work capacity and its limiting factors at high altitude. In: The Physiological Effects of High Altitude, W.H. Weine (ed). New York: Macmillan, pp. 233-247.

Barcroft, J., C.A. Binger, A.V. Bock, J.H. Doggart, H.S. Forbes, G. Harrop, J.C. Meakins and A.C. Redfield (1923). Observations upon the effect of high altitude on the physiological processes of the human body carried out in the Peruvian Andes chiefly at Cerro de Pasco. Phil. Trans. Royal Soc. London Ser B. 211:351-480.

Bencowitz, H.Z., P.D. Wagner and J.B. West (1982). Effect of change in P50 on exercise tolerance at high altitude: a theoretical study. J. Appl. Physiol. 53:1487-1495.

Bidani, A. and E.D. Crandall (1978). Slow postcapillary changes in blood pH in vivo: titration with acetazolamide. J. Appl. Physiol. 45:565-573.

Bidani, A., E.E. Crandall and R.E. Forster (1978). Analysis of post-capillary pH changes in blood in vivo after gas exchange. J. Appl. Physiol. 44:770-781.

Brown, E.G., R.W. Krouskop, F.E. McDonnell, C.C. Monge, F. Sarnquist, N.J. Winslow and R.M. Winslow (1985). Simultaneous and continuous measurement of breath-by-breath oxygen uptake and arterial-venous oxygen content. J. Appl. Physiol. 58:1138-1139.

Buick, F.J., N. Gledhill, A.B. Froese, L. Spriet and E.C. Meyers (1980). Effect of induced erythrocythemia on aerobic work capacity. J. Appl. Physiol. 48:636-642.

DeGraff, A.C., R.F. Grover, R.L. Johnson, J.W. Hammond and J.M. Miller (1970). Increased diffusing capacity of the lung in persons native to 3,100 in North America. J. Appl. Physiol. 29:71-76.

Dempsey, J.A., P.G. Hanson and K.S. Henderson (1984). Exercise-induced arterial hypoxemia in healthy human subjects at sea level. J. Physiol. 355:161-175.

Eaton, J.W., T.D. Skelton and E. Berger (1974). Survival at extreme altitude: Protective effect of increased hemoglobin-oxygen affinity. Science 185:743-744.

Erslev, A. (1981). Erythroid adaptation to altitude. Blood Cells 7:495-508.

Guyton, A.C., C.E. Jones and T.G. Coleman (1973). Cardiac Output and Its Regulation. 2nd edition, Philadelphia: Saunders.

Hebbel, R.P., J.W. Eaton, R.S. Kronenberg, E.D. Zanjani, L.G. Moore and E.M. Berger (1978). Human llamas: Adaptation to altitude in subjects with high hemoglobin oxygen affinity. J. Clin. Invest. 62:593-600.

Hill, E.P., G.G. Power and R.D. Gilbert (1977). Rate of pH changes in blood plasma in vitro and in vivo. J. Appl. Physiol. 42:928-934.

Jones, N.L., E.J.M. Campbell, R.H.T. Edwards and D.G. Robertson (1975). Clinical Exercise Testing. Philadelphia: W.B. Saunders.

Kelman, G.R. (1966). Calculation of certain indices of cardio-pulmonary function using a digital computer. Respir. Physiol. 1:335-343.

Kelman, G.R. (1967). Digital computer procedure for the conversion of PCO_2 into blood CO_2 content. Respir. Physiol. 3:111-115.

Neville, J.R. (1976). Altered heme-heme interaction and tissue-oxygen supply: A theoretical analysis. Br. J. Haem. 34:387-395.

Pugh, L.G.C.E. (1964). Cardiac output in muscular exercise at 5800 m (19,000 feet). J. Appl. Physiol. 19:441-447.

Roughton, F.J.W. and R.E. Forster (1957). Relative importance of diffusion and chemical reaction rates in determining rate of exchange of gases in the human with special reference to true diffusing capacity of pulmonary membrane and volume of blood in the lung capillaries. J. Appl. Physiol. 11:290-302.

Staub, N.C., J.M. Bishop and R.E. Forster (1962). Importance of diffusion and chemical reaction rates in O_2 uptake in the lung. J. Appl. Physiol. 17:21-27.

Thomas, L.J. (1972). Algorithms for selected blood acid-base and blood calculations. J. Appl. Physiol. 33:154-158.

Vandegriff, K.D. and J.S. Olson (1985). Morphological and physiological factors affecting oxygen uptake and release by red blood cells. J. Biol. Chem. 259:12619-12627.

VanderElst, C. and J. Kreukniet (1982). Some aspects of the oxygen transport in patients with chronic obstructive lung diseases and respiratory insufficiency. Respiration 43:336-343.

Wagner, P.D. (1977). Diffusion and chemical reaction in pulmonary gas exchange. Physiol. Rev. 57:257-312.

West, J.B., P.H. Hackett, K.H. Maret, J.S. Milledge, R.M. Peters, C.J. Pizzo and R.M. Winslow (1983). Pulmonary gas exchange on the summit of Mt. Everest. J. Appl. Physiol. 55:678-687.

Winslow, R.M. (1986). A model for red cell O_2 uptake. International J. Clin. Monitor. Comput. 2:81-93.

Winslow, R.M. and W.F. Anderson (1982). The hemoglobinopathies. In: The Metabolic Basis of Inherited Disease, J.B. Stanbury (ed): 5th edition, pp. 1666-1710.

Winslow, R.M. and C.C. Monge (1986). Hypoxia, Polycythemia, and Chronic Mountain Sickness. Baltimore: Johns Hopkins University Press.

Winslow, R.M., C.C. Monge, E.G. Brown, H.G. Klein, F. Sarnquist and N.J. Winslow (1985). The effect of hemodilution on O_2 transport in high-altitude polycythemia. J. Appl. Physiol. 59:1495-1502.

Winslow, R.M., C.C. Monge, N.J. Statham, C.G. Gibson, S. Charache, J. Whittembury, O. Moran and R.L. Berger (1981). Variability of oxygen affinity of blood: human subjects native to high altitude. J. Appl. Physiol. 51:1411-1416.

Winslow, R.M., M. Samaja and J.B. West (1984). Red cell function at extreme altitude on Mount Everest. J. Appl. Physiol. 56:109-116.

Winslow, R.M., M. Samaja, N.J. Winslow, L. Rossi-Bernardi and R.I. Shrager (1983). Simulation of the continuous O_2 equilibrium curve over the physiologic range of pH, 2,3-diphosphoglycerate, and pCO_2. J. Appl. Physiol. 54:524-529.

Winslow, R.M., M.L. Swenberg, R.L. Berger, R.I. Shrager, M. Luzzana, M. Samaja and L. Rossi-Bernardi (1977). Oxygen equilibrium curve of normal human blood and its evaluation by Adair's equation. J. Biol. Chem. 252:2331-2337.

CAN METABOLITES CONTRIBUTE IN REGULATING BLOOD OXYGEN AFFINITY?

Russell E. Isaacks

Research Laboratories, V. A. Medical Center and
Department of Medicine, University of Miami
Miami, FL 33125

There are a number of interrelated factors contributing to the regulation of blood oxygen affinity in animals. The red blood cell in mammals has an active glycolytic system and a relationship exists between metabolism and function of the red cell in oxygen transport during hypoxia. The relationship in red cells of most mammals involves binding of 2,3-bisphosphoglycerate (2,3-P_2-glycerate), a glycolytic intermediate, with deoxyhemoglobin, regulating hemoglobin function and enhancing oxygen delivery to the tissues (Benesch and Benesch, 1967; Chanutin and Curnish, 1967). The in vivo level of 2,3-P_2-glycerate in human erythrocytes increases above normal during hypoxia from altitude or from certain pathological conditions, but no satisfactory way is known to actually manipulate the in vivo concentration of this compound. In human erythrocytes, the levels of 2,3-P_2-glycerate and ATP, which also influences hemoglobin binding, can be depleted after in vitro incubation at 37° for 12-24 hours (Lian et al., 1971); red cells incubated with glycolate also lose 2,3-P_2-glycerate rapidly without changes in ATP concentrations (Rose, 1976). The level of 2,3-P_2-glycerate can be enriched several-fold in human red cells by incubating with inosine, pyruvate, and inorganic phosphate (Lian et al., 1971). Importantly, during the depletion and enrichment of the cells 2,3-P_2-glycerate levels, there is a corresponding increase and decrease in blood oxygen affinity (Lian et al., 1971). Further, the allosteric properties of hemoglobins are such that increases in [H^+], CO_2 and temperature, features which are all present in tissue capillary beds, also facilitate the release of oxygen.

The erythrocytes of the chick embryo and sea turtle embryo contain 2,3-P_2-glycerate as the major organic phosphate (4-6 mM) during the last stages of embryonic development (Isaacks et al., 1975; 1980). Its rapid disappearance from the circulating red blood cells shortly after hatching suggests that it may function as a transitory modulator of hemoglobin function during this period (Isaacks and Harkness, 1980). As 2,3-P_2-glycerate disappears from the erythrocytes, inositol pentakisphosphate (inositol-P_5) and ATP concentrations begin to accumulate with corresponding marked decreases in blood oxygen affinity. Inositol-P_5 binds to both avian and human hemoglobin, reducing oxygen affinity and is, in fact, the most effective hemoglobin modulatory occurring in animal red blood cells (Isaacks and Harkness, 1980; Vandercasserie et al., 1973). Presumably it functions in bird erythrocytes to modulate hemoglobin function, although neither it nor 2,3-P_2-glycerate influences hemoglobin function in adult sea turtle red

137

cells, consequently its exact function is still unclear. The concentration of inositol-P$_5$ changes little in the mature erythrocyte and is apparently not metabolically regulated in response to hypoxia in an analogous manner to changes in 2,3-P$_2$-glycertate in erythrocytes of mammals. Recent data indicated that carbon dioxide and hydrogen ions in concert with inositol-P$_5$ may play a physiological role in regulating oxygen delivery during oxygen deficit in birds and sea turtles (Isaacks et al., 1982; Isaacks et al., 1986). Additionally, metabolites produced in response to hypoxia and known to be substrates of erythrocytes may also be physiologically significant in regulating blood oxygen affinity in these species. We have recently found that erythrocytes from chickens of various ages can be stimulated to accumulate 2,3-P$_2$-glycerate after in vitro incubation with inosine and pyruvate with corresponding decreases in blood oxygen affinity (Isaacks et al., in prep).

In this communication, data are given to indicate that chicken red cells incubated with adenosine and pyruvate also accumulate 2,3-P$_2$-glycerate with corresponding decreases in blood oxygen affinity. Further, red cells from 14 month green (Chelonia mydas mydas) and loggerhead (Caretta caretta) sea turtles incubated with inosine and pyruvate accumulate 2,3-P$_2$-glycerate to a lesser extent than bird erythrocytes and with only modest changes in oxygen affinity. The erythrocytes from mature green sea turtles did not accumulate 2,3-P$_2$-glycerate.

MATERIALS AND METHODS

Samples of freshly-drawn blood from cockerels at ages 13- or 18-weeks were filtered through glass wool to remove any small clots, and the plasma and buffy coat were removed by aspiration after centrifugation at 300 x g for 30 minutes. The erythrocytes were washed twice with Krebs-Ringer's buffer, adjusted to pH 7.45, and supplemented with 17 mM inorganic phosphate and 13.9 mM glucose. The ability of the chicken erythrocyte to accumulate 2,3-P$_2$-glycerate and its effect on the oxygen affinity of the cell suspensions were determined.

The washed chicken erythrocytes were resuspended to a hematocrit of about 20% in the same buffer, which has been supplemented with 20 mM of adenosine and 20 mM of pyruvate, and incubated at 39-40°C, pH 7.45, in a shaking water bath in an atmosphere of 95% oxygen and 5% carbon dioxide. Aliquants were removed and placed in an ice bath after 0, 1, 2, 4, and 6 hours incubation. The cell suspensions were centrifuged at 300 x g, the buffer medium removed, and the erythrocytes washed three times by repeated resuspension in cold 0.154 M NaCl and centrifugation. The aqueous-soluble phosphates were extracted from the washed red cells (0.3-0.4 cm^3 RBC) with 18% perchloric acid as previously described (Isaacks et al., 1983). Total phosphates (TPi) of each extract were determined on 0.1 ml aliquants in triplicate. The concentration of 2,3-P$_2$-glycerate in the PCA-extracts was estimated enzymatically. Inorganic phosphate (Pi), ATP, and inositol-P$_5$ were separated in the PCA-extracts by their respective elution with 100 ml each of 0.05, 0.3, and 1.0 N HCl from mini-columns (0.7 x 10.0 cm columns; Bio-Rad Laboratories) with a 7.0 cm resin bed volume of Dowex 1-X8 anion exchange resin in the chloride form (Isaacks et al., 1983). These compounds were quantitated by wet-ash phosphate analysis. Some of the saline-washed erythrocyte suspension was retained for measuring packed cell volume by the microcapillary method for calculation of phosphate concentration per cm^3 RBC. The P$_{50}$ (partial pressure of oxygen in mm Hg, or torr, at which 50% of hemoglobin is saturated) was determined on each sample of saline-washed erythrocytes by a modification of the Longmuir and Chow (1970) method as described (Lian et al, 1971).

Table I. Incubation of Chicken Erythrocytes with Adenosine and Pyruvate
         (20 mM)

Age[1]	Incub. Time	μmole/ml RBC						
		TPi	Pi	ATP	DPG	IP_5	P_{50}	n
13 wks	0	32.52 ±.24	6.03 ±.13	2.29 ±.11	0.05 ±.04	4.36 ±.56	47.8	2.98
	1	36.17 ±4.11	9.05 ±1.17	2.98 ±.14	0.58 ±.02	4.62 ±.26	52.0	2.94
	2	44.72 ±.26	10.97 ±.54	3.80 ±.09	1.15 ±.01	4.78 ±.08	52.8	3.20
	4	50.82 ±2.54	11.12 ±.16	3.64 ±.09	2.20 ±.04	4.84 ±.30	54.4	2.64
	6	49.10 ±11.35	11.67 ±1.13	3.45 ±1.48	3.54 ±1.08	4.61 ±.83	72.5	3.64
18 wks	0	36.41 ±.24	6.35 ±.05	2.16 ±.42	0	5.34 ±1.30	42.7	3.5
	1	49.41 ±2.69	13.04 ±1.87	3.67 ±2.08	0.16 ±.05	5.64 ±1.85	47.7	3.5
	2	42.45 ±.32	12.79 ±2.13	3.44 ±1.95	0.62 ±0	5.40 ±1.19	49.8	3.6
	4	57.66 ±.45	19.17 ±.56	3.42 ±0.03	1.37 ±.10	5.61 ±.0	55.8	3.1
	6	67.31 ±2.34	22.16 ±1.41	4.26 ±.92	1.90 ±.31	5.48 ±.42	53.3	2.8

[1] 13-wks and 18-wks are ages of birds in weeks; values are means
plus or minus standard deviation; DPG is $2,3-P_2$-glycerate; IP_5
is inositol-P_5; n is Hill coefficient.

Blood samples were also taken from yearling green and loggerhead sea
turtles (14 months) and from an adult green sea turtle for incubation
studies. The erythrocytes were collected and washed in the same manner as
above but incubated in Krebs-Ringer's supplemented with 20 mM inosine and 20
mM pyruvate. RBC phosphate composition and oxygen affinity of the cell
suspension were determined as described above.

RESULTS

Incubation of erythrocytes from 13- and 18-week old chickens with the
Krebs-Ringer buffer supplemented with 20 mM adenosine and 20 mM pyruvate for
6 hours resulted in an increase in $2,3-P_2$-glycerate concentration of the red
cells from 0.05 mM to 3.5 mM and 0 to 1.9 mM, respectively (Table I). The
total cell phosphate content per ml RBC from the 13 and 18-week old birds
increased 16.6 and 30.9 mM, respectively, after 6 hours incubation. Most of
the increased phosphate content comes primarily from expected increases in

Pi with incubation but also from the stimulated synthesis of $2,3-P_2$-glycerate. ATP concentrations were not only maintained but actually increased by 51 and 97 per cent, respectively, in cells from the 13- and 18-week chickens after 6 hours incubation, whereas inositol-P_5 levels changed little (Table I). Interestingly, as the cells are enriched in $2,3-P_2$-glycerate, there is a marked increase in P_{50} of the cell suspensions (Table I). These results are similar to previous studies on chicken erythrocytes from birds of various ages incubated with inosine and pyruvate (unpublished observations).

Incubation of erythrocytes from 14 month old green sea turtles and loggerhead sea turtles with the same buffer but supplemented with 20 mM inosine and 20 mM pyruvate resulted in increases of $2,3-P_2$-glycerate from 0 to 0.2 mM and 0.37 to 0.83 mM, respectively, after 6 hours incubation (Table II). The increases in total cell phosphate content per ml of cells from 8.5 to 14.5 mM, respectively, were due primarily to increases in Pi (Table II). ATP concentration decreased 12 per cent in the cells from the green sea turtle but increased 58 per cent in the red cells from the yearling loggerhead sea turtle after 6 hours incubation (Table II). Inositol-P_5 concentrations are low but unchanged during incubation of the cells from both species. The P_{50} of the cell suspensions of these two species at this age increased very modestly after 6 hours incubation (Table II).

When the erythrocytes from an adult green sea turtle (>14 years) were incubated with the same concentrations of inosine and pyruvate, no accumulation of $2,3-P_2$-glycerate was detected. Total cell phosphate increased 9.9 mM because of increased uptake of Pi and a 38 per cent increase in concentration of ATP was observed (Table II). Inositol-P_5 concentration did not change.

DISCUSSION

Inositol-P_5 is present in rather large amounts (3-5 mM) in red cells of mature birds and is known to be a potent modulator of hemoglobin. However, its role in regulating blood oxygen affinity in response to hypoxia is still uncertain, primarily because its concentration in vivo changes little, particularly once it reaches maximum levels found in the mature erythrocyte. Perhaps its function is to establish and maintain bird blood at a much lower oxygen affinity in keeping with a generally higher metabolic rate than most mammals. Perhaps other factors in concert with inositol-P_5 participate in regulating blood oxygen affinity in birds, if indeed it is required. Previous studies revealed that increasing $[H^+]$ and CO_2 markedly affected oxygen affinity of bird blood (Isaacks et al., 1986), which could have physiological significance. In addition, inosine, adenosine and pyruvate produced in response to hypoxia may influence blood oxygen affinity. The data presented here, as well as previous studies with inosine and pyruvate, show that adenosine and pyruvate are red cell substrates which stimulate accumulation of $2,3-P_2$-glycerate with corresponding decreases in oxygen affinity. It should be noted that red cells from chickens 13- and 18-weeks of age do not normally contain $2,3-P_2$-glycerate. The concentration of metabolites produced in response to hypoxia would not likely be as much as we have used in these in vitro experiments. Whether they would be sufficient to result in synthesis and accumulation of $2,3-P_2$glycerate under in vivo physiological conditions is not known.

Data from previous studies revealed that hemoglobin oxygen affinity in the adult loggerhead and green sea turtles was not regulated by organic phosphates as it was in the embryos of each specie. The lack of regulation of hemoglobin oxygen affinity in the adults was apparent because oxygen dissociation curves of the phosphate-free hemoglobins changed very little

Table II. Incubation of Sea Turtle Erythrocytes with Inosine and Pyruvate (20 mM)

| | | μmole/ml RBC | | | | | | |
| | | Green Sea Turtle | | | | | | |
Age[1]	Incub. Time	TPi	Pi	ATP	DPG	IP$_5$	P$_{50}$	n
14 mo.	0	21.19 ±.40	8.20 ±.84	4.28 ±.23	0	0.22 ±.01	32.5	2.72
	1	26.43 ±1.21	12.49 ±1.43	4.40 ±.72	0.29 ±.21	0.21 ±.01	33.9	2.71
	2	29.85 ±.05	14.10 ±.44	4.70 ±.25	0.14 ±.03	0.21 ±.005	34.9	2.73
	4	29.40 ±.20	17.26 ±.98	3.90 ±.20	–	0.21 ±.002	37.2	2.74
	6	29.64 ±.36	18.22 ±.90	3.76 ±.21	0.20	0.20 ±.01	35.6	2.6
Adult	0	11.87 ±.15	4.88 ±.18	2.30 ±.05	–	0.26 ±.02		
	1	16.34 ±.72	7.64 ±.50	2.75 ±.09	–	0.31 ±.07		
	2	17.27 ±2.34	9.52 ±.51	2.99 ±.22	–	0.30 ±.08		
	4	20.26 ±.97	11.0 ±.34	2.93 ±.09	–	0.36 ±.06		
	6	21.74 ±.74	12.27 ±.42	3.18 ±.09	–	0.22 ±.03		
		Loggerhead						
14 mo.	0	22.30	8.88	3.77	0.37	0.43	45.1	3.03
	1	22.2	9.49	3.48	0.54	0.38	45.1	3.13
	2	29.07	15.30	4.30	0.69	0.44	47.3	3.05
	4	35.26 ±.44	17.04 ±.08	5.69 ±.10	0.17 ±.02	0.40 ±.03	48.8	3.14
	6	36.79 ±.55	17.21 ±.85	5.96 ±.29	0.83 ±.15	0.39 ±.06	49.9	3.00

[1] 14 mo. is age of yearling turtles in months; values are mean plus or minus standard deviation; DPG is 2,3-P_2-glycerate; IP$_5$ is inositol-P_5; n is Hill coefficient.

in the presence of organic phosphates and too the combined physiological concentration of $2,3-P_2$-glycerate and inositol-P_5 are probably too low in vivo to modulate hemoglobin function (Isaacks et al., 1982). The calculated in vivo molar ratio of organic phosphate ($2,3-P_2$-glycerate plus inositol-P_5): hemoglobin was 0.15 and 0.125 for the adult loggerhead and green sea turtles, respectively, as compared to ratios of about 0.65 to 1.0 for most birds and mammals, respectively. Data on other studies have shown that hemoglobin oxygen affinity of juvenile loggerhead and green sea turtles (8-9 months post-hatch) responded to organic phosphates even though their hemoglobin electrophoretic composition was already quite similar, if not identical, to that of the adult. Other data indicated that blood oxygen affinity of these two marine species changes dramatically with increasing concentration of CO_2 and to a lesser extent with decreasing pH. The data presented here again point up a difference between erythrocytes of the young turtle to synthesize and accumulate at least some $2,3-P_2$-glycerate from inosine and pyruvate with some changes in oxygen affinity and those of the mature green sea turtle which did not accumulate $2,3-P_2$-glycerate.

These results show that bird red cells are responsive to metabolites and are capable of turning-on synthesis of $2,3-P_2$-glycerate with corresponding increases in P_{50}. Our results further indicate that the ability to synthesize this compound decreases with age of the bird. The data indicate that at least young sea turtle erythrocytes are able to respond to inosine and pyruvate but to a lesser extent than that observed in bird cells. Whether these metabolites produced during hypoxia actually participate in regulating blood oxygen affinity under physiological situations has yet to be determined.

ACKNOWLEDGEMENT

This work was supported by a National Science Foundation Grant DMB-8413681.

REFERENCES

Benesch, R. and R.E. Benesch (1967). The effect of organic phosphates from the human erythrocyte on the allosteric properties of hemoglobin. Biochem. Biophys. Res. Commun. 26:162-167.

Chanutin, A. and R.R. Curnish (1967). Effect of organic and inorganic phosphate on the oxygen equilibrium of human erythrocytes. Arch. Biochem. Biophys. 121:96-102.

Isaacks, R., P. Goldman and C. Kim (1986). Studies on avian erythrocyte metabolism. XIV. Effect of CO_2 and pH on P_{50} in the chicken. Am. J. Physiol. 250:R260-R266.

Isaacks, R.E. and D.R. Harkness (1975). 2,3-Diphosphoglycerate in erythrocytes of chick embryos. Science 189:393-394.

Isaacks, R.E., D.R. Harkness (1980). Erythrocyte organic phosphates and hemoglobin function in birds, reptiles and fishes. Am. Zool. 20:115-129.

Isaacks, R.E., D.R. Harkness and J.R. White (1982). Regulation of hemoglobin function and whole blood oxygen affinity by carbon dioxide and pH in the loggerhead (Caretta caretta) and green sea turtle (Chelonia mydas mydas). Hemoglobin 6:549-568.

Isaacks, R., C. Kim, H.L. Liu, P. Goldman, A. Johnson, Jr. and D.R. Harkness (1983). Studies on avian erythrocyte metabolism. XIII. Changing organic phosphate composition in age-dependent density populations of chicken erythrocytes. Poultry Sci. 62:1639-1646.

Lian, C.Y., S. Roth and D.R. Harkness (1971). The effect of alteration of intracellular 2,3-DPG concentration upon oxygen binding of intact erythrocytes containing normal and mutant hemoglobins. Biochem. Biophys. Res. Commun. 45:151-158.

Longmuir, I.S. and J. Chow (1970). Rapid method for determining effects of agents on oxyhemoglobin dissociation curves. J. Appl. Physiol. 28:343-345.

Rose, Z.B. (1976). A procedure for decreasing the level of 2,3-bisphospho-glycerate in red cells in vitro. Biochem. Biophys. Res. Commun. 73:1011-1017.

THE ROLE OF BETA-ADRENERGIC RECEPTORS IN THE CARDIAC OUTPUT

RESPONSE DURING CARBON MONOXIDE HYPOXIA

C.K. Chapler, M.J. Melinyshyn, S.M. Villeneuve and S.M. Cain*

Department of Physiology
Queen's University
Kingston, Ontario K7L 3N6, Canada
and
*Department of Physiology and Biophysics
University of Alabama at Birmingham
Birmingham, Alabama 35294, USA

INTRODUCTION

Cardiac output increases during carbon monoxide hypoxia (COH) in anesthetized dogs when the level of carboxyhemoglobin exceeds 40% (Einzig et al., 1980; King et al., 1984; King et al., 1985; Sylvester et al., 1979). This compensatory response partially offsets the decrease in whole body oxygen delivery which results from the reduced oxygen content; oxygen uptake is maintained in spontaneously breathing anesthetized dogs at both 50 and 65% carboxyhemoglobin (King et al., 1984). The mechanisms underlying the cardiac output response during COH are not fully understood. It has been shown that nonselective β_1 and β_2-adrenergic blockade with propranolol resulted in lower values for cardiac output at 30 minutes of COH than in unblocked animals (King et al., 1985; Villeneuve et al., 1985). The effect of propranolol could have resulted from blockade of β_1, β_2 or a combination of the β_1 and β_2-adrenergic receptor sub-types. In the present study, the effects of the selective β_2 blocker ICI 118,551 on cardiac output and whole body oxygen uptake responses were observed during severe COH (62% decrease in arterial oxygen concentration) in anesthetized, spontaneously breathing dogs. The data were compared to our earlier results obtained during COH in dogs treated with propranolol.

METHODS

The experiments were carried out using spontaneously breathing male mongrel dogs anesthetized with pentobarbital sodium (30 mg/kg i.v.); additional doses of 30-60 mg were given as required. A cuffed endotracheal tube was inserted to insure a patent airway and a heating pad was placed around the trunk of the animal to help maintain body temperature, which was measured using a rectal thermistor. The brachial artery was catheterized for measurement of arterial blood presure and for obtaining arterial blood samples. A second catheter was inserted in the right external jugular vein and advanced into the right ventricle to obtain mixed venous blood samples. Catheters were also placed in the right femoral artery and vein for use during induction of COH.

The animals were divided into two groups. The β_2-adrenergic receptors were blocked in one group (n=7) using the highly selective β_2 antagonist ICI 118,551 (100 µg/kg i.v.) (O'Donnell and Wanstall, 1980; Bilski et al., 1983) administered 30 minutes prior to collection of two control samples. A booster dose of 60 µg/kg i.v. was given one hour later. The second group (n=6) served as a time control (no β-blockade). Two pairs of mixed venous and arterial blood samples were collected simultaneously within a 10 min interval prior to the induction of COH. The values from these samples were averaged to obtain a mean control value.

Immediately following the second control sample, COH was induced using an in situ dialysis method (King et al., 1984). Briefly, this method involved pumping blood from the right femoral artery through a Gambro fiber dialyzer and returning it to the animal via the right femoral vein. Inflow into the dialyzer was maintained at 52 ml/min while 100% CO gas was passed through the dialysate compartment. The tubing and dialyzer (volume approximately 125 ml) were filled with 6% dextran (molecular weight 70,000) in Tyrode solution. Arterial oxygen concentration was reduced by an average of 62% (from 16.8 to 6.4 vol%) by this method.

A third pair of blood samples was obtained from both the time control and β_2 block groups at 30 min of COH. Arterial and venous blood samples were analyzed for oxygen concentration, PO_2, PCO_2, and pH. Blood gases and pH were measured using a Corning 165 blood gas analyzer. The values for oxygen concentration were derived from values of O_2 saturation of Hb obtained using a CO-Oximeter. Whole body oxygen uptake was recorded on a Godart Pulmotest and cardiac output was calculated using the Fick equation. Total peripheral resistance and whole body O_2 delivery were calculated as the ratio of mean arterial pressure to cardiac output and the product of cardiac output and arterial O_2 concentration, respectively. The data from the present study were then compared to our earlier results obtained using a comparable experimental protocol and a similar level of COH (65% reduction in arterial oxygen content) in dogs treated with propranolol (1 mg/kg) which produced a β_1 and β_2 adrenergic blockade (King et al, 1985).

The statistical comparisons were made using a repeated measures design analysis of variance. Statistically significant differences were accepted at $p<0.05$. Individual comparisons were made using Duncan's multiple range test (Bruning and Kintz, 1977).

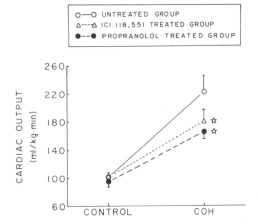

Fig. 1. Cardiac output values prior to (CONTROL) and at 30 min of carbon monoxide hypoxia (COH). Mean values ± SEM.
* Significant difference (p<0.05) between untreated and treated group(s).

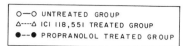

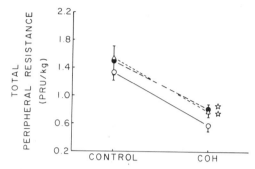

Fig. 2. Total peripheral resistance values prior to (CONTROL) and at 30 min of carbon monoxide hypoxia (COH). Mean values ± SEM.
 * Significant difference (p<0.05) between untreated and treated group(s).

RESULTS

Cardiac output values are shown in Fig. 1. There were no significant differences between the groups at the control sampling period, while cardiac output was significantly increased (p<0.05) in all groups at 30 min of COH. The value for cardiac output at 30 min of COH in the untreated group (no beta-blockade) was significantly higher (p<0.05) than that observed in either ICI 118,551 or propranolol groups. Cardiac output in the ICI 118,551 treated group rose to a level that was not different from that observed in the propranolol treated group at 30 min of COH. Total peripheral resistance fell during COH in all groups (p<0.05) with the values being significantly higher in both treated groups than in the untreated group at 30 min of COH (Fig. 2). The values in the treated groups were not different from each other during either the control or COH periods.

The data for heart rate, stroke volume and mean arterial pressure are shown in Fig. 3. At the control sampling period, the heart rates of all three groups were significantly different from each other; in order of increasing values these were propranolol-treated, ICI 118,551 treated, and the untreated group, respectively. Heart rate increased significantly in the untreated (24 bpm) and ICI 118,551 (20 bpm) treated groups during COH; but no significant change was observed in the propranolol treated group. The heart rate values at 30 minutes of COH remained significantly different between the three groups with the same rank order as seen during the control sampling period. Stroke volume was significantly increased in all three groups during COH but the values were not different between the groups at either control or COH sampling times. Mean arterial pressure was significantly lower in the untreated group at control than in either treated group. A significant decrease (p<0.05) in MAP occurred in all three groups during COH with values being significantly higher in the ICI 118,551 treated group than the untreated group at 30 min of COH.

Whole body oxygen transport values fell during COH (Fig. 4) in all three groups with values being significantly lower in both treated groups than in the untreated group. The whole body oxygen extraction ratio (a-$\bar{v}O_2$/CaO_2; Fig. 5) was increased during COH in all three groups. The values in the treated groups were significantly higher than those in the

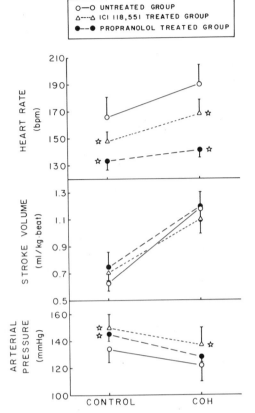

Fig. 3. Heart rate, stroke volume and mean arterial pressure values prior to (CONTROL) and at 30 min of carbon monoxide hypoxia (COH). Mean values ± SEM.
    * Significant difference (p<0.05) between untreated and treated group(s).

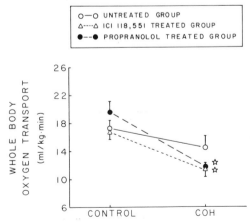

Fig. 4. Whole body oxygen transport values prior to (CONTROL) and at 30 min of carbon monoxide hypoxia (COH). Mean values ± SEM.
    * Significant difference (p<0.05) between untreated and treated group(s).

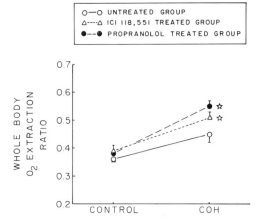

Fig. 5. Whole body oxygen extraction ratio values prior to (CONTROL) and at 30 min of carbon monoxide hypoxia (COH). Mean values ± SEM.
    * Significant difference (p<0.05) between untreated and treated group(s).

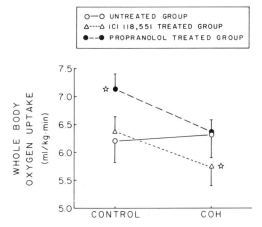

Fig. 6. Whole body oxygen uptake values prior to (CONTROL) and at 30 min of carbon monoxide hypoxia (COH). Mean values ± SEM.
    * Significant difference (p<0.05) between untreated and treated group(s).

untreated group at 30 min of COH. The data for whole body oxygen uptake are shown in Fig. 6. At the control sampling period, values for the propranolol treated group were significantly higher than either the ICI 118,551 or the untreated group. Whole body oxygen uptake at 30 min of COH remained at the prehypoxic level in the untreated group while values were significantly decreased from their respective control values in both ICI 118,551 and propranolol treated groups.

CONCLUSIONS

    The results shown in this paper demonstrate that the highly selective β_2-adrenergic blocking agent ICI 118,551 and the combined β_1 and β_2 blocking drug propranolol both produce a similar reduction in the cardiac output response during severe COH. Further, total peripheral resistance was significantly greater during COH in both beta receptor-blocked groups than that seen in the untreated (no beta-block) series. These findings provide evidence that about 35% of the increase in cardiac output that occurred at 30 min of COH (approximately a 65% reduction in arterial oxygen content) depended on activation of β_2-adrenergic receptors while β_1 receptors appeared to have little or no role in this response. The physiological importance of the β_2 receptors in the cardiovascular adjustments to severe COH was made evident by the failure of whole body oxygen uptake to be maintained during the hypoxic insult.

ACKNOWLEDGEMENTS

Research supported by grants from the Medical Research Council of Canada and National Institutes of Health (Grant No. HL-14693). ICI 118,551 was donated by Imperial Chemical Industries, Pharmaceuticals Division. The authors express their gratitude to R. Leggo for her technical assistance.

REFERENCES

Bilski, A.J., S.E. Halliday, J.D. Fitzgerald and J.L. Wale (1983). The Pharma-cology of a β_2-Selective Adrenoceptor Antagonist (ICI 118,551). J. Cardiovasc. Pharmacol. 5:430-437.
Bruning, J.L. and B.L. Kintz (1977). Supplemental computations for analysis of variance. In Computational handbook of statistics. 2nd ed. Edited by F. Scott, Foresmann, Glenview, IL.
Einzig, S., D.W. Nickoloff and R.V. Lucas (1980). Myocardial perfusion abnormalities in carbon monoxide poisoned dogs. Can. J. Physiol. Pharmacol. 58:396-405.
King, C.E., S.M. Cain and C.K. Chapler (1984). Whole body and hindlimb cardiovascular responses of the anesthetized dog during CO hypoxia. Can. J. Physiol. Pharmacol. 62:769-774.
King, C.E., S.M. Cain and C.K. Chapler (1985). O_2 transport and uptake in dogs during CO hypoxia with and without β-block. Oxygen Transport to Tissue - VI. 591-598, Plenum Publishing Corp.
O'Donnell, S.R. and J.L. Wanstall (1980). Evidence that ICI 118,551 is a potent, highly beta$_2$-selective adrenoceptor antagonist and can be used to characterize beta-adrenoceptor populations in tissues. Life Sci. 27:671-677.
Sylvester, J.T., S.M. Scharf, R.D. Gilbert, R.S. Fitzgerald and R.J. Traystman (1979). Hypoxic and CO hypoxia in dogs: hemodynamics, carotid reflexes, and catecholamines. Am. J. Physiol. 236:H22-H28.
Villeneuve, S.M., C.K. Chapler, C.E. King and S.M. Cain (1985). The role of beta-adrenergic receptors in carbon monoxide hypoxia. Proc. Can. Fed. Biol. Soc. (Abs) 28:77.

ANALYSIS OF GAS EXCHANGE IN THE OPHTHALMIC RETE OF PIGEONS

Marvin H. Bernstein

Department of Biology
New Mexico State University
Las Cruces, NM 88003

Many birds have ophthalmic retia (OR), paired vascular networks in the head, that cool arterial blood flowing toward the brain. They thus independently regulate brain and core body temperatures (Bernstein et al., 1979; Kilgore et al., 1979). The system works because evaporation from nasal and oral membranes cools venous blood before it flows through the OR. This blood then takes up heat from warmer arterial blood flowing countercurrent to it (Midtgård, 1983; Clair, 1985). The vascular and neural anatomy of the OR suggests that the sympathetic nervous system controls both total blood flow and the proportioning of flow between heat-exchange and shunt vessels (Midtgård, 1985). Birds routinely tolerate body temperatures of 45°C, indicating an extreme capacity for heat storage. This and the ability to cool the brain serves during heat exposure and exercise by reducing the need for evaporative cooling.

Recent observations of high cerebrospinal-fluid (csf) O_2 tension in pigeons (114 Torr), compared with an arterial O_2 tension (PaO_2) of 82 Torr, have led to the hypothesis that the cephalic mucosa and the OR may function as gas exchangers as well as heat exchangers (Bernstein et al., 1984). Mucosal blood is thought to release CO_2 and take up O_2 from air, increasing PO_2 and decreasing PCO_2 in venous blood flowing to the OR. The high PvO_2 and low $PvCO_2$ allow the arteries and veins in the OR to exchange gases, increasing PaO_2 and perhaps hemoglobin O_2 saturation. During gas exchange in the OR, this could compensate for the increase in O_2 affinity that accompanies reduced temperature. On the other hand, if OR arteries and veins exchange gases without a prior increase in venous O_2 tension (PvO_2), O_2 tension and content could fall markedly in the cooled arterial blood reaching the brain (Pinshow et al., 1985).

In support of O_2 enhancement, Bernstein et al., (1984) observed that the PO_2 difference between csf and systemic arterial blood decreased when they prevented contact between air and the oronasal mucosa. Pinshow et al. (1985) have analyzed the effects of temperature and PCO_2 on O_2 combining properties in pigeon blood and have shown that O_2 transport to and O_2 diffusion in the brain would indeed be enhanced if the proposed mechanism really operated. In experiments on pigeons using H_2 in normoxic or hypoxic inspired gas, the tracer appeared in the carotid artery and the brain in quick succession, whereas H_2 administered via the nasal cavity alone appeared only in the brain. In the latter protocol, greatly diluted H_2 appeared in the blood in the jugular vein but appeared to be excreted in one pass through the lungs (Pierce, et al., unpublished observation). Local application of a vasoconstrictor to the nasal membranes prevented nasally administered H_2 from appearing in the brain, whereas local vasodilation enhanced brain H_2. An increased body-brain temperature difference, indicating more heat exchange in the OR, was accompanied by an increase in the appearance of nasally administered H_2 in the brain (Pierce, et al., unpublished observation). All these results support the idea of extrapulmonary enhancement of brain oxygen.

As pointed out previously (Bernstein, et al., 1984; Pinshow et al., 1985), for the system to function as proposed it is necessary that gases be exchanged across oronasal mucosa and between OR arteries and veins. The only analysis of OR function has been that presented by Midtgård (1983) for heat exchange. The present contribution attempts to extend the approach in Midtgård's analysis to the potential for gas exchange in the OR of pigeons.

Calculations

Heat Transfer in the Ophthalmic Rete

In Midtgård's (1983) analysis, the rate of heat transfer from the arteries to the veins depends on the resistance to heat flux between blood and its respective vessel walls, and through the walls themselves. The sum of these resistances is the inverse of the total conductance (U). Midtgård (1983) measured arterial dimensions in the OR of 40 species of birds over a wide range of body weights and presented a series of allometric expressions for these in passerine and nonpasserine species. He also estimated the allometric expression for arterial blood flow in the OR from arterial flow information in mammals. He used these relationships to calculate allometric expressions for U in passerine and nonpasserine birds of 4.6×10^4 $m^{-0.18}$ and 4.6×10^4 $m^{-0.17}$, respectively, where m is body mass in grams. From the theory for ideal countercurrent heat exchange, Midtgård (1983) also derived an expression for the temperature change in arterial blood (ΔT) during its traverse from the inlet to the outlet of the OR:

$$\Delta Ta = (Tai - Tvi)/[1 + m \ c/(As \ U)] \tag{1}$$

where Tai and Tvi are the temperatures of arterial and venous blood, respectively, as they arrive at their respective inlets to the OR. Tai and Tvi are here assumed to be 41 and $36°C$, based on direct measurements in cormorants (Bernstein and Hudson, unpublished observation). In equation (1) m is the total mass-flow rate of OR arterial blood, c is the specific heat of blood, and As is the total arterial surface area for heat exchange. It was assumed that m and c in the arteries were equal to the corresponding parameters in veins, and m was taken to be the product of blood flow (Vb) and density. It is unlikely that all the assumptions involved are wholly accurate; still equation (1) provides a reasonable approximation of the temperature change in arterial blood (Clair, 1985). The actual temperature of the arterial blood leaving the OR was taken as Tai - ΔTa. For this analysis I have applied Midtgård's allometric equations for OR arteries and U to a hypothetical 400 g pigeon.

Gas Transfer in the Ophthalmic Rete

Gas transfer coefficients. I estimated gas transfer coefficients (G) for O_2 and CO_2 in OR arteries by means analogous to those used by Midtgård for U. The total resistance to gas transfer (1/G) was taken to be the sum of the individual resistances in the arterial and venous blood and in the arterial walls: (1/Ga + 1/Gv + 1/Gw), where Ga and Gv are the arterial and venous blood gas transfer coefficients and Gw is the gas transfer coefficient for arterial walls. Assuming Ga = Gv = Gb, the expression for 1/G becomes 2/Gb + 1/Gw. Gb is determined as Sh Kb/D_1, where Sh is the Sherwood number for convective mass transport (Vogel, 1981), Kb is Krogh's diffusion constant for blood, and D_1 is the luminal diameter of the arteries. Sh, a dimensionless number, is calculated in terms of two other dimensionless numbers, the Reynolds number (Re) and the Schmidt number (Sc) (Welty, et al., 1984) which require knowledge of velocity and gas capacitance (βg) of blood. The rate of gas diffusion across vessel walls is Kw As Pg/E, where Pg is the gas tension difference across a wall of thickness E and Kw is Krogh's diffusion constant for the wall. Thus, Gw can be defined as Kw/E, and the expression for total resistance to gas transfer becomes:

$$1/G = 2D_1/(Sh\ Kb) + E/Kw \tag{2}$$

Since in Midtgård's morphometric analysis of the OR, E was 0.2 D_1, the expression can be rewritten:

$$1/G = 0.2\ D_1\ [10/(Sh\ Kb) + 1/Kw] \tag{3}$$

where G is in $mmol/(cm^2\ s\ Torr)$, D_1 is calculated from Midtgård's (1983) allometric equations, and the Krogh constants for blood and arterial wall are assumed to be independent of temperature and equal to those for water and muscle, respectively.

Changes in OR arterial gas tensions. By assuming that the OR functions as an ideal countercurrent exchanger, I derived equation (4) to describe the change in arterial gas tension (ΔPag) along the length of the OR:

$$\Delta Pag = (Paig - Pvig)/[1 + \dot{Vb}\ \beta b\ g/(As\ Gb)] \tag{4}$$

where Paig and Pvig are respectively the gas tensions in the inlets to the arteries and veins of the OR, at the corresponding blood temperatures (Tai and Tvi); the other symbols are as defined above.

Calculations of ΔPaO_2 were carried out for a hypothetical 400-g pigeon using a range of values for $PaiO_2$ (60-100 Torr) and $PviO_2$ (40-120 Torr) in multiple combinations. Blood flow and exchange surface area values were calculated from the allometric equations of Midtgård (1983) for nonpasserine birds. For each combination of conditions, βbO_2 was calculated as the ratio of the difference between total blood O_2 content in the OR's arterial and venous inlets ($CaiO_2 - CviO_2$), to the corresponding O_2 tensions (Paig - Pvig). βbCO_2 was calculated similarly. $CaiO_2$ and $CviO_2$ were calculated for the corresponding PO_2, PCO_2 and temperature values by the equation for pigeon blood presented by Pinshow et al., (1985), assuming an O_2 carrying capacity of 16.4 volume % (Weinstein et al., 1985). Total CO_2 content at both sites was calculated from PCO_2 and O_2 saturation by the equation of Weinstein et al., (1985) for CO_2-combining properties of pigeon blood in relation to the Haldane effect. The PO_2 in the arterial blood flowing out of the OR, ($PaiO_2 - \Delta PaO_2$), thus reflected the effects of temperature, PCO_2, and pH changes along the arteries, as well as any transfer of O_2 between venous and arterial blood that may have occurred.

RESULTS AND DISCUSSION

Table I shows the results of the calculations of rete dimensions from allometric equations. These data were employed, along with thermal properties (Midtgård, 1983) and gas-combining properties (Weinstein et al., 1985) of pigeon blood, to calculate the arterial changes in temperature and gas tension along the OR.

Table I. Dimensions of one ophthalmic rete in a 400-gram hypothetical pigeon, calculated from allometric equations of Midtgård (1983).

Mean arterial length, mm	4.3
Mean arterial wall thickness, μm	10.5
Mean arterial luminal diameter, mm	0.053
Total cross-sectional area, cm^2	0.0011
Total arterial exchange surface, cm^2	0.29
Total arterial blood flow, cm^3/min	0.76

Heat Transfer

The estimates of arterial temperature change in the OR, calculated from equation (1), indicate that heat exchange is nearly complete, arterial outlet temperature (Tao) closely approximating Tvi. This agrees with estimates by Clair (1985) from a more complete vascular heat exchange model developed by Mitchell and Myers (1968).

Gas Transfer

As Table II shows, gas transfer coefficients for CO_2 and O_2 were extremely low, amounting to 4×10^{-7} and 1.2×10^{-8} mmol/(cm^2 s Torr), respectively, and they were unaffected by temperature or gas tensions in arterial and venous blood. The relatively high saturation values make facilitated diffusion negligible. The arterial blood gases were calculated to decrease during flow through the OR; PO_2 fell by 9-10 Torr and PCO_2 fell by 1 Torr. These results can be explained entirely by the blood temperature change. This was true over the entire ranges of PCO_2 and PO_2 values used in the calculations. Even under conditions of simulated high altitude in which blood temperature and gas tensions are greatly reduced and βb for both gases may change markedly (Weinstein et al., 1985), arteriovenous gas transfer appeared not to occur in the OR.

The conclusion seems to be that vascular geometry in the OR precludes arteriovenous gas transfer. For arterial PO_2 not to decrease during blood's traverse of the OR under the arterial and venous inlet conditions given below the "Extrapulmonary Gas Exchange" heading in Table II, the estimates of arterial flow, arterial wall thickness, or βbO_2 would have to be reduced, the exchange surface would have to be increased, or these alterations would have to occur in combination.

Nevertheless, the experimental evidence mandates cautious interpretation of the present calculations. The data from the experiments on PO_2 in csf and from the H_2 tracer studies described above seem to support unequivocally a nonsystemic pathway from oronasal mucosa to brain via the OR. Furthermore, as pointed out previously (Bernstein, in press), it is reasonable to use H_2 as an O_2 analog in the tracer studies mentioned above, since the diffusion rates and Krogh constants, as well as the solubilities, are similar for the two gases.

Given the uncertainties associated with the use of allometric equations to calculate values for individual species, the present analysis must be viewed as preliminary. To reconcile it with the experimental results, additional information about vascular dimensions and retial blood flow will be required. To settle finally the question of brain O_2 enhancement, direct measurement of PO_2 in blood just before and just after it traverses the OR will be needed, a feat whose technical requirements have so far thwarted all attempts to achieve it.

Table II. Calculated gas transfer parameters at arterial and venous inlet and arterial outlet sites in the ophthalmic rete (OR) of a hypothetical 400-gram pigeon, using data in Table I, equations (3) and (4), and blood gas-combining properties reported by Pinshow et al. (1985) and Weinstein et al. (1985). Conventional data assume only cephalic blood-tissue gas exchange, whereas Extrapulmonary Gas Exchange data were calculated assuming CO_2 and O_2 exchange in oronasal mucosa and in OR.

	Conventional			Extrapulmonary Gas Exchange		
	Venous Inlet	Arterial Inlet	Arterial Outlet	Venous Inlet	Arterial Inlet	Arterial Outlet
Carbon Dioxide						
[1] βbCO_2, mmol/(l Torr)	0.22			0.23		
[1] GCO_2, mmol/(cm^2 s Torr)	4.4×10^{-7}			4.5×10^{-7}		
PCO_2, Torr	40	33	33	18	33	32
CO_2 Content, mmol/l	23	21.4	21.4	17.9	21.4	21.4
Oxygen						
[1] βbO_2, mmol/(l Torr)	0.092			0.029		
[1] GO_2, mmol/(cm^2 s Torr)	1.3×10^{-8}			1.2×10^{-8}		
PO_2, Torr	40	80	71	99	80	71
O_2 Content, mmol/l	3.5	6.4	6.4	7.2	6.4	6.4
P_{50}, Torr	41	43	37	29	43	43

[1] Blood gas capacitance (βb) and gas transfer coefficients (G) calculated for conditions of arterial and venous blood at inlets to OR.

REFERENCES

Bernstein, M.H. (in press) Temperature and oxygen supply in avian brain. In: <u>Comparative Pulmonary Physiology: Current Concepts</u>, edited by S.C. Wood. New York, Dekker.

Bernstein, M.H., I. Sandoval, M.B. Curtis and D.M. Hudson (1979). Brain temperature in pigeons: Effects of anterior respiratory bypass. J. Comp. Physiol. 129:115-118.

Bernstein, M.H., H.L. Duran and B. Pinshow (1984). Extrapulmonary gas exchange enhances brain oxygen in pigeons. Science 226:564-566.

Clair, P.M. (1985). The Rete Mirabile Ophthalmicum of the Double-crested Cormorant (<u>Phalacrocorax</u> <u>auritus</u>): Its Form and Function. M.S. Thesis, New Mexico State University, Las Cruces, New Mexico.

Kilgore, D.L., Jr., D.F. Boggs and G.F. Birchard (1979). Role of the rete mirabile ophthalmicum in maintaining the body to brain temperature difference in pigeons. J. Comp. Physiol. 129:119-122.

Midtgård, U. (1983). Scaling of the brain and the eye cooling system in birds: A morphometric analysis of the rete ophthalmicum. J. Exp. Zool. 225:197-207.

Midtgård, U. (1985). Innervation of the avian ophthalmic rete. Fortschritte der Zoologie 30:401-404.

Mitchell, J.W. and G.E. Myers (1968). An analytical model of the countercurrent heat exchange phenomena. Biophys. J. 8:897-911.

Pinshow, B., M.H. Bernstein and Z. Arad (1985). Effects of temperature and PCO_2 on O_2 affinity of pigeon blood: implications for brain O_2 supply. Am. J. Physiol. 249:R758-R764.

Vogel, S. (1981). <u>Life in Moving Fluids, the Physical Biology of Flow</u>. Boston, Willard Grant Press.

Weinstein, Y., M.H. Bernstein, P.E. Bickler, D.V. Gonzales, F.C. Samaniego and M.A. Escobedo (1985). Blood respiratory properties in pigeons at high altitudes: effects of acclimation. Am. J. Physiol. 249:R765-775.

Welty, J.R., C.E. Wicks and R.E. Wilson (1984). <u>Fundamentals of Momentum, Heat, and Mass Transfer</u>. New York, Wiley.

OXYGEN TRANSFER IN THE TISSUES

MATCHING O_2 DELIVERY TO O_2 DEMAND IN MUSCLE: I. ADAPTIVE VARIATION

Ewald Weibel and Susan R. Kayar

Department of Anatomy
University of Berne
Buehlstrasse 26, CH-3000 Berne 9
Switzerland

THE PROBLEM

This paper and the following one by C.R. Taylor et al. report an attempt to look at the conditions for O_2 delivery from the atmosphere to the mitochondria in muscle cells in an integrated approach, and this in a double sense: 1) we look at all steps in the transfer pathway that leads from the lung through the circulation and the microvasculature to the O_2 consumption site in the mitochondria; 2) we consider both the structural design and the functional performance at each step. In addition, we are interested mostly in what is happening at very high levels of O_2 consumption, first because under these conditions 90-95% of $\dot{V}O_2$ occurs in the muscles, so we know the target for O_2 entering the system at the lung, and second because the limit to $\dot{V}O_2$ is of particular interest in this context. We would like to know at what level in the respiratory system is this limit set and for what reasons, physiological or structural.

THE MODEL FOR STRUCTURE-FUNCTION CORRELATION

Fig. 1 shows a model of the mammalian respiratory system set up in the form of a cascade, similar to the model used in previous studies (Weibel and Taylor, 1981). The pathway is broken into four steps: the pulmonary gas exchanger, the circulatory transport of O_2 by the blood and heart, the tissue gas exchanger in the muscle microvasculature, and the mitochondria as O_2 consumers. Under steady state conditions, O_2 flow is the same for each step. At each level, $\dot{V}O_2$ is determined by the product of two terms, the first being essentially functional, whereas the second is strongly affected by structural parameters.

The goal of our studies is to obtain estimates of all the variables shown in this model and to relate them to the limit of aerobic work, i.e., to $\dot{V}O_2$max. We focus our attention particularly on the structural parameters as primary candidates as potential limiting factors of physiological regulation, the reason being that regulating structure to functional demand is a slow and costly process which requires morphogenesis. On the basis of these arguments, we have formulated the hypothesis of symmorphosis, defining it as a state of structural design commensurate to functional needs resulting from regulated morphogenesis (Taylor and Weibel, 1981). This hypothesis predicts that each part of the respiratory system is matched to all others

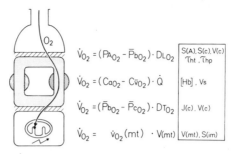

$$\dot{V}_{O_2} = (P_{A_{O_2}} - \bar{P}_{b_{O_2}}) \cdot D_{L_{O_2}} \quad \boxed{\begin{array}{c} S(A), S(c), V(c) \\ \dot{T}ht, \dot{T}hp \end{array}}$$

$$\dot{V}_{O_2} = (C_{a_{O_2}} - C_{\bar{v}_{O_2}}) \cdot \dot{Q} \quad \boxed{[Hb], Vs}$$

$$\dot{V}_{O_2} = (\bar{P}_{b_{O_2}} - \bar{P}_{c_{O_2}}) \cdot D_{T_{O_2}} \quad \boxed{J(c), V(c)}$$

$$\dot{V}_{O_2} = \dot{v}_{O_2}(mt) \cdot V(mt) \quad \boxed{V(mt), S(im)}$$

Fig. 1. Schematic representation of the pathway for oxygen in the mammalian respiratory system from the lung (top) through the circulatory system (stippled portion in center), to the mitochondria in skeletal muscle (bottom). The pathway has been divided into four steps, with the oxygen transfer ($\dot{V}_{O_2}$) of each step defined as the product of two terms. The first term is essentially functional, while the second term is strongly affected by the structural parameters listed at the right.

and to the limit of aerobic performance of the entire system. Our studies are designed to test this hypothesis.

THE APPROACH

The experimental design of all our studies involves basically four stages: (1) After a period of conditioning to the experimental procedures the animals are made to run on a treadmill at different speeds up to the speed that elicits $\dot{V}_{O_2}$max; each run lasts for 5-7 min so as to reach steady state levels of performance. (2) During these runs all the necessary physiological measurements are made at steady state. (3) When all these measurements are made, the animal is sacrificed in the course of a terminal experiment; at this time, we obtain tissue samples of all relevant organs (lung, heart, skeletal muscles) following sampling strategies which allow us to obtain morphometric information pertaining to the whole body. (4) The tissue samples thus obtained are used for a thorough morphometric analysis using stereological methods (Weibel, 1979). This approach, performed on the same animals, allows us then to work out structure-function correlations on the basis of the integrated model presented above (Fig. 1).

In order to test the hypothesis of symmorphosis, we need to vary the basic global reference parameter, $\dot{V}_{O_2}$max. For this we use three different strategies (Fig. 2): (1) Allometric variation where mass-specific $\dot{V}_{O_2}$max of small mammals, such as mice, is 4-7 times higher than in large mammals, such as cows; (2) Adaptive variation, where "normal" animals are compared with "athletic" species of the same size, which achieve 2-3 times higher $\dot{V}_{O_2}$max; (3) Induced variation, i.e., training to higher levels of aerobic performance, by which one can achieve differences in $\dot{V}_{O_2}$max of about 1.3-fold.

In this paper we are looking at the modifications in the parameters determining $\dot{V}_{O_2}$max occurring with adaptive variation (Taylor et al., 1987), whereas the following paper by Taylor et al. considers allometric variation (Weibel and Taylor, 1981).

Adaptive Variation

Most mammalian species, like healthy man, can increase their rate of O_2 consumption by about 10-fold between rest and maximal levels of exercise.

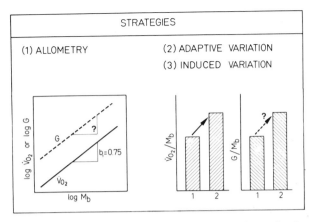

Fig. 2. Three primary strategies for testing the hypothesis of symmorphosis. Strategy 1: Allometry examines differences in mass-specific oxygen uptake rate ($\dot{V}O_2$) of animals over a range of body mass, and calculates this difference as the slope of a regression. This slope is then compared to the slope of the regression for the conductance (G) of some other parameter. Strategy 2: Adaptive Variation examines differences in $\dot{V}O_2$ of "normal" versus "athletic" animals. Strategy 3: Induced Variation examines differences in the conductance (G) of some parameter experimentally induced to reach a higher level, for example by exercise training.

But some "athletic" species, like human athletes, achieve an up to 20-fold increase of $\dot{V}O_2$ (Prosser, 1973). Accordingly, $\dot{V}O_2$max of the athletic species is 2-3 times larger than that of the more normal species. Table I shows recent data obtained on two adaptive pairs, namely dogs compared to goats and ponies to calves of similar body mass, where $\dot{V}O_2$max of the athletic species was 2.5 times higher (Taylor et al., 1987). We shall exploit this difference in asking in what way the functional and structural parameters which determine $\dot{V}O_2$ (Fig. 1) are modified to allow this higher $\dot{V}O_2$max. The model of Fig. 1 will be followed from bottom up, beginning with the mitochondria as the O_2 sink and ending with the lung.

The Mitochondria

Oxidative phosphorylation occurs at the respiratory chain enzymes, which are densely packed into the inner mitochondrial membranes (Schwerz-

Table I. Oxygen uptake rate per unit body mass ($\dot{V}O_2$/Mb) at rest and at maximal aerobic exercise, for the "athletic" dog and pony versus the "normal" goat and calf.

STRATEGY 2: ADAPTIVE VARIATION

COMPARE:	$\dot{V}_{O2}$ / M_b		
	rest	max	max / rest
DOGS (28 kg)	0.17	2.29	13 x
GOATS (30 kg)	0.14	0.95	7 x
PONIES (170 kg)	0.07	1.48	21 x
CALVES (140 kg)	0.08	0.61	8 x

$$mlO_2.sec^{-1}.kg^{-1}$$

161

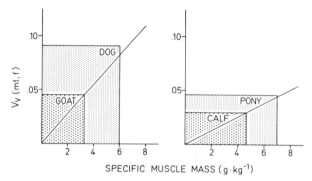

Fig. 3. Greater muscle mass and greater volume density of
mitochondria (V_v(mt,f)) contribute approximately equally to
the greater total volume of mitochondria in representative
locomotor muscles (M. semitendinosus and M. vastus medialis)
of "athletic" dog and pony versus "normal" goat and calf.

mann et al., 1986). The rate of O_2 consumption is limited by the number of
such functional units and by the rate at which they can operate. Since the
density of inner membranes is constant for skeletal muscle the amount of
enzymes available for oxidative phosphorylation can be estimated by the
mitochondrial volume, V(mt), whereas the maximal rate at which they can
operate is reflected by the rate of O_2 consumption of the unit mitochondrial
volume, $\dot{V}O_2$ (mt), at $\dot{V}O_2$max. The question we need to address is whether the
higher $\dot{V}O_2$max of the athletic species (dogs and ponies) is due to an
increased rate of enzyme activity or to a larger mitochondrial volume, or to
some combination of both factors.

Fig. 3 shows that the total mass of mitochondria in skeletal muscle is
larger in the athletic species; this is due to a larger muscle mass combined
with a higher volume density of mitochondria (Hoppeler et al., 1987). This
increase in total mitochondrial volume is proportional to $\dot{V}O_2$max; accordingly, the rate of O_2 consumption by the unit mitochondrial volume at $\dot{V}O_2$max
is invariant at 3.5-4.5 ml O_2/min (Table II). This confirms the observation
of Hoppeler and Lindstedt (1985) that the maximal rate of $\dot{V}O_2$ by mitochondria is invariant.

Table II. The maximal oxygen uptake rate per unit volume of
mitochondria ($\dot{V}O_2$max/Vmt) is the same in dog, goat, pony and
calf. This value is the quotient of the oxygen uptake rate per
unit body mass ($\dot{V}O_2$/Mb) and the total volume of mitochondria
per unit body mass (Vmt/Mb).

ADAPTIVE VARIATION: at $\dot{V}_{02}$max

O_2 consumption in locomotor muscle mitochondria

	($\dot{V}_{02}$max/M_b)	/	(Vmt/M_b)	=	$\dot{V}_{02}$max/Vmt
DOG	2.28		40.6		3.37
GOAT	.90		13.8		3.90
D/G	2.5		2.94		.86
PONY	1.48		19.5		4.55
CALF	0.61		9.1		3.98
P/C	2.4		2.13		1.14
	ml.sec^{-1}.kg^{-1}		ml.kg^{-1}		ml.min^{-1}.ml^{-1}

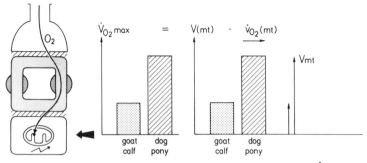

Fig. 4. The greater maximal oxygen uptake rate ($\dot{V}_{O_2}$max) of dog and pony versus goat and calf is due entirely to a greater volume of mitochondria (V(mt)), with the maximal oxygen uptake rate of a unit volume of mitochondria ($\dot{V}_{O_2}$(mt)) constant.

We therefore conclude that, at the level of muscle mitochondria, adaptive variation is achieved simply by building more mitochondria of the same kind (Fig. 4). Adaptation is thus achieved totally by structural means. It is interesting to note that this also holds for myocardium in these species (Karas et al., 1987a).

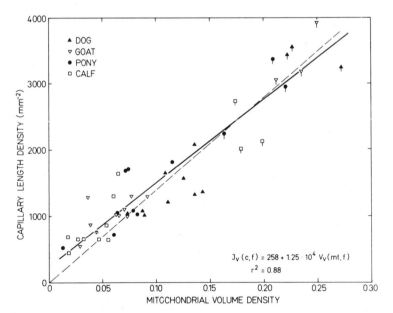

Fig. 5. Capillary length density versus mitochondrial volume density for selected skeletal muscles (diaphragm, M. semitendinosus and M. vastus medialis) and heart (marked with vertical bars) of dog, goat, pony and calf. The solid line indicates the least-square regression of the data. The dotted line indicates the regression expected if there were a direct proportionality between capillaries and mitochondria.

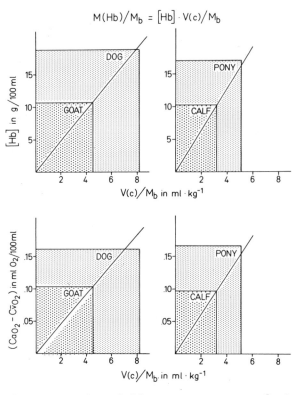

$$M(Hb)/M_b = [Hb] \cdot V(c)/M_b$$

Fig. 6. A greater hemoglobin concentration ([Hb]) and a greater mass-specific volume of capillaries (V(c)/Mb) contribute approximately equally to the greater total mass-specific amount of hemoglobin (M(Hb)/Mb) present in the blood of dog and pony versus goat and calf. The greater arterio-venous oxygen concentration difference (CaO$_2$ - C$\bar{v}$O$_2$) and greater mass-specific capillary volume contribute approximately equally to the higher oxygen delivery by the blood in the dog and pony versus goat and calf.

The Capillary Supply of Muscle

The rate at which capillaries can deliver O$_2$ to the muscle cells is partly determined by the amount of capillary blood available for O$_2$ delivery. This is related to the length density of capillaries (Hoppeler et al., 1981) which is obtained from the counts of capillary density multiplied by a tortuosity factor. The first question we can ask is whether the capillary length density is proportional to the mitochondrial volume density. Fig. 5 shows that this is the case, at least in first approximation (Conley et al., 1987). However, a closer look at the data shows that the athletic species have a somewhat lower capillary-to-mitochondria ratio; indeed, it turns out that the total muscle capillary volume is only 1.7 times greater in the athletic species and, therefore, does not match the differences in VO$_2$max. As shown in Fig. 6, the 2.5-fold greater rate of O$_2$ delivery from capillaries depends also on a 1.6-times larger arterio-venous difference in O$_2$ concentration; this is explained by the 1.6-fold higher hemglobin concentration in the blood of the athletic species.

We therefore conclude that the higher rate of O$_2$ delivery to muscle cells required in the athletic species is achieved by a combination of two structural adaptations (Fig. 7): an increase in the capillary density and an

Table III. The greater maximal oxygen uptake rate ($\dot{V}O_2$max) of "athletic" dog and pony versus "normal" goat and calf is the product of a greater arterio-venous blood oxygen concentration difference ($(CaO_2 - C\bar{v}O_2)$)max and a greater cardiac output ($\dot{Q}$max) at maximal exercise.

ADAPTIVE VARIATION: at $\dot{V}_{O2}$max

O_2 transport by circulation and blood

	$\dot{V}_{O2}$max	=	$(Ca_{O2} - C\bar{v}_{O2})$max	$\dot{Q}$max
DOG	2.29		0.16	14.4
GOAT	0.95		0.10	9.3
D / G	2.41		1.55	1.56
PONY	1.48		0.17	8.9
CALF	0.61		0.10	6.3
P / C	2.43		1.42	1.71
	$ml.sec^{-1}.kg^{-1}$		$ml.ml^{-1}$	$ml.sec^{-1}.kg^{-1}$

increase in the number of erythrocytes contained in the unit volume of blood (Conley et al., 1987). The functional parameters, such as mean capillary transit time, are invariant, however.

The Circulation of Blood

The transport of O_2 by the circulation is determined by the Fick equation (Fig. 1, second line), that is, by the product of cardiac output, $\dot{Q}$, and the arterio-venous difference in O_2 concentration, $(CaO_2 - C\bar{v}O_2)$. We have observed above (Fig. 6), that the latter parameter is larger in the athletic species; accordingly, the 2.5-fold higher level of $\dot{V}O_2$max can be achieved by increasing $\dot{Q}$ by 1.6-fold, as shown in Table III (Karas et al., 1987a). This increase in cardiac output is achieved totally by increasing stroke volume, and thus the size of the ventricle, whereas the functional parameter frequency is invariant within each pair (Fig. 8); maximal heart frequency is determined by body size.

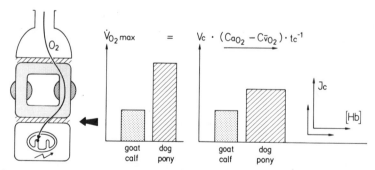

Fig. 7. A higher arterio-venous oxygen concentration difference in capillary blood (CaO_2-$C\bar{v}O_2$) and a greater capillary volume (V(c)) contribute approximately equally to the higher maximal oxygen uptake rate ($\dot{V}O_2$max) of dog and pony versus goat and calf. The greater oxygen supply from the capillary blood of the dog and pony is provided by a longer total length of capillaries (J(c)) and a greater hemoglobin concentration ([Hb]) in the blood. Mean minimal transit time for blood in capillaries (t_c) is constant within each pair of animals.

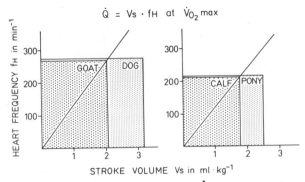

$$\dot{Q} = V_s \cdot f_H \quad \text{at} \quad \dot{V}_{O_2} \text{max}$$

Fig. 8. The greater cardiac output ($\dot{Q}$) of dog and pony versus goat and calf is due entirely to a greater stroke volume (V_s), with heart frequency (f_H) constant for each pair of animals exercising at their maximal aerobic limit ($\dot{V}_{O_2}$max).

We therefore conclude that in adaptive variation, higher rates of O_2 transport by the circulation are achieved totally by structural adaptation, but the two contributing components, heart and blood, share the adaptive effort to about equal parts (Fig. 9).

The Pulmonary Gas Exchanger

Oxygen uptake in the lung is driven by the P_{O_2} difference between alveolar air and capillary blood, two functional parameters affected by ventilation and perfusion. On the basis of the Bohr equation (Fig. 1), the conductance is the pulmonary diffusing capacity, DL_{O_2}. The structural parameters that determine the diffusing capacity of the lung are the alveolar and capillary surface area, the capillary blood volume, and the harmonic mean thickness of the tissue and plasma barriers; the model developed by Weibel (1970/71), therefore, allows one to estimate the effect of differences in structure on different rates of O_2 uptake. Previous studies had suggested that the different levels of $\dot{V}_{O_2}$ that occur with adaptive variation could be matched by a proportionately larger DL_{O_2}; but these results were inconclusive because they were not based on estimations of $\dot{V}_{O_2}$max (Weibel, 1979).

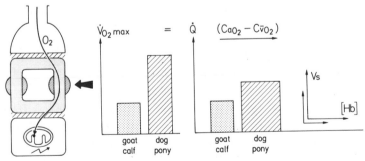

Fig. 9. A higher cardiac output ($\dot{Q}$) and a greater arterio-venous blood oxygen concentration difference ($Ca_{O_2} - C\bar{v}_{O_2}$) contribute approximately equally to the greater maximal oxygen uptake rate of dog and pony versus goat and calf. A greater stroke volume (V_s) and greater hemoglobin concentration ([Hb]) share equally in this effort.

Table IV. The greater maximal oxygen uptake rate ($\dot{V}O_2$max) of "athletic" dog and pony versus "normal" goat and calf is the product of a greater oxygen diffusing capacity of the lung (DLO_2) and a greater oxygen pressure gradient from alveoli to lung capillaries ($PAO_2-\overline{Pb}O_2$).

ADAPTIVE VARIATION: lung morphometry
Pulmonary O_2 uptake

	$\dot{V}_{O2}$max	=	DL_{O2}	·	$(PA_{O2} - \overline{Pb}_{O2})$
DOG	2.27		.118		19.2
GOAT	0.90		.080		11.3
D / G	**2.52**		**1.46**		**1.70**
PONY	1.48		.079		18.9
CALF	0.61		.050		12.1
P / C	**2.44**		**1.57**		**1.56**

$$ml.sec^{-1}.kg^{-1} \qquad\qquad mmHg$$

When we estimated DLO_2 from morphometric data collected on the animals of this study, we found that DLO_2 was only 1.5 times larger in the athletic species (Table IV), and part of this was even contributed by the higher hemoglobin concentration (Weibel et al., 1987). Accordingly, we had to conclude that the alveolar-capillary PO_2 difference was larger in the athletic species by a factor of 1.6 (Table IV). In order to look for the reasons behind this difference in driving force, we calculated estimates of PAO_2 and performed a Bohr integration of the capillary PO_2 profile on the basis of available physiological data (Karas et al., 1987b). The result of this analysis is shown in Fig. 10: goats and calves achieve O_2 equilibration of their capillary blood with alveolar air after less than half the capillary transit time, whereas the athletic dog and pony need 3/4 of the transit time for equilibration.

This leads to the conclusion that about half of the path length of the pulmonary capillary is redundant in the "normal" species. The athletic species use part of this redundancy to achieve a higher rate of O_2 uptake, and thus establish a larger PO_2 difference; but they also need to increase

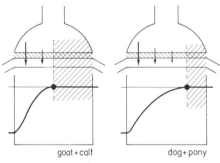

goat + calf   dog + pony

Fig. 10. Calculation of the pulmonary capillary PO_2 profile by Bohr integration indicates that oxygen equilibration between alveolar air and capillary blood is achieved in 1/2 the transit time of blood in pulmonary capillaries of goat and calf, and in 3/4 the transit time in dog and pony.

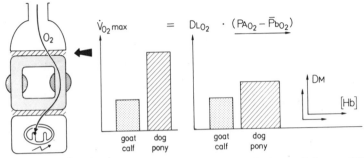

Fig. 11. A greater oxygen diffusing capacity of lung (DLO_2) and greater alveolar-blood oxygen pressure gradient ($PAO_2 - \overline{P}bO_2$) contribute approximately equally to the greater maximal oxygen uptake rate ($\dot{V}O_2$max) of dog and pony versus goat and calf. This greater oxygen diffusing capacity is generated in equal parts by a greater hemoglobin concentration ([Hb]) in pulmonary blood and more lung structure (DM) in the dog and pony.

DLO_2 to some extent, partly by making more lung structure, but also by exploiting the larger hemoglobin concentration of the blood (Fig. 11).

CONCLUSIONS

In studying the modifications in the structural parameters supporting O_2 flow through the various steps of the O_2 pathway as they occur with adaptive variation, we discovered three patterns (Fig. 12): 1) At the last step of the pathway the mitochondrial volume is quantitatively adapted to $\dot{V}O_2$max; only one parameter is involved so that the hypothesis of symmorphosis holds in its simplest form. 2) The transport of O_2 by the circulation, as well as the delivery of O_2 from the capillaries, are adapted in the athletic animals by a shared adaptive effort on the part of the blood and the heart, each contributing about equally to overall adaptation; symmorphosis also holds, as adaptation is based on modifying structures, but it now involves two components. 3) In adapting pulmonary gas exchange to higher levels of $\dot{V}O_2$max athletic animals appear to first exploit some of the

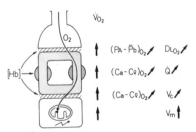

Fig. 12. The pathway for oxygen in the mammalian respiratory system is modified to achieve a greater maximal oxygen uptake rate ($\dot{V}O_2$max) in "active" versus "normal" animals. At the level of the lung, circulatory system and muscle microcirculation, the modifications involve an adaptive effort shared equally between one primarily functional parameter and one parameter that is strongly affected by structure. Within the muscles, $\dot{V}O_2$max is increased solely by increasing the volume of mitochondria (Vmt).

redundancy of their gas exchange apparatus, but they also augment their diffusing capacity by both enlarging lung structure and the hemoglobin concentration of the blood. On the basis of current evidence, we must, therefore, conclude that symmorphosis does not - or only partially - hold in the lung, since functional changes contribute half of the adaptive effort (Taylor et al., 1987).

ACKNOWLEDGEMENTS

The illustrations are reproduced from Taylor, Karas, Weibel and Hoppeler, Respir. Physiol. 69:1-127, 1987. This work was supported by grant 3.036.84 from the Swiss National Science Foundation.

REFERENCES

Conley, K.E., S.R. Kayar, K. Roesler, H. Hoppeler, E.R. Weibel and C.R. Taylor (1987) Adaptive variation in the mammalian respiratory system in relation to energetic demand: IV. Capillaries and their relationship to oxidative capacity. Respir. Physiol. (69:47-64).

Hoppeler, H., S.R. Kayar, H. Claassen, E. Uhlmann and R.H. Karas (1987) Adaptive variation in the mammalian respiratory system in relation to energetic demand: III. Skeletal muscles: setting the demand for oxygen. Respir. Physiol. (69:27-46).

Hoppeler, H. and S.L. Lindstedt (1985) Malleability of skeletal muscle tissue in overcoming limitations: Structural elements. J. Exp. Biol. 115:355-364.

Hoppeler, H., O. Mathieu, E.R. Weibel, R. Krauer, S.L. Taylor and C.R. Lindstedt (1981) Design of the mammalian respiratory system. VIII. Capillaries in skeletal muscles. Respir. Physiol. 44:129-150.

Karas, R.H., C.R. Taylor, J.H. Jones, R.B. Reeves and E.R. Weibel (1987) Adaptive variation in the mammalian respiratory system in relation to energetic demand: VII. Flow of oxygen across the pulmonary gas exchanger. Respir. Physiol. (69:101-115).

Karas, R.H., C.R. Taylor, K. Roesler and H. Hoppeler (1987) Adaptive variation in the mammalian respiratory system in relation to energetic demand: V. Limits to oxygen transport by the circulation. Respir. Physiol. (69:65-79).

Prosser, C.L. (1973) Comparative Animal Physiology, Philadelphia: W.B. Saunders Co., 966 pp.

Schwerzmann, K., L.M. Cruz-Orive, R. Eggmann, A. Saenger and E.R. Weibel (1986) Molecular architecture of the inner membrane of mitochondria from rat liver: A combined biochemical and stereological study. J. Cell Biol. 102:97-103.

Taylor, C.R., K.E. Longworth, and H. Hoppeler (1987) Matching O_2 delivery to O_2 demand in muscle: II. Allometric variation in energy demand. This symposium.

Taylor, C.R. and E.R. Weibel (1981) Design of the mammalian respiratory system: I. Problem and strategy. Respir. Physiol. 44:1-10.

Weibel, E.R. (1979) Stereological Methods. Vol. I: Practical Methods for Biological Morphometry. New York: Academic Press, chapter 4 and 6.

Weibel, E.R., L.B. Marques, M. Constantinopol, F. Doffey, P. Gehr and C.R. Taylor (1987) Adaptive variation in the mammalian respiratory system in relation to energetic demand: VI. The pulmonary gas exchanger. Respir. Physiol. (69:81-100)

Weibel, E.R. and C.R. Taylor (1981) Design of the mammalian respiratory system. Respir. Physiol. 44:1-164.

MATCHING O_2 DELIVERY TO O_2 DEMAND IN MUSCLE:

II. ALLOMETRIC VARIATION IN ENERGY DEMAND

C. Richard Taylor, Kim E. Longworth and Hans Hoppeler

C.F.S., Museum of Comparative Zoology
Harvard University
Old Causeway Road
Bedford, MA 01730

SYMMORPHOSIS: THE HYPOTHESIS

This paper and the preceding paper by Weibel and Kayar address the question of whether, or to what extent, the structures involved in O_2 consumption and delivery are matched to each other. This match between structure and function is a "common sense" prediction of economic design. We have given this principle the name symmorphosis, defining it as a state of structural design commensurate to functional needs resulting from regulated morphogenesis (Taylor and Weibel, 1981).

Symmorphosis predicts that the design of each part of the respiratory system is matched to that of all others. This hypothesis can only be tested if the entire respiratory system is considered, because the transfer of O_2 in this system involves a series of linked steps (Fig. 1). This pathway for O_2 (Weibel, 1984) consists of: ventilation of the lung; diffusion between the air in the lung and the blood in the pulmonary capillaries; convective transfer from the lung to the muscle; diffusion between the muscle capillaries and the mitochondria; and consumption by the mitochondria. The flow through each of the steps must be the same under steady state conditions. Under rate limiting conditions of maximal oxygen uptake, $\dot{V}O_2max$, if all of the available structure for O_2 transfer or uptake is utilized at each step, limits would be matched and reached simultaneously at all steps.

The transfer or uptake of oxygen at each level is determined by a combination of structural and functional parameters (Dejours, 1981; Weibel, 1984), and both must be taken into account to evaluate symmorphosis. The approach of both this and the previous paper is to consider how both structural and functional parameters change at each level in the pathway for oxygen in order to bring about the large differences in $\dot{V}O_2max/Mb$ that are found among different species in nature. The previous paper considered the 2-3 fold differences that occur as the result of adaptive variation among animals of the same size. This paper considers the 4-10 fold differences that occur as the result of allometry, i.e., differences in body size.

The Mitochondria

Under conditions of $\dot{V}O_2max$, oxygen is consumed by the mitochondria of a 30 g mouse at about 4 times the rate as in a 300 kg cow (Fig. 2). Under

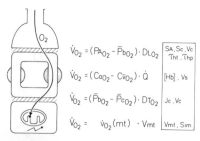

$$\dot{V}_{O_2} = (P_{A_{O_2}} - \bar{P}_{b_{O_2}}) \cdot D_{L_{O_2}} \quad \boxed{\begin{array}{c} \text{SA,Sc,Vc} \\ \text{Tht,Thp} \end{array}}$$

$$\dot{V}_{O_2} = (C_{a_{O_2}} - C_{\bar{v}_{O_2}}) \cdot \dot{Q} \quad \boxed{[\text{Hb}] \text{ , Vs}}$$

$$\dot{V}_{O_2} = (\bar{P}_{b_{O_2}} - \bar{P}_{c_{O_2}}) \cdot D_{T_{O_2}} \quad \boxed{\text{Jc , Vc}}$$

$$\dot{V}_{O_2} = \dot{v}_{O_2}(\text{mt}) \cdot V_{mt} \quad \boxed{\text{Vmt , Sim}}$$

Fig. 1. Model of the respiratory system indicating O_2 flow rates and the structural and functional parameters determining these rates. The box at the right lists some of the structural parameters which have been quantified at each step. At steady state rates of oxygen consumption, $\dot{V}_{O_2}$, the flow of O_2 through all steps in the pathway must be equal, and is determined by the rate of O_2 consumption of the muscle mitochondria. Reproduced with permission from: Weibel, E.R., C.R. Taylor, H. Hoppeler and R.H. Karas (1987). Adaptive variation in the mammalian respiratory system in relation to energetic demand. I. Introduction to problem and strategy. Respir. Physiol. 69:1-6.

these conditions, the oxygen is directed primarily to mitochondria of skeletal muscle (Folkow and Neil, 1971). These mitochondria are producing ATP by oxidative phosphorylation to meet the demands imposed by the contractile machinery. Phosphorylation is carried out by respiratory chain enzymes which are built into the mitochondria's multiply infolded inner membrane, the cristae. The concentration of these enzymes in muscle mitochondria is a valid estimate of the quantity of respiratory chain enzymes in the muscle cell (Hoppeler et al., 1987b).

The 4 fold higher rate of O_2 uptake by the mouse when compared to the cow could be accomplished structurally by building more mitochondria, functionally by increasing the rate at which phosphorylation occurs within a given volume of mitochondria, or by some combination of changes of both structural and functional parameters. A morphometric technique for estimating the volume of mitochondria, $V(\text{mt})$, in the entire skeletal musculature has been developed (Hoppeler et al., 1984). The total volume of muscle mitochondria increases with body size with approximately the same scaling factor as $\dot{V}_{O_2}$max (Fig. 2a). Thus it appears to be primarily a structural component that increases to meet the higher demand for ATP of each gram of muscle in smaller animals.

Fig. 2b plots $\dot{V}_{O_2}$max/$V(\text{mt})$ against Mb, the rate at which each ml of mitochondria consumes oxygen at $\dot{V}_{O_2}$max. The graph shows that this rate does not change with body size and each ml of mitochondria of both large and small animals consume approx 2.5-5 ml O_2 each minute despite a 4-fold difference in $\dot{V}_{O_2}$max/Mb. This constancy of metabolic activity was also observed in the previous paper by Weibel, et al., in the situation in which maximal rates of O_2 uptake differ by 2-3 fold with adaptive variation. This observation was first reported by Hoppeler and Lindstedt (1985) and is now supported by mounting evidence from a variety of sources (Hoppeler et al., 1981b; Hoppeler, et al., 1984, 1987a; Taylor, 1987).

The Muscle Capillaries

Under conditions of $\dot{V}_{O_2}$max the 4 fold higher rate of O_2 consumption by the mitochondria of the 30 g mouse compared to the 300 kg cow must be

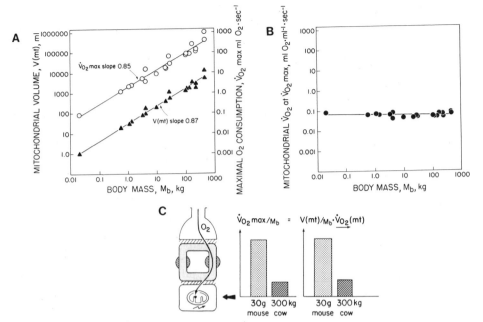

Fig. 2. A, $\dot{V}O_2$max and whole body muscle mitochondrial volume, V(mt), plotted as a function of body mass, Mb, on logarithmic coordinates have nearly equal slopes. The equation for these functions, calculated by the method of least squares, are:

$$\dot{V}O_2\text{max} = 1.79 \ \text{Mb}^{0.854}$$

where $\dot{V}O_2$max has the units of ml $O_2 \cdot \text{sec}^{-1}$ and Mb is in kg; 95% confidence intervals: intercept 1.30, 2.48; slope 0.772, 0.936

$$V(mt) = 30.12 \ \text{Mb}^{0.865}$$

where V(mt) has the unit ml and Mb is in kg; 95% confidence intervals: intercept 23.71, 38.77; slope 0.804, 0.926.

Only data where both measurements were made on the same animal are used. V(mt) is calculated or measured for the total skeletal muscle mass of 19 species using data from Hoppeler et al., (1981); Hoppeler et al., (1984), Hoppeler et al., (1987) and personal communications from H. Hoppeler, S.R. Kayar and S.L. Lindstedt.

B, mitochondrial O_2 consumption at $\dot{V}O_2$max, $\dot{V}O_2$(mt), plotted as a function of Mb on logarithmic coordinates is independent of Mb. $\dot{V}O_2$(mt) is calculated according to Hoppeler and Lindstedt (1985) as $\dot{V}O_2$max/V(mt), using data from Fig. 2A. The slope of the regression is not significantly different from zero. The mean $\dot{V}O_2$(mt) for all other species is 0.050 ml $O_2 \cdot \text{ml}^{-1} \cdot \text{sec}^{-1} \pm$ S.E. 0.003 (3.54 ml $O_2 \cdot \text{ml}^{-1} \cdot \text{min}^{-1}$).

C, the left pair of histograms show $\dot{V}O_2$max of each gram of a 30 g mouse consumes O_2 at four times the rate of each gram of a 300 kg cow (when $\dot{V}O_2$max is calculated using the allometric equation of Fig. 2A. The right pair of histograms depict $\dot{V}O_2$max as an area, the product of the volume of mitochondria, V(mt)/Mb (height), and the rate at which the mitochondria consume O_2 at $\dot{V}O_2$max, $\dot{V}O_2$(mt) (width). The increase in $\dot{V}O_2$max/Mb from cow to mouse is accomplished almost entirely by an increase in V(mt) while $\dot{V}O_2$(mt) is independent of body size.

matched by a 4 fold higher rate of O_2 diffusion from the capillaries to the mitochondria. The O_2 diffusion in the muscles can be described by the equation (Conley et al., 1987):

$$\dot{V}O_2max = (\bar{P}bO_2 - \bar{P}cO_2) \cdot DTO_2 \qquad (1)$$

where $\bar{P}bO_2$ is the mean capillary O_2 pressure, $\bar{P}cO_2$ is the mean intracellular O_2 pressure and DTO_2 is the diffusive conductance of the tissue gas exchanger. Unfortunately, a well-founded model expressing DTO_2 in terms of biophysical and structural parameters is not yet available. However, the length and volume of capillaries per unit volume of tissue are, as a first approximation, basic descriptors of the capillary network design bearing on the structural parameters determining DTO_2. Using the morphometric technique of Hoppeler et al., (1984), the mean capillary density and total skeletal muscle capillary length and volume for an individual animal can be estimated. With these calculations, we can consider the relative structural and functional contributions to O_2 flow at the capillary level using equation 2:

$$\dot{V}O_2max = V(c) \cdot \dot{V}O_2(c) \qquad (2)$$

where $V(c)$ is the volume of muscle capillary in ml and $\dot{V}O_2(c)$ is the rate at which O_2 flows from the capillary to the mitochondria at $\dot{V}O_2max$ in ml $O_2 \cdot ml^{-1} sec^{-1}$.

The length and volume of capillaries in skeletal muscles increases with a slope of 0.94 with body mass, which is greater than the slope of 0.86 for $\dot{V}O_2max$ (Fig. 3a). Thus, small animals have a relatively larger capillary volume per unit body mass and the increase in this structural component contributes to the higher O_2 delivery at this step. Using measured $\dot{V}O_2max$ and $V(c)$, we then calculate $\dot{V}O_2(c)$, the rate at which O_2 flows through each ml of capillary: it is slightly higher in smaller animals (Fig. 3b), and contributes to their higher mass specific O_2 consumption (Fig. 3c).

A higher $\dot{V}O_2(c)$ in smaller animals requires a faster rate of delivery of oxygen to and/or a greater extraction of O_2 from the capillary. A faster rate of delivery could be accomplished by a greater blood flow per unit time, a higher concentration of hemoglobin in the blood, and/or a more complete extraction of oxygen as it passes through the capillary. The concentration of hemoglobin in, and the extraction of oxygen from the blood is reportedly independent of body size (Schmidt-Nielsen, 1984). Therefore, the blood flow through the capillaries must be more rapid in the smaller animals. Average muscle transit time, tc, has been found to decrease from about 0.9 sec in 500 kg horse and cow (Kayar et al., 1987), to about 0.6 sec in the 20-30 kg goat and dog (Conley et al., 1987) to about 0.3 sec in the 4 kg fox (J.E.P.W. Bicudo personal communication).

The higher $\dot{V}O_2(c)$ must also be associated with a faster rate of diffusion of O_2 from the capillaries to the mitochondria. This could be accomplished by a larger DTO_2 and/or a larger pressure head for diffusion according to equation 1. The average diffusion distance between capillaries and mitochondria will decrease with decreasing body size as a result of the increase in volume density of both capillaries and mitochondria. Other things being equal, the shorter diffusion path will result in a larger DTO_2. However, it is possible to find pairs of animals, such as goats and cows (Conley et al., 1987; Kayar et al., 1987), that have similar estimates of mean muscle mitochondrial and capillary density (and, therefore, presumably similar mean O_2 diffusion distances), but different estimates of tc. Thus, diffusion distance alone cannot account for all differences in mass-specific oxygen delivery rates.

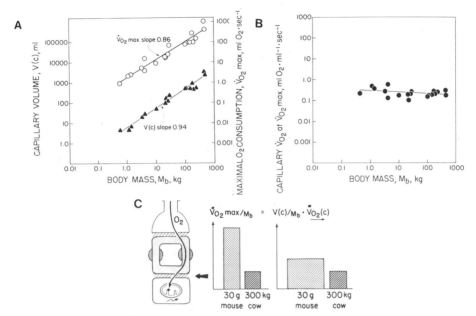

Fig. 3.   A, $\dot{V}O_2$max and whole body muscle capillary volume, V(c), plotted as a function of body mass, Mb, on logarithmic coordinates, have different slopes. The equations for these functions, calculated by the method of least squares, are:

$$\dot{V}O_2\text{max} = 1.79 \text{ Mb}^{0.854}$$

where $\dot{V}O_2$max has the units of ml $O_2 \cdot \text{sec}^{-1}$ and Mb is in kg; 95% confidence intervals: intercept 1.15, 2.79; slope 0.748, 0.980

$$V(c) = 5.88 \text{ Mb}^{0.942}$$

where V(c) has the unit ml and Mb is in kg; 95% confidence intervals: intercept 4.30, 8.04; slope 0.861, 1.023.

Only data where both measurements were made on the same animal are used. V(c) is calculated or measured for the total skeletal muscle mass of 19 species using data from Hoppeler et al., 1981; Hoppeler et al., 1987 and personal communications from J.E.P.W. Bicudo, H. Hoppeler, S.R. Kayar and S.L. Lindstedt.

B, oxygen delivery per ml of muscle capillary, $\dot{V}O_2$(c), increases slightly with a decrease in Mb. Lower values of Mb extrapolate to 0.353 ml $O_2 \cdot \text{ml}^{-1} \cdot \text{sec}^{-1}$ for a 30 g mouse compared to the value of 0.199 ml $O_2 \cdot \text{ml}^{-1} \cdot \text{sec}^{-1}$ predicted for a 300 kg cow. $\dot{V}O_2$(c) is calculated as $\dot{V}O_2$max/V(c) using data from Fig. 3A. The equation for the regression is:

$$\dot{V}O_2(c) = 0.284 \text{ Mb}^{-0.062}$$

where $\dot{V}O_2$(c) has the units ml $O_2 \cdot \text{ml}^{-1} \cdot \text{sec}^{-1}$ and Mb is in kg; 95% confidence intervals: intercept 0.194, 0.416; slope - 0.161, 0.038.

C, the left pair of histograms show $\dot{V}O_2$max of each gram of a 30 g mouse consumes O_2 at 3.5 times the rate of each gram of a 300 kg cow (when $\dot{V}O_2$max is calculated using the allometric equations of Fig. 3A). The right pair of histograms depict $\dot{V}O_2$max/Mb as an area, the product of the volume of capillaries, V(c)/Mb (height), and the rate at which the capillaries deliver O_2 at $\dot{V}O_2$max, $\dot{V}O_2$(c) (width). The increase in $\dot{V}O_2$max/Mb from cow to mouse is accomplished by a 1.7 fold increase in V(c)/Mb and a 2.1 fold increase in $\dot{V}O_2$(c).

It is also possible that a larger pressure gradient for O_2 might contribute to a more rapid diffusion in smaller animals. Since $\dot{V}(mt)$ is directly proportional to $\dot{V}O_2max$ (Fig. 2B) and $\dot{V}O_2max/V(c)$ decreases slightly with body mass (Fig. 3b), then $V(mt)/V(c)^2$ must also decrease with increasing body mass. However, it is not clear if this could provide enough of an O_2 gradient difference to account for the shorter tc in smaller animals.

Thus, with allometric variation we find a shared effort: two structural parameters, a larger $V(c)$ and shorter diffusion path length, and two functional parameters, tc and O_2 pressure gradient, are involved in meeting the higher O_2 delivery at this step. It is interesting that with adaptive variation (discussed in the previous paper), there is a shared adaptive effort between structural parameters ($V(c)$ and hemoglobin concentration) while the functional parameter of tc is unchanged.

The Circulation

Oxygen must be transported from the capillaries in the lung to each gram of muscle 4 times faster in the mouse than in the steer. This could be accomplished by a higher cardiac output, $(\dot{Q})$ a larger difference in arteriovenous O_2 concentration, $CaO_2 - C\bar{v}O_2$, or some combination of the two, as seen from the Fick equation:

$$\dot{V}O_2max = \dot{Q} \cdot (CaO_2 - C\bar{v}O_2). \tag{3}$$

Fig. 4a shows that cardiac output increases with body size in the same manner as $\dot{V}O_2max$, accounting for nearly all of the increase, whereas $CaO_2 - C\bar{v}O_2$ is nearly independent of body size (Fig. 4b). In contrast, the previous paper reported that both parameters contributed equally to the higher O_2 delivery of the athletic animals with adaptive variation.

Different mechanisms account for the larger $\dot{Q}/Mb$ with allometric (personal communication from J.E.P.W. Bicudo, J.H. Jones and K.E. Longworth) and with adaptive variation (Karas et al., 1987b). $\dot{Q}/Mb$ is the product of a functional parameter, heart rate, f_H, and a structural parameter, stroke volume (V_S/Mb):

$$\dot{Q}/Mb = f_H \cdot V_S/Mb \tag{4}$$

Higher frequencies account for about half of the difference in $\dot{Q}max/Mb$ with body size, thus a larger VS/Mb must account for the other half according to eq. 4 (Fig. 5). There are only a few measurements of these parameters at $\dot{V}O_2max$ and they span a relatively narrow range of Mb; thus, our conclusions must remain tentative. Adding to the uncertainty is the fact that heart mass and stroke volume at rest are a constant proportion of body mass (Schmidt-Nielsen, 1984; Stahl, 1967).

The Lung

As is the case with each of the other steps in the respiratory system, both a structural parameter, diffusing capacity of the lung for O_2 (DLO_2), and a functional parameter, the pressure head for diffusion between alveolar air and capillary blood ($PAO_2 - \bar{Pb}O_2$) determine the flow of O_2 in the lung. This can be expressed quantitatively using the Bohr (1909) equation:

$$\dot{V}O_2max = (PAO_2 - \bar{Pb}O_2) \cdot DLO_2. \tag{5}$$

DLO_2 scales with a slope of about 1.1 with body mass (Fig. 6A) (Gehr et al., 1981; Stahl, 1967), thus the structural parameter, DLO_2/Mb, is invari-

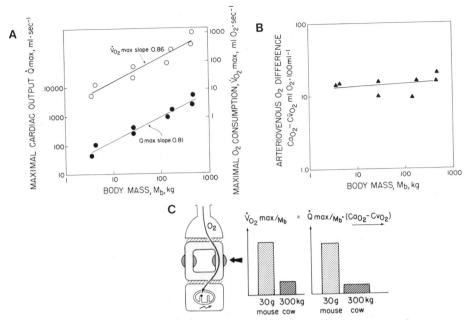

Fig. 4. A, plots of $\dot{V}O_2$max and maximal cardiac output, $\dot{Q}$max, as a function of Mb on logarithmic coordinates have nearly equal slopes. The equations for these functions calculated by the method of least squares are:

$$\dot{V}O_2max = 2.60\ Mb^{0.857}$$

where $\dot{V}O_2$max has the units of ml $O_2 \cdot sec^{-1}$ and Mb is in kg; 95% confidence intervals: intercept 0.86, 8.37; slope 0.589, 1.124.

$$\dot{Q}max = 21.90\ Mb^{0.810}$$

where $\dot{Q}$max has the units ml$\cdot sec^{-1}$ and body mass is in kg; 95% confidence intervals: intercept 10.80, 44.41; slope 0.648, 0.972.

Only data where both measurements were made on the same animal are used: cardiac output is calculated from the Fick equation as the ratio of $\dot{V}O_2$max to arteriovenous O_2 concentration difference, $CaO_2 - C\bar{v}O_2$. Data from Hoppeler et al., (1987) and personal communication from J.E.P.W. Bicudo, J.H. Jones and K.E. Longworth.

B, arteriovenous O_2 concentration difference, $CaO_2 - C\bar{v}O_2$, plotted as a function of Mb on logarithmic coordinates is nearly independent of body mass. Data are from measurements made on 8 species of mammals (Hoppeler et al., 1987, Taylor et al., 1987, personal communication from J.E.P.W. Bicudo, J.H. Jones and K.E. Longworth). The slope of the regression is not significantly different from zero. The mean $CaO_2 - C\bar{v}O_2$ is 14.9 ml $O_2 \cdot 100$ ml^{-1} ±S.E. 1.3.

C, the left pair of histograms shows that at $\dot{V}O_2$max each gram of a 30 g mouse consumes O_2 at 4 times the rate of each gram of a 300 g cow (when $\dot{V}O_2$max is calculated using the allometric equations of Fig. 4A). The right pair of histograms depict $\dot{V}O_2$max/Mb as an area, the product of maximal cardiac output, $\dot{Q}$max (height), and the arteriovenous oxygen concentration

(Continued)

177

difference, $CaO_2 - C\bar{v}O_2$ (width). The increase in $\dot{V}O_2max/Mb$ from cow to mouse is accomplished entirely by an increase in $\dot{Q}max$, while $CaO_2 - C\bar{v}O_2$ remains independent of body size, or even decreases slightly.

---

ant with size while the functional pressure head for diffusion must increase to account for the higher $\dot{V}O_2$max of the mouse compared to the cow (Weibel et al., 1981; Weibel, 1987).

In the previous paper, Weibel et al. showed that about half of the 2-3 fold difference in maximal O_2 flow across the lung with adaptive variation was due to differences in the amount of structure and about half to the differences in functional pressure head (Weibel et al., 1987). However, the higher pressure head in the athletic species was the simple consequence of utilizing a larger fraction of the transit time for gas exchange (Karas et al., 1987a).

The higher pressure head in smaller animals could also result from simply utilizing a greater fraction of the lung for gas exchange as body size decreases. Recent measurements on 4 kg foxes (Longworth et al., 1986) enable us to calculate the time course of the change in PO_2 and O_2 concentration as the blood transits their lungs. O_2 exchange appears to be complete after the blood has transited only 2/3 of their lungs. We have calculated that a 4 g shrew would have adequate lung structure to meet its higher O_2 demands if it used all of its lung (Fig. 6B). This greater utilization of the lung leads to a much lower mean capillary PO_2, $\bar{Pb}O_2$, and this might well explain how the pressure head is increased in small animals. However, other explanations are also possible and a definitive answer awaits comparable measurements on smaller animals.

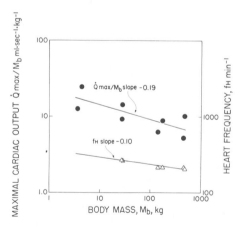

Fig. 5. Allometric plots of mass specific cardiac output at $\dot{V}O_2max$, $\dot{Q}max/Mb$, and heart frequency at $\dot{V}O_2max$, f_H, illustrates that these two functions scale differently with Mb:

$$\dot{Q}max/Mb = 21.90 \ Mb^{-0.190}$$

95% confidence intervals: intercept 10.80, 44.41; slope - 0.028, 0.352.

$$f_H = 368.3 \ Mb^{-0.097}$$

where f_H has the unit of min^{-1} and Mb is in kg; 95% confidence intervals: intercept 285.9, 474.5; slope - 0.046, 0.148.

Data from Taylor et al., (1987) and personal communication J.E.P.W. Bicudo, J.H. Jones and K.E. Longworth.

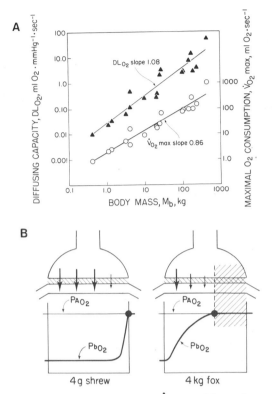

Fig. 6. A, allometric plots of $\dot{V}O_2$max/Mb and mass specific diffusing capacity of the lung for O_2 (DLO_2/Mb) illustrate that these two functions scale differently with Mb:

$$DLO_2 = 0.0258 \text{ Mb}^{1.084}$$

where DLO_2 has the units ml $O_2 \cdot \text{mmHg}^{-1} \cdot \text{sec}^{-1}$ and Mb is in kg; 95% confidence intervals: intercept 0.0129, 0.0517; slope 0.895, 1.272.

$$\dot{V}O_2\text{max} = 1.78 \text{ Mb}^{0.864}$$

where $\dot{V}O_2$max has the units of ml $O_2 \cdot \text{sec}^{-1}$ and Mb is in kg; 95% confidence intervals: intercept 1.15, 2.75 slope; 0.752–0.980.

The pressure head for diffusion, ΔPO_2, is equal to the ratio $\dot{V}O_2$max/DLO_2, therefore it increases with a decrease in Mb. Data from Gehr et al., 1981; Longworth et al., 1987, and personal communication from J.H. Jones.

B, by using a Bohr integration procedure to calculate changes in O_2 content as blood transits the lung (Karas et al., 1987), it can be shown that the DLO_2/Mb of the fox is adequate to meet the O_2 delivery needs of a 4 g shrew at $\dot{V}O_2$max. A higher $\dot{Q}$max/Mb and a shorter transit time in the shrew results in a lower mean capillary PO_2 ($\overline{P}bO_2$) and, therefore, a larger pressure head between alveolar gas (PAO_2) and $\overline{P}bO_2$.

CONCLUSIONS

The principle of symmorphosis clearly applies to the design of the respiratory system at several steps:

(1) It applies in its simplest form with respect to the mitochondria where a single structural parameter, the volume of mitochondria $(V(mt))$, varies directly in proportion to $\dot{V}O_2max$ (Fig. 2C).

(2) At the level of the capillaries, symmorphosis applies and involves changes in both the structural parameters $(V(c)$ and $DTO_2)$ and the functional parameter, tc (Fig. 3C).

(3) At the level of the circulation symmorphosis also applies. $\dot{Q}max$ accounts for all of the change in $\dot{V}O_2max$ while $CaO_2 - C\bar{v}O_2$ is constant (Fig. 4C). The change in $\dot{Q}max$ involves changes in both structural (V_S) and functional (f_H) parameters.

At the level of the lung, the principle of symmorphosis does not appear to apply with respect to O_2 transfer. It may apply with respect to some other functional requirement that may play a dominant role in determining the design of the lung.

ACKNOWLEDGEMENTS

The preparation of this manuscript was supported by grants from the U.S. National Science Foundation (DCB-8612294), The Swiss National Science Foundation (3.036.84) and a U.S. National Science Foundation Pre-doctoral Fellowship to Kim E. Longworth.

REFERENCES

Bohr, C. (1909). Ueber die spezifische Tätigkeit der Lungen bei der respiratorischen Gasaufnahme. Scand. Arch. Physiol. 22:221.

Conley, K.E., S.R. Kayar, K. Rösler, H. Hoppeler, E.R. Weibel and C.R. Taylor (1987). Adaptive variation in the mammalian respiratory system in relation to energetic demand: IV. Capillaries and their relationship to oxidative capacity. Respir. Physiol. 69:47-64.

Dejours, P. (1981). Principles of Comparative Physiology. Second edition. Elsevier/North-Holland, Amsterdam, p. 265

Folkow, B. and E. Neil (1971). Circulation. Oxford University Press, London, p. 593.

Gehr, P., D.K. Mwangi, A. Ammann, S. Sehovic, G.M.O. Maloiy, C.R. Taylor and E.R. Weibel (1981). Design of the mammalian respiratory system. V. Scaling morphometric pulmonary diffusing capacity to body mass: wild and domestic mammals. Respir. Physiol. 44:61-86.

Hoppeler, H., O. Mathieu, R. Krauer, H. Claassen, R.B. Armstrong and E.R. Weibel (1981a). Design of the mammalian respiratory system. VI. Distribution of mitochondria and capillaries in various muscles. Respir. Physiol. 44:87-111.

Hoppeler, H., O. Mathieu, E.R. Weibel, R. Krauer, S.L. Lindstedt and C.R. Taylor (1981b). Design of the mammalian respiratory system. VIII. Capillaries in skeletal muscles. Respir. Physiol. 44:129-150.

Hoppeler, H., S.L. Lindstedt, E. Uhlmann, A. Niesel, L.M. Cruz-Orive and E.R. Weibel (1984). Oxygen consumption and the composition of skeletal muscle tissue after training and inactivation in the European woodmouse (Apodemus sylvaticus). J. Comp. Physiol. B, 155:51-61.

Hoppeler, H. and S.L. Lindstedt (1985). Malleability of skeletal muscle tissue in overcoming limitation: structural elements. J. Exp. Biol. 115:355-364.

Hoppeler, H., J.H. Jones, S.L. Lindstedt, C.R. Taylor, E.R. Weibel, K.E. Longworth, R. Straub and A. Lindholm (1987a). Relating VO_2max to skeletal muscle mitochondria in horses. In: Proceedings of the International Conference on Equine Exercise Physiology (In press).

Hoppeler, H., S.R. Kayar, H. Claassen, E. Uhlmann and R.H. Karas (1987b). Adaptive variation in the mammalian respiratory system in relation to energetic demand: III. Skeletal muscles: setting the demand for oxygen. Respir. Physiol. 69:27-46.

Karas, R.H., C.R. Taylor, J.H. Jones, R.B. Reeves and E.R. Weibel (1987a). Adaptive variation in the mammalian respiratory system in relation to energetic demand: VII. Flow of oxygen across the pulmonary gas exchanger. Respir. Physiol. 69:101-105.

Karas, R.J., C.R. Taylor, K. Rosler and H. Hoppeler (1987b). Adaptive variation in the mammalian respiratory system in relation to energetic demand: V. Limits to oxygen transport by the circulation. Respir. Physiol. 69:65-79.

Kayar, S.R., H. Hoppeler, R.B. Armstrong, S.L. Lindstedt and J.H. Jones (1987). Minimal blood transit time in muscle and capillaries. Federation Proc. 46:352.

Longworth, K.E., J.H. Jones, J.E.P.W. Bicudo, R.H. Karas and C.R. Taylor (1986). Scaling pulmonary diffusion with body mass in mammals. Physiologist 29:142.

Schmidt-Nielsen, K. (1984). Scaling: Why is Animal Size so Important? Cambridge University Press, Cambridge, p. 241

Schwerzmann, K., L.M. Cruz-Orive, R. Eggmann, A. Sanger and E.R. Weibel (1986). Molecular architecture of the inner membrane of mitochondria from rat liver: A combined biochemical and sterological study. J. Cell. Biol. 102:97-103.

Stahl, W.R. (1967). Scaling of respiratory variables in mammals. J. Appl. Physiol. 22: 453-460.

Taylor, C.R. and E.R. Weibel (1981). Design of the mammalian respiratory system. I. Problem and strategy. Respir. Physiol. 44:1-10.

Taylor, C.R. (1987). Structural and functional limits to oxidative metabolism: insights from scaling. Ann. Rev. Physiol. 49:135-146.

Taylor, C.R., R.H. Karas, E.R. Weibel and H. Hoppeler (1987). Adaptive variation in the mammalian respiratory system in relation to energetic demand. II. Reaching the limits to oxygen flow. Respir. Physiol. 69:7-26.

Weibel, E.R., C.R. Taylor, P. Gehr, H. Hoppeler, O. Mathieu and G.M.O. Maloiy (1981). Design of the mammalian respiratory system. IX. Functional and structural limits for oxygen flow. Respir. Physiol. 44: 151-164.

Weibel, E.R. (1984). The Pathway for Oxygen: Structure and Function in the Mammalian Respiratory System. Harvard University Press, Cambridge, p. 425.

Weibel, E.R. (1987). Scaling of structure and functional variables in the respiratory system. Ann. Rev. Physiol. 49:147-159.

Weibel, E.R., L.B. Marques, M. Constantinopol, F. Doffey, P. Gehr and C.R. Taylor (1987). Adaptive variation in the mammalian respiratory system in relation to energetic demand: VI. The pulmonary gas exchanger. Respir. Physiol. 69:81-100.

THE RELATIONSHIP OF TISSUE OXYGENATION TO CELLULAR BIOENERGETICS

Guillermo Gutierrez

Department of Internal Medicine
University of Texas Health Science Center
6431 Fannin
Houston, TX   77030

INTRODUCTION

As primeval cellular organisms coalesced into complex entities, the diffusion distance from atmosphere to the energy producing organelles increased. This resulted in the evolution of an elaborate mass transport system with the dual purpose of carrying oxygen to the mitochondria and removing one of the biproducts of metabolism, carbon dioxide. This gas transport system has allowed man and other living creatures to thrive under a wide range of environmental conditions. However, the delivery of O_2 to the tissues may be compromised when the limits of adaptation are reached. This may occur in high altitude, space travel, or when disease is present. Under those conditions the resulting tissue hypoxia results in a series of microvascular responses which help to preserve the flow of energy to the mechanisms responsible for maintaining the integrity of the cell membrane (Duling and Klitzman, 1980; Kontos 1986).

I will examine several aspects of the relationship of O_2 transport to the cellular production of energy. Particular attention will be given to the hypoxic condition.

THE GENERATION OF ENERGY

Cellular mechanisms for the production of energy can be broadly divided into two categories, anaerobic and aerobic. The ultimate aim of these processes is the production of adenosine triphosphate (ATP). The energy derived from the hydrolysis of the high energy bonds of ATP is used for muscle contraction, protein synthesis, and ion transport across the cellular membrane. The hydrolysis of ATP results in the following reaction:

$$ATP + H_2O \xrightarrow[\text{ATPase}]{} ADP + Pi + H^+$$

(1)

Where ATPase is any enzyme capable of causing the hydrolysis of ATP.

The generation of energy, anaerobic, as well as aerobic, begins with the sequential breakdown of one of the initial metabolic fuels: glucose, fat or protein. In the case of glycolysis, the fermentation of one mole of glucose or its stored equivalent, glycogen, results in the production of two

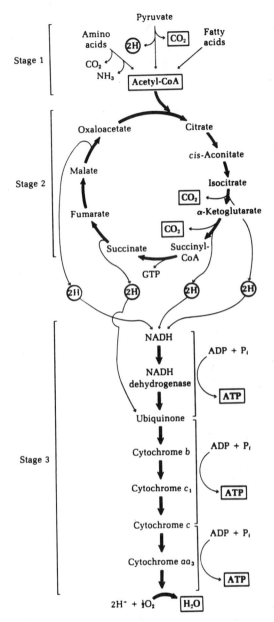

Fig. 1. The tricarboxylic acid cycle and oxidative phosphorylation. From Lehninger (1982), reprinted with permission.

ATPs, two reducing equivalents in the form of reduced nicotinamide adenine
dinucleotide (NADH), and two moles of pyruvate. To allow the process of
glycolysis to continue, NADH must be re-oxydized to NAD. In the absence of
oxygen this takes place in the cytosol by the lactate dehydrogenase
reaction, as pyruvate is converted to lactate. Therefore, glycolysis is an
anaerobic process capable of generating ATP producing lactate:

$$\text{glucose} + 2Pi + 2ADP \longrightarrow 2 \text{ lactate}^- + 2ATP + 2H_2O \qquad (2)$$

It should be noted that lactate production by hypoxic cells is not the
source of acidosis seen in anaerobic glycolysis. Hydrogen ions are

generated when the anaerobically formed ATP is hydroxylized to ADP and Pi, as shown in equation (1) (Gevers, 1977)

Aerobic organisms have a more efficient method to dispose of the NADH generated during glycolysis: oxidative phosphorylation, which takes place in the inner mitochondrial membrane (Fig. 1). Since NADH does not cross the mitochondrial membrane, the electrons are transported by the glycerol phosphate shuttle into the mitochondrion, where series of electron carriers are located in its inner membrane. These electron carriers undergo successive oxidation-reduction reactions and transfer electrons from NADH to oxygen. At least three of these reactions are coupled with the phosphorylation of adenosine diphosphate (ADP) producing ATP. The net yield of oxidative phosphorylation is the formation of 3 moles of ATP for each NADH, or two moles of ATP in the case of another form of reducing equivalent, flavin adenine dinucleotide, FADH.

During the aerobic generation of energy, pyruvate is converted to acetyl CoA in the mitochondrium and enters the tricarboxylic acid cycle. An entire revolution of this cycle results in the additional generation of 11 ATP per mole of pyruvate, so that a total of 36 moles of ATP are formed by aerobic metabolism from one mole of glucose metabolized. This compares quite favorably with the two moles of ATP generated by glycolysis, and thus it is not surprising that the bulk of the production of ATP in man and other vertebrates is through oxidative phosphorylation.

The anaerobic production of ATP can also take place by the creatine kinase reaction:

$$PCr + ADP + H^+ \quad <\!-\!-\!-\!-\!-> \quad ATP + creatine \qquad\qquad (3)$$

This reaction constitutes a ready source of energy, and, as we shall see, it appears to be very important in the hypoxic response of muscle, both cardiac and skeletal, and brain.

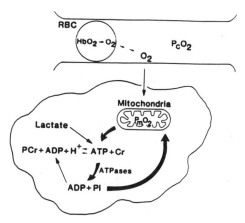

Fig. 2. Capillary O_2 transport and the markers of cellular bioenergetics. Oxygen released from the red blood cells diffuses into the mitochondria where ATP is produced by oxidative phosphorylation. The aerobic energy cycle, as shown by the heavy arrows, consists of the breakdown of ATP by the ATPases into ADP and Pi, with the subsequent resynthesis of ATP in the mitochondria. Also shown are anaerobic sources of ATP, the creatine kinase reaction and glycolysis with the formation of lactate.

THE AEROBIC ENERGY CYCLE

Fig. 2 is a schematic representation of the relationship of cellular O_2 delivery to the aerobic and anaerobic cellular processes that generate ATP. As blood flows in the capillaries, O_2 dissociates from hemoglobin in the red blood cells and travels down a diffusional gradient into the tissues. Among the major determinants of this diffusion process are the oxyhemoglobin dissociation reaction, the effective diffusion capacity of the tissues, the thickness of the plasma layer around the red blood cells (RBC), and the difference in O_2 tension between capillary (PcO_2) and mitochondria (PmO_2) (Krogh, 1919). In the mitochondria, O_2 serves as the final electron acceptor in the cytochrome chain and ATP is produced by oxidative phosphorylation, initiating the aerobic energy cycle shown by the heavy arrows in Fig. 2. This ATP is freely available to the rest of the cell, either by diffusion across the mitochondrial membrane, or by the hypothesized creatine-phosphocreatine energy shuttle (Bessman and Geiger, 1981). This theory maintains that ATP and ADP are compartmentalized in the mitochondria and in the sites of energy utilization, such as the cell membrane and myofibrils. PCr serves to shuttle energy phosphates between the mitochondria and the various energy consuming portions of the cell, where ATP is hydroxylized to ADP and Pi by the ATPases. With an adequate supply of oxygen and fuel substrate, ADP and Pi are converted back into ATP, closing the aerobic energy cycle.

With reductions in the amount of oxygen available, the cell may turn to the creatine kinase reaction and glycolysis as anaerobic methods of ATP production. Anaerobic glycolysis, which results in the production of lactate and protons as ATP is hydrolyzed, does not appear to be the preferred defense strategy against hypoxia in mammalian tissues (Hochachka, 1985). On the other hand, the creatine kinase reaction is a readily available source of ATP, and it has the added benefit of mopping up one H^+ during the generation of ATP. The creatine kinase reaction has been shown to occur in many tissues, including cardiac and skeletal muscle, and brain. To summarize, with reductions in O_2, there may be: 1) increases in Pi, since it cannot be converted by the mitochondria to ATP; 2) decreases in PCr, as the creatine kinase reaction is used to generate ATP; 3) increases in lactate; and 4) decreases in intracellular pH.

The importance of the creatine kinase reaction in the defense against hypoxia has not been fully appreciated until recent times. With the advent of 31-Phosphorus magnetic resonance spectroscopy (MRS) it is possible to sequentially monitor changes in PCr, Pi, and ATP during hypoxia in an accurate and non-invasive manner. More will be said about MRS later.

THE RELATIONSHIP OF OXYGEN TRANSPORT TO OXYGEN CONSUMPTION

The consumption of oxygen by the tissues is roughly equivalent to the production of ATP by the mitochondria. There are two reasons for the lack of a one-to-one correspondence between $\dot{V}O_2$ and ATP production. In the first place, approximately 2% of the consumable oxygen is used by cellular processes, other than oxidative phosphorylation, during the formation of reactive O_2 metabolites, such as superoxide anion, hydrogen peroxide and hydroxyl radicals (Grisham and McCord, 1986). Another factor that determines the number of ATPs per mole of O_2 consumed is the fuel used to produce the acetyl CoA entering the TCA cycle. As opposed to fat or protein, glucose makes the most effective use of the available oxygen, since two extra ATPs are produced during glycolysis prior to the formation of acetyl CoA. However, for the purposes of this discussion, and in fact, for most clinical or physiological considerations, one can assume that $\dot{V}O_2$ represents the aerobic production of ATP.

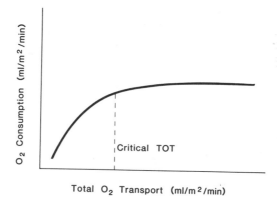

Fig. 3. Schematic representation of the biphasic relationship of O_2 consumption to the total O_2 transport, as found in animal experiments. $\dot{V}O_2$ remains constant until the critical O_2 transport is reached, where it begins to decline.

The relationship between the transport of oxygen by the cardiovascular system ($\dot{T}O_2$), and $\dot{V}O_2$ was first explored in 1964 by Stainsby and Otis in their classical paper. Their results are schematically illustrated in Fig. 3, where $\dot{V}O_2$ is shown as a function of $\dot{T}O_2$. Initially, as $\dot{T}O_2$ decreases, $\dot{V}O_2$ remains more or less unchanged. The early constancy of the $\dot{V}O_2$ may be the result of microvascular adaptations that reduce the distance that O_2 must diffuse across, from hemogloin to mitochondria (Duling and Klitzman, 1980). Therefore, the O_2 extraction ratio (O_2ER), rises. With further declines in $\dot{T}O_2$ these microvascular adaptations reach their limit and the number of O_2 molecules reaching the mitochondria falls below that required to maintain the uninterrupted production of ATP by the anerobic energy cycle. When $\dot{T}O_2$ can not sustain aerobic ATP production [critical $\dot{T}O_2$ ($\dot{T}O_2$crit)], the cell must either lower its energy requirements or find alternate sources of energy to supplement the reduced production of ATP by oxidative phosphorylation. In either case, $\dot{V}O_2$ will fall.

Lowering cellular energy requirements by some form of partial metabolic arrest is an adaptive respose to hypoxia. This is seen in the goldfish in anoxic water, the freshwater turtle and to an extreme in brine shrimp embryos (Hochachka, 1986). Except for diving mammals who use the diving reflex to apportion their O_2 storages to the most hypoxia sensitive organs, mammalian cells have no known mechanism to help them reduce their energy requirements (Hochachka, 1986). Therefore, the anaerobic sources of ATP described previously must be utilized by the tissues below the $\dot{T}O_2$crit.

The biphasic nature of the $\dot{T}O_2$ - $\dot{V}O_2$ relationship has been amply confirmed in numerous animal experiments. Of particular interest is the study by Cain (1977) comparing a decrease in $\dot{T}O_2$ produced by hypoxemia to another produced by isovolemic anemia. Both experiments resulted in a similar $\dot{T}O_2$crit of 10 ml/min·kg, but the anemic group had a significantly higher mixed venous PO_2 at the critical point. This data suggested that it was the $\dot{T}O_2$ and not the capillary PO_2, as reflected by the mixed venous PO_2, that regulated tissue $\dot{V}O_2$. In other words, these data implied that the O_2 transport system was limited by the bulk flow of O_2 carried to the capillaries, and not by diffusion. Analysis of this question using a mathematical model of tissue oxygenation suggests that the difference in venous PO_2 may be the result of time-dependent factors governing the release of O_2 from the RBC and that, in fact, the system is diffusion-limited, since the calculated capillary PO_2 were similar in both cases (Gutierrez, 1986).

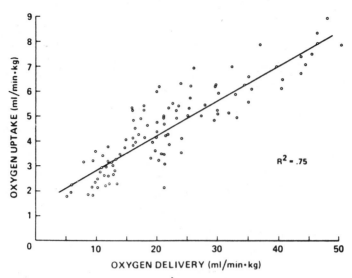

Fig. 4. Grouped data for $\dot{T}O_2$ (denoted in the fig. as O_2 delivery) and $\dot{V}O_2$ in patients with ARDS (n=20). From Danek et al., (1980), reprinted with permission.

Other investigators also have found a biphasic $\dot{T}O_2$ - $\dot{V}O_2$ relationship using anesthetized, mechanically ventilated dogs, including studies where $\dot{V}O_2$ was measured independent of the cardiac output, using the gas exchange method (Pepe and Culver, 1985). These studies also found a $\dot{T}O_2$crit in the neighborhood of 10 ml/min·kg, lending support to the theory that the $\dot{T}O_2$crit is a fixed quantity which could be of value as an index of failure of the O_2 transport system. However, studies on dogs whose basal $\dot{V}O_2$ was decreased by hypothermia and then exposed to hypoxemia, revealed a $\dot{T}O_2$crit of 6 ml/min·kg, a significantly lower value when compared to normothermic control (Gutierrez et al, 1986). In addition, it appears that increasing the affinity of hemoglobin for oxygen reduces the ability of the tissues to extract O_2 from blood and increases the $\dot{T}O_2$crit. This increase in $\dot{T}O_2$crit has been demonstrated in hypoxemic dogs chronically fed with sodium cyanate, which shifts the oxyhemoglobin dissociation curve (ODC) to the left (Warley and Gutierrez, in press). These results suggest that the $\dot{T}O_2$crit is not a fixed quantity, but that it varies with the metabolic needs for oxygen and also with factors that affect the oxygen transport system.

A different story emerges when data obtained from critically ill patients are examined. With the exception of the work of Shibutani et al., (1983), who studied 58 patients undergoing cardiac surgery, all the clinical studies show a linear relationship between $\dot{T}O_2$ and $\dot{V}O_2$. Danek et al., (1980) were the first to describe this relationship in patients with the adult respiratory disease syndrome (ARDS). These investigators measured $\dot{T}O_2$, $\dot{V}O_2$ and mixed venous PO_2 in 20 patients with ARDS and 12 others with various diseases. In all but one of the patients, the $\dot{T}O_2$ -$\dot{V}O_2$ relationship was linear, even in those regions where the $\dot{T}O_2$ rose to approximately 50 ml/min·kg, three times that of normal subjects at rest (Fig. 4). In the patients without ARDS, $\dot{V}O_2$ appeared to be unrelated to $\dot{T}O_2$. This finding was confirmed by Mohsenifar et al., (1983) and Nishimura (1984). The latter concluded that patients behave as oxygen conformers, increasing their $\dot{V}O_2$ in response to an increment in $\dot{T}O_2$.

Gutierrez and Pohil (1986) performed a prospective study where thirty mechanically ventilated patients were studied throughout the length of their illness. At least eight $\dot{T}O_2$ - $\dot{V}O_2$ determinations per patient were

188

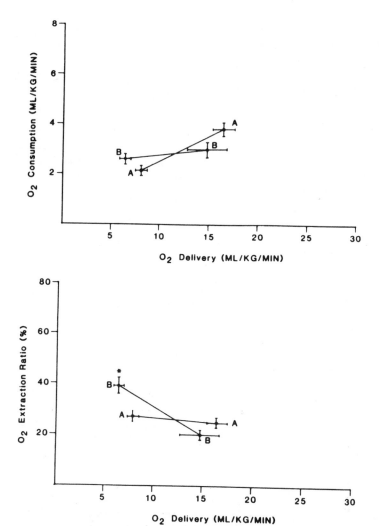

Fig. 5. Relationship of O_2 transport to $\dot{V}O_2$ and O_2 extraction ratio in critically ill patients (n=30). See text for details. From Gutierrez and Pohil (1986) reprinted with permission.

performed. The patients presented various clinical conditions, including ARDS, gastrointestinal hemorrhage, congestive heart failure and sepsis. Six different types of curves, including polynomials of orders one through three, exponential, logarithmic and power curves, were fitted to the individual data using regression analysis. It was found that the coefficient of correlation for a straight line was not different from that of the other functions. It was concluded that a linear function accurately portrayed the $\dot{T}O_2$ - $\dot{V}O_2$ relationship in acutely ill patients, regardless of their clinical conditions. A retrospective analysis of the data showed that the patients could be divided into two groups, depending upon the statistical strength of the $\dot{T}O_2$ - $\dot{V}O_2$ relationship (Fig. 5). In one group (group A), the $\dot{V}O_2$ was tightly dependent on changes in $\dot{T}O_2$, and in the other (group B) the $\dot{V}O_2$ varied little with alterations in $\dot{T}O_2$. As one would expect, the correlation between $\dot{T}O_2$ and $\dot{V}O_2$ was dependent upon the ability of the tissues to extract oxygen from blood. Interestingly, the group whose $\dot{V}O_2$ depended on the $\dot{T}O_2$ (group A) had a 70% mortality rate. These patients who could not increase O_2ER in response to a decrease in $\dot{T}O_2$, had conditions capable of disrupting microvascular control, peripheral blood distribution, or tissue gas exchange. These diseases included septic shock, ARDS, and severe gastrointestinal bleeding. The other group of patients could increase their O_2ER with decreases in $\dot{T}O_2$ and maintained their $\dot{V}O_2$ relatively constant. This group, which had a mortality rate of only 30%, included patients with heart failure and decreased cardiac output, although some of these patients had ARDS.

Several theories have been proposed to explain the difference between the animal experiments displaying a biphasic $\dot{T}O_2 - \dot{V}O_2$ relationship and the clinical studies showing a linear function (Dantzker, 1987). Perhaps the most plausible one is that advanced by Kreuzer and Cain (1985), who proposed that the biphasic curve is the true physiological behavior of the O_2 transport system. The linear dependency of $\dot{V}O_2$ on $\dot{T}O_2$ was explained on the basis of an increased basal $\dot{V}O_2$ in patients with ARDS, combined with a decreased ability to extract O_2 from blood. Therefore, unless supranormal levels of $\dot{T}O_2$ are achieved by the cardio-pulmonary system, ARDS patients cannot achieve that portion of the curve where $\dot{V}O_2$ is independent of $\dot{T}O_2$. They called this situation a pathological supply dependency. This theory, attractive as it is, does not explain the recent data of Annat et al., (1986) who measured $\dot{V}O_2$ in eight mechanically ventilated patients with ARDS. They acutely lowered $\dot{T}O_2$ using positive end-expiratory pressure and found no change in $\dot{V}O_2$ or increases in blood lactate, suggesting that these patients were capable of increasing O_2 extraction to meet cellular O_2 requirements. Furthermore, the basal $\dot{V}O_2$ of these patients was not different to that found in other mechanically ventilated patients.

It is possible that the differences between experimental animal and patient data are related to the temporal characteristics of the data collection. In the case of the experimental data, $\dot{V}O_2$ is the dependent variable, while $\dot{T}O_2$ is the independent variable. A healthy animal preparation, whose energy requirements are relatively constant throughout the length of the experiment, is exposed for a relatively short time to reductions in $\dot{T}O_2$ produced by decreases in O_2 saturation, cardiac output, or hemoglobin concentration. In addition, the animals usually are anesthetized, paralyzed and mechanically ventilated, which may result in a pharmacologically induced tachycardia and a lower than normal $\dot{V}O_2$. Under these conditions $\dot{T}O_2$ could be far greater than that required for the aerobic production of ATP and $\dot{V}O_2$ would remain constant for a wide range of $\dot{T}O_2$s above the $\dot{T}O_2$crit. Finally, the microcirculatory control mechanisms of these animal preparations are intact, allowing them to maximize diffusion distances and the fraction of O_2 extracted from blood.

The above conditions are different in the clinical setting where data are collected from patients over several days. These patients experience fluctuations in their $\dot{V}O_2$ and may even have pathological alterations in microcirculatory control (Kariman and Burns, 1985), which may vary during the length of the study. Therefore, it is possible that in critically ill patients $\dot{T}O_2$ is at times below that required to sustain basal energy requirements and $\dot{V}O_2$ changes with alterations in $\dot{T}O_2$. At other times the oxygen supply to the tissues may be adequate, but there may be increases in metabolic rate, as the result of fever, increased work of breathing, etc, which results in a higher $\dot{V}O_2$ and compensatory increases in cardiac output and $\dot{T}O_2$. In the former case the $\dot{V}O_2$ depends on the $\dot{T}O_2$, while in the latter the $\dot{T}O_2$ rises to meet tissue oxygen needs, thus becoming a function of the $\dot{V}O_2$. This will produce a linear $\dot{T}O_2 - \dot{V}O_2$ relationship throughout the whole range of $\dot{T}O_2$.

The preceding argument implies that a linear relationship is the normal state of affairs, since above the $\dot{T}O_2$crit it provides the tissues with the required $\dot{T}O_2$ to satisfy energy demands, which has the advantage of minimizing cardiac work. Therefore, it is possible that only when energy requirements remain unchanged and lower than normal, as in anesthetized, mechanically ventilated dogs, that a biphasic relationship becomes evident as $\dot{T}O_2$ is decreased.

It may be argued that above the $\dot{T}O_2$crit a linear $\dot{T}O_2 - \dot{V}O_2$ relationship is energy efficient since it matches cellular O_2 needs with capillary O_2 transport. However, below the $\dot{T}O_2$crit it could eliminate the safety factor provided by the flat portion of the biphasic function. In other words, a linear $\dot{T}O_2 - \dot{V}O_2$ relationship has a limited ability to raise the O_2ER. The latter can be graphically visualized as the slope of a line connecting the origin with the $\dot{T}O_2 - \dot{V}O_2$ relationship. For example, in the extreme case of a line passing through the origin, the O_2ER is fixed and a fall in $\dot{T}O_2$ results in an immediate decrease in $\dot{V}O_2$. However, it may be that increases in O_2ER produced by microvascular adjustments result in different slopes and different linear operating curves. Advancing this argument one step further, one may conclude that a single $\dot{T}O_2 - \dot{V}O_2$ relationship may not exist. Instead, it could be hypothesized that there is a region of action in the $\dot{T}O_2 - \dot{V}O_2$ plane which is bounded by the physical constraints of the system, such as the ability to increase cardiac output, the transport characteristics of hemoglobin, including its concentration and the position of the oxygen dissociation curve (ODC), the degree of microvascular control, etc. This region of action allows the oxygen transport system a certain latitude in optimizing the transport of O_2 to match cellular needs. How these constraints affect the system at a particular time will determine the shape of the $\dot{T}O_2 - \dot{V}O_2$ relationship within the region of action.

The information we need to better analyze clinical data are the energy requirements of the patient. Thus, we could define the $\dot{T}O_2$crit as the point where the tissue ATP requirements are greater than the ATP production by oxidative phosphorylation, as reflected by the $\dot{V}O_2$.

CONCEPTUAL FEEDBACK MODEL OF TISSUE OXYGENATION AND ENERGY PRODUCTION

The feedback system shown in Fig. 6 has been proposed as a possible analogy of the oxygen delivery and tissue oxygenation process (Gutierrez and Pohil, 1986). This is a complex feedback model where the dynamics of RBC deoxygenation (Gutierrez, 1986; Vandegriff and Olson, 1984), capillary recruitment (Honig et al., 1980), capillary transit time (Honig and Odoroff, 1984), and the various feedback loops provided by the cardiovascular system are in continuous interplay to assure an adequate supply of O_2 to the mitochondria.

The middle of the diagram shows $\dot{T}O_2$, defined as the product of the arterial O_2 concentration (CaO_2) and the cardiac output ($\dot{Q}$). As blood enters the systemic capillaries the RBCs release their oxygen. The rate of O_2 release in the capillaries is shown as a function of $\dot{T}O_2$, the capillary transit time, the kinetics of RBC deoxygenation, and the intracapillary PO_2. The transfer function defining capillary O_2 release is poorly defined as the result of the complexity of the microcirculation. This transfer function must account for changes in microvascular hematocrit (Desjardins and Duling, 1987), the effect of stagnant layers around the RBCs (Sinha, 1983), capillary anatomy (Pittman, 1986) and the space between RBCs (Federspiel and Sarelius, 1981). Subtracting the rate of O_2 consumption by the tissues from the rate of O_2 release in the capillaries yields the rate of change of the intracapillary PO_2. Integration of this term results in the mean capillary PO_2, PcO_2. This term is a very useful abstraction that permits the linkage of the vascular O_2 delivery process with another abstraction, the mean tissue PO_2, PtO_2, or if one prefers it, the intramito-

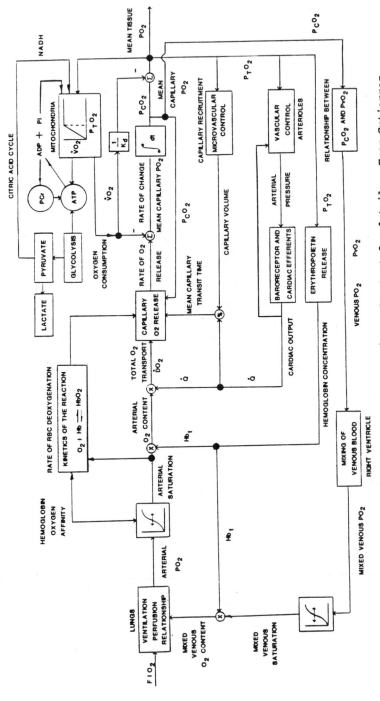

Fig. 6. Feedback model of tissue oxygenation. See text for details. From Gutierrez and Pohil (1986) reprinted with permission.

chondrial PO_2, PmO_2 which in turn directly affects $\dot{V}O_2$. PtO_2 has been chosen in this model as the controlled variable of the system, and its value will affect the degree of control exerted by the various feedback systems on the O_2 delivery process.

As a first order approximation, $\dot{V}O_2$, PcO_2 and PtO_2 may be related by the assumption of passive diffusion of O_2 from the capillaries to the tissues, $\dot{V}O_2 = Kd \times (PcO_2 - PtO_2)$, where Kd is defined as an average effective diffusion capacity of the tissues (Krogh, 1919). From the above equation: $PtO_2 = PcO_2 - \dot{V}O_2 /Kd$. This equation is represented in the model by subtracting the product of the $\dot{V}O_2$ and the term $- 1/Kd$ from PcO_2.

The relationship between PtO_2 and $\dot{V}O_2$ is represented by a biphasic function, with $\dot{V}O_2$ as the dependent variable (Sugano et al., 1974). This relationship assumes a $\dot{V}O_2$ independent of PtO_2 until the latter reaches levels at which the mitochondria cannot sustain oxidative phosphorylation. At that point $\dot{V}O_2$ decreases as a function of PtO_2. With adequate PtO_2 levels, $\dot{V}O_2$ becomes independent of PtO_2 and varies only as a function of the concentration of utilizable substrates, and the cellular energy needs, as expressed by the phosphate potential [ATP]/[ADP][Pi] (Chance and Hollunger, 1963). The level of PtO_2 where $\dot{V}O_2$ begins to decline has been defined by Sugano et al., (1974) as the "critical oxygen concentration", where the kinetics of mitochondrial respiration make the transition from zero-order to first order. The mitochondrial PO_2 at this point is extremely low, less than 0.5 Torr (Chance, 1982).

The upper right corner of the model shows that ATP is metabolized to ADP and Pi by the cellular ATPases. These products of metabolism, along with NADH produced in the tricarboxylic acid cycle, also serve as input to the mitochondrial transfer function. Shown in this section of the model are two other sources of ATP generation: the use of phosphocreatine and anaerobic glycolysis. This portion of the model provides the link between the physiological changes in O_2 delivery and consumption with cellular bioenergetics.

The rest of the model shows how the PtO_2 serves as the controlled variable for multiple physiological feedback loops, which in turn affect TO_2. It should be remembered that each transfer function in this model represents processes of vast complexity and their description are beyond the scope of this chapter.

This model represents an initial attempt at correlating the physiological variables of tissue oxygenation with cellular energy production and utilization. It is by no means complete, but it does point out some of the difficulties in interpreting the relationship of $\dot{V}O_2$ to $\dot{T}O_2$. As shown in the model, $\dot{V}O_2$ and $\dot{T}O_2$ are not directly related by a transfer function, but are influenced by changes in the rate of O_2 release in the capillaries, which is a complex process. It is, therefore, unlikely that a constant $\dot{T}O_2$ $-\dot{V}O_2$ relationship exists.

RELATIONSHIP OF CELLULAR OXYGEN TENSION TO MITOCHONDRIAL ENERGY PRODUCTION

As shown in the feedback model, the production of ATP and $\dot{V}O_2$ are directly related to PmO_2. Therefore, it is important to determine the relationship between capillary PO_2, intracellular PO_2 and mitochondrial energy production, since by monitoring these O_2 tensions a clear picture of the state of cellular bioenergetics may emerge.

It has been determined that cytochrome oxidase has a very high affinity for O_2. Sugano et al., (1974) determined that the PO_2 required for the half maximal reduction of cytochrome c (Km) was between 0.2 to 0.02 torr. The existence of steep diffusion gradients of O_2 from the cell membrane to the mitochondria have been postulated to explain the dependency of oxidative phosphorylation on the relatively high capillary PO_2. According to this theory, very steep radial and longitudinal oxygen diffusion gradients are created as the RBCs travel through the capillaries. As a result of these gradients, the portion of the tissue nearest the arterial side is well supplied with oxygen. Therefore, in hypoxia there will be two regions, a well oxygenated one closer to the arterial side, and an anoxic portion towards the venous side. The transition between these regions has been calculated to be tenths of a micron, so that in the oxygenated region oxidative phosphorylation proceeds at a normal rate, while in the anoxic region the production of ATP has stopped.

On the other hand, Nuutinen et al., (1982, 1983) studied the regulation of coronary blood flow in isolated rat hearts and found that decreases in the [ATP]/[ADP][Pi] ratio were associated with increases in coronary flow, regardless of the method used to produce coronary vasodilation. They concluded that mitochondrial oxidative phosphorylation is the sensor that links cellular O_2 consumption to coronary blood flow and serves to regulate local oxygen delivery. Furthermore, Wilson et al., (1977,1979) maintain that the phosphate potential is a function of the PO_2 throughout the physiologic range. Using cultured kidney cells they obtained simultaneous measurements of the mitochondrial [NAD+]/[NADH], the cytoplasmic [ATP]/[ADP][Pi], and the respiratory rate as the tissue PO_2 was reduced. They found a decrease in the [ATP]/[ADP][Pi] which was accompanied by a progressive increase in the reduction of cytochrome c. The changes in these parameters allowed the rate of O_2 consumption and ATP production to remain unchanged down to a PO_2 of 5 torr. They concluded that limited oxygen supply affects cellular metabolism at much higher concentrations than the Km of cellular respiration, but $\dot{V}O_2$ remains relatively unchanged due to compensatory changes in the [ATP]/[ADP][Pi] ratio and the progressive reduction of cytochrome c. The oxygen dependence of the cytochrome c oxidase reaction at near physiological oxygen tensions forms the basis for oxidative phosphorylation as an oxygen sensing mechanism. This theory fits very well with data showing a uniform distribution of O_2 within the cytosol (Gayeski and Honig, 1986).

Which theory is correct, that which postulates steep O_2 gradients from cytosol to mitochondria, or that stating that oxidative phosphorylation is capable of sensing changes in PO_2 at relatively high concentrations? This is an important question whose answer is yet to be determined. Fortunately, newer technological advances are making it possible for us to probe deeper into the mechanisms of cellular metabolism and function. Among these techniques are myoglobin cryospectrophotometry, positron emission tomography, and magnetic resonance spectroscopy. The following section reviews the basic principles and application of magnetic resonance spectroscopy.

MAGNETIC RESONANCE SPECTROSCOPY - A NON-INVASIVE WINDOW INTO CELLULAR BIOENERGETICS

In simple terms, a magnetic resonance spectrometer consists of a radio transmitter capable of sending high frequency radio waves into a sample placed in a strong and homogeneous magnetic field. These radiofrequency waves are absorbed by certain atomic nuclei in the magnetic field, altering their spatial orientation. When the radio transmitter is turned off, these nuclei return to their original position, generating electrical signals which can then be used to develop a frequency spectrum. Therefore, although MRS is a nuclear phenomenon, the generation of measurable signals does not involve the emission of ionizing radiation. Reviews on the basic principles and applications of MRS to the hypoxic condition have been published recently (Gadian, 1982; Gutierrez and Andry, 1987; Radda, 1986).

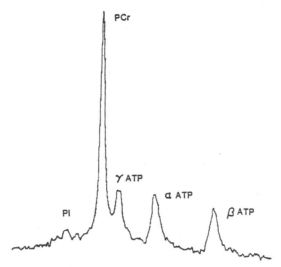

Fig. 7. A typical 31-Phosphorus spectrum obtained from rabbit skeletal muscle. The areas under the peaks are proportional to the concentration of the compounds shown. The intra-cellular pH can be determined from the separation of the resonant frequencies of Pi and PCr (the chemical shift).

The MRS spectrum contains several peaks which correspond to the resonant frequencies of the chemical groups containing a particular isotope nuclei. This separation of resonant frequencies provides the means whereby the spectrum can be used to identify a molecule. Furthermore, since the peak area is the sum of the resonanting nuclei, this area can be used to quantify the concentration of the compound under investigation.

Fig. 7 shows a typical 31-Phosphorus MRS spectrum from skeletal muscle. 31-P is a naturally occurring isotope that forms part of several important molecules that participate in the transfer of energy within the cell, ATP, ADP, AMP, Pi and PCr. This spectrum was obtained by applying a radiofrequency pulse of 32.17 megahertz which is the resonant frequency of 31-phosphorus in a static magnetic field of 2 tesla (1 tesla = unit of magnetic flux equal to 10000 gauss). Since the phosphorus atoms form part of different molecules, the local magnetic effects produce a spectrum with several peaks, rather than a single peak. These peaks correspond to the resonance of Pi, PCr and the gamma, alpha, and beta phosphates of ATP. Since the gamma and alpha peaks contain contributions from the phosphates of ADP and ATP, the beta peak is used to quantify the concentration of ATP. Additional information that can be derived from this spectrum is the intra-cellular pH by measuring the separation between the Pi and PCr peaks (Moon and Richards, 1973). This separation is also called the chemical shift and it is expressed in arbitrary units of parts per million (p.p.m.). Changes in the peak areas and chemical shift reflect in vivo interactions in cellular energy metabolism.

METABOLIC STUDIES WITH 31-PHOSPHORUS NMR

When the rate of ATP hydrolysis becomes greater than the rate of ATP production by oxidative phosphorylation, Pi begins to accumulate. Simul-taneously, the use of the creatine kinase reaction and anaerobic glycolysis to produce ATP result in a decrease in PCr and an increase in pHi respec-tively. These changes can be sequentially measured with 31-P MRS providing

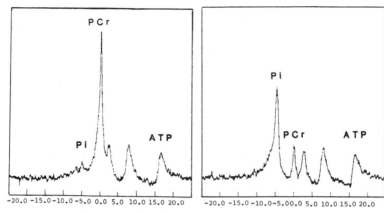

LIVE RABBIT   HYPOXIA

Fig. 8. Typical changes inthe 31-P spectrum of rabbit skeletal muscle during hypoxia. There is a reduction in the PCr peak with an increase in Pi. The ATP concentration appears unchanged.

a dynamic picture of the cellular energy state, as shown in Fig. 8, where 31-P spectra of rabbit skeletal muscle are shown prior to and during hypoxia. The hypoxic spectrum shows an increase in Pi and a decrease in PCr with no change in ATP concentration.

The decrease in PCr and the concomitant increase in Pi described above, provide a convenient index of tissue hypoxia, the PCr/Pi ratio. This index has been used to characterize the status of cell bioenergetics in hypoxic skeletal muscle, heart, and brain (Chance et al., 1985a). Under most conditions, it appears that the PCr/Pi ratio is equivalent to the [ATP]/[ADP][Pi] ratio, (Chance et al, 1985a). Changes in the rate of mitochondrial ATP production are reflected in the PCr/Pi ratio for tissues containing creatine kinase (Gyulai et al, 1985), including heart, skeletal muscle, and brain.

MRS has certain limitations: the signals derived from an MRS experiment represent a spatial and temporal average, and therefore, the technique is inherently poor in spatial localization (Chance et al, 1985b). The sensitivity of the MRS measurement is strongly dependent upon the signal to noise ratio of the signal, which in turn depends on the concentration of the nuclei under study and the length of time during which the spectra are accumulated. MRS can detect metabolites found in concentrations of approximately 1 mM, which include 1-H, 31-P, and enriched 13-C, but cannot measure those present in small or trace amounts.

APPLICATIONS OF MAGNETIC RESONANCE SPECTROMETRY TO THE STUDY OF HYPOXIA

Studies on the response to hypoxia of skeletal muscle have shown a pattern similar to that observed in cardiac muscle and brain, that is, a progressive decrease in the PCr/Pi ratio with preservation of ATP levels until the PCr stores are nearly depleted. Sapega et al., (1985) studied the effect of ischemia on canine hindlimbs. During three hours of total ischemia there was a decrease in PCr which was accompanied by a rise in Pi. Reperfusion after three hours of ischemia resulted in a rapid and complete reversal of these changes. Phosphocreatine resynthesis was detected within ten seconds after reperfusion, and complete recovery required approximately one minute for every preceding hour of ischemia.

There is little work correlating alterations in cellular bioenergetics with measurements of changes in oxygen delivery and consumption. Idstrom et al (1985) perfused an isolated rat hindlimb preparation with Krebs-Henseleit bicarbonate buffer without erythrocytes, equilibrated with 5% CO_2 and fractions of O_2 varying from 95% to 20%. 31-P spectra were obtained from the rat leg at different levels of $\dot{T}O_2$ with the limb at rest and during and after stimulation. The relationship between O_2 and $\dot{V}O_2$ obtained in these experiments is best described by a linear function. Likewise, the authors found a linear relationship between PCr/Pi and $\dot{V}O_2$. The dependency of $\dot{V}O_2$ on $\dot{T}O_2$ at constant flow was interpreted to mean that both the flow rate and the oxygen concentration are important in maintaining adequate tissue oxygenation. In other words, it is essential to maintain a gradient of O_2 from blood to tissue, as well as an adequate blood flow. In these experiments the PCr recovery was dependent on the oxygen delivery, in contrast to the data reported by Sapega et al., (1985) who found a rapid resynthesis of PCr following total ischemia. The in vivo rat hindlimb shows a complete recovery of PCr/Pi within 3 minutes of cessation of stimulation, which agrees with the results of Kushmerick and Meyer (1985). However, the perfused preparation showed a longer time for PCr/Pi recovery at a $\dot{T}O_2$ of 60 µmol/h·g and no recovery at all at $\dot{T}O_2$ of 30 and 20 µmol/h·g. There were no changes in ATP concentration during stimulation for any of the groups.

The next section illustrates the use of 31-P MRS to correlate physiological parameters of tissue oxygenation with changes in the markers of cellular bioenergetics.

THE EFFECT OF HYPOXEMIA AND ISCHEMIA ON SKELETAL MUSCLE BIOENERGETICS

It is known that totally ischemic brain tissue suffers less damage than if exposed to incomplete ischemia (Rehncrona et al., 1979). It also appears that hypoxemia can be more harmful to the brain than total ischemia (Myers, 1979; Siesjo, 1984). The greater injurious effects of hypoxemia and partial ischemia have been explained on the basis of an increased tissue level of lactate. It is hypothesized that in total ischemia the glycolytic pathways are capable of producing lactic acid only to the degree of available glycogen in the tissues. On the other hand, partially ischemic or hypoxemic tissues receive a constant supply of glucose which may result in enhanced lactic acidosis and cellular death (Ljunggren et al., 1974). This has been described as an "acidotic failure" (Siesjo, 1984).

A different view had been proposed by Chance et al., (1985a), who believe that oxidative metabolism has an instrinsic property that makes hypoxic tissues more sensitive to damage when lactic acidosis is also present. In other words, there may be cellular death in hypoxemia or ischemia with minimal acidosis whenever the production of ATP by the mitochondria is not sufficient to maintain the needs of cellular ATPase. According to Chance et al., (1985a), a reduction in oxygen delivery to the mitochondria results in a change in the transfer function for ADP control of oxidative phosphorylation. For a given cellular energy requirement, a decrease in tissue PO_2 will shift the operating point of the ADP control system to an increasingly unstable position. The relative position of the ADP control operating point has been characterized by the PCr/Pi ratio. It is thought that a decrease of this ratio to a value less than one may result in the instability of the ADP control system, which in turn may lead to cellular death. This has been described as a "kinetic failure".

To test the hypothesis that hypoxemia and ischemia may elicit different cellular responses to O_2 deprivation, we exposed a blood-perfused rabbit hindlimb preparation to total ischemia or to severe hypoxemia (arterial PO_2 less than 10 torr) and evaluated the effect of these interven-

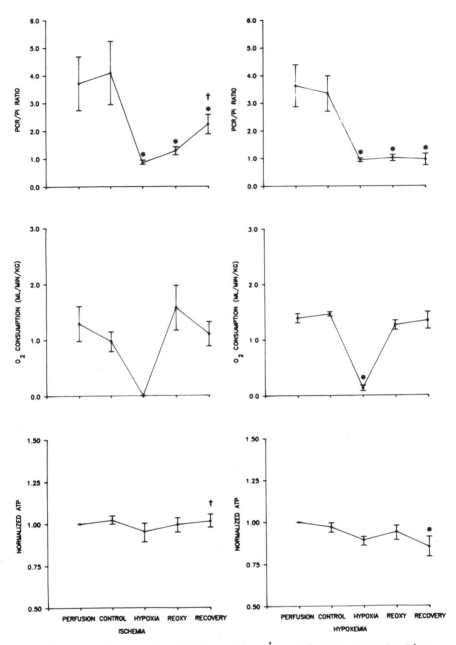

Fig. 9. Changes in PCr/Pi ratio, $\dot{V}O_2$ and ATP concentration measured with 31-P MRS during ischemia and hypoxemia (n=4 each).

tions on cellular bioenergetics. $\dot{V}O_2$, $\dot{T}O_2$ and venous lactate concentration were measured, except during the period of total ischemia. Changes in the relative concentrations of ATP, PCr/Pi and pHi were monitored with 31-P MRS. Following a one hour control period, one group (n=4) was exposed to total ischemia, while the arterial PO_2 of the other group (n=4) was decreased to 5±2 torr. In both groups the PCr/Pi ratio was allowed to decrease to a value less than one, at which point the preparation was reoxygenated at control levels of $\dot{T}O_2$. The time required for the PCr/Pi ratio to fall below one was 56.0±10.0 minutes and 63.8±2.5 minutes for ischemia and hypoxemia,

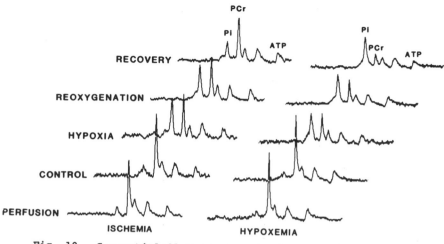

Fig. 10. Sequential 31-P spectra obtained during an ischemic and a hypoxemic experiment. The legend to the left of the spectrum denotes the experimental period. See text for details.

respectively (N.S.). This was followed by a 30 minute recovery period at control levels of $\dot{T}O_2$.

As shown in Fig. 9, the major difference between the ischemic and hypoxemic tissues lies in the response of the PCr/Pi ratio to reoxygenation. The data are presented in five different phases, corresponding to measurements taken at the initiation of perfusion (PERFUSION), the end of the control period (CONTROL), the end of the hypoxic period (HYPOXIA), following the resumption of perfusion with oxygenated blood (REOXY), and at the end of the recovery period (RECOVERY). Shown in Fig. 10 are a series of 31-P MRS spectra obtained from an ischemic and a hypoxemic experiment.

Following the period of ischemia, $\dot{V}O_2$ returned to control, ATP concentration remained constant, PCr was resynthesized, and the concentration of Pi decreased. These changes indicate that the bioenergetic machinery of ischemic tissues was functioning well, despite having been deprived of O_2 to the degree that resulted in a PCr/Pi ratio well below one. It appears that the ischemic tissues maintained the necessary creatine kinase flux in the direction of ATP production, and thus were able to maintain the required cellular energy output constant.

The hypoxemic tissues behaved differently during reoxygenation. While O_2 consumption returned to control, PCr was not resynthesized and the Pi concentration remained high. Furthermore, the ATP concentration was significantly lower than control at the end of the recovery period. Therefore, the capability of the tissues to conserve energy appears to have been impaired, since neither the [ATP]/[ADP][Pi] nor the PCr/Pi ratio, (Gyulai et al., 1985), did recover with reoxygenation. These findings suggest that the metabolic machinery had been damaged and not enough ATP was being produced by oxidative phosphorylation to satisfy cellular needs or to resynthesize PCr.

During the period of total ischemia there appears to have been little or no anaerobic glycolysis, judging from the lack of change in venous blood lactate concentration (Fig. 11). Conversely, the blood lactate concentration rose during and following the period of hypoxemia, despite a return of TO_2 to control. This rise in lactate concentration produced a significant

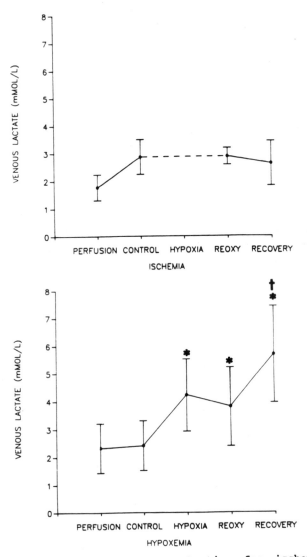

Fig. 11. Venous lactate concentration for ischemia and hypoxemia (n=4 in each group; mean ± SE). There was a significant rise in venous lactate during hypoxemia which continued to increase during the recovery period (*, $p < 0.05$). The venous lactate concentration of the hypoxemic group was significantly greater than that of ischemia at the end of the recovery period (+, $p < 0.05$).

decrease in the intracellular pH of the hypoxemic tissues when compared to the control value (Fig. 12). Therefore, it is appealing to speculate that a greater lactate accumulation in response to an uninterrupted supply of glucose substrate was responsible for the lack of PCr resynthesis following the hypoxemic period.

However, at the end of the hypoxic period, the mean intracellular pH of the hypoxemic group was not statistically different from that of the ischemic group. This lack of substantial acidosis during the hypoxemic period disputes the "acidotic failure" mechanism, and raises the possibility that factors other than an increase in lactate production may be responsible for the different responses to ischemia and hypoxemia.

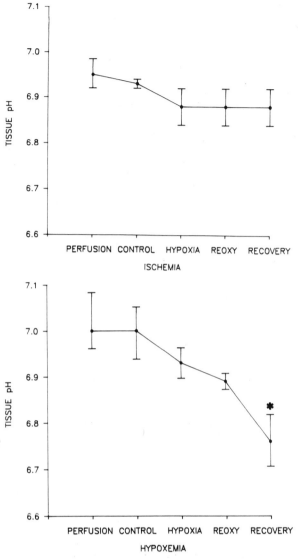

Fig. 12. Tissue pH calculated from the chemical shift between
the PCr and Pi peaks in the 31-P spectra using the equation
developed by Moon and Richards (1973) (n=4 in each group; mean
± SE). There were no changes in pH_i with ischemia. There was
a decline in pHi in the hypoxemic group during the recovery
period (p<0.05). There were no differences in pHi between the
groups.

Another possibility to consider is that the combination of a low PCr/Pi
ratio and an increase in blood lactate resulted in the failure of the
hypoxemic group to resynthesize PCr. This appears unlikely, since the
PCr/Pi decreased to values less than one in both experimental groups, but
the PCr/Pi of the ischemic group recovered following reperfusion, suggesting that the ischemic tissues were spared a "kinetic failure" despite having
similar pHis as the hypoxemic group. In addition, electrically stimulated
skeletal muscle, well perfused and oxygenated, has been shown to be
resilient to low PCr/Pi ratios, lactate accumulation and decreases in
intracellular pH (Kushmeric and Meyer, 1985).

A more likely mechanism may be that of superoxide radical production during hypoxemia. Both the ischemic and hypoxemic preparations were exposed to hypoxic insults of similar magnitude in terms of time, decreases in O_2 transport and the level of PCr/Pi attained during O_2 deprivation. However, we were unable to completely abolish $\overset{.}{T}O_2$ in hypoxemia. This was the result of difficulties encountered in completely desaturating hemoglobin, since any small leak in the system introduced enough O_2 to maintain a hemoglobin saturation of a few percentage points. Therefore, the hypoxemic group continued to acquire and consume O_2 during the hypoxic period, albeit to a minimal degree. This may be of significance, since enough O_2 may have been allowed to flow into the hypoxemic preparations to induce the activation of autolytic systems resulting in superoxide-induced membrane damage.

These experiments show that, under the conditions tested, skeletal muscle is unable to resynthesize PCr following a period of severe hypoxemia. It is also apparent from these data that the $\overset{.}{V}O_2$ fails to portray the true state of cellular homeostasis following a severe hypoxemic insult. Further work remains to be done to determine the role played by the above mentioned mechanisms in the pathogenesis of cellular damage during hypoxia.

CONCLUSION

I have attempted to give an overview of the factors that influence the generation of energy by the tissues, and how tese energy-producing processes may interact with physiologic measures of tissue oxygenation. Our present understanding of these interactions is incomplete but thanks to the application of new technology we can anticipate that rapid advances will take place in this field in the near future.

REFERENCES

Annat, G., J.P. Viale, C. Percival, M. Froment and J. Motin (1986). Oxygen delivery and uptake in the adult respiratory distress syndrome. Am. Rev. Respir. Dis. 133:999-1001.

Bessman, S.P. and P.J. Geiger (1981). The transport of energy in muscle: The phosphorylcreatine shuttle. Science 211:448-452.

Cain, S.M. (1977). Oxygen delivery and uptake in dogs during anemic and hypoxic hypoxia. J. Appl. Physiol. 42:228-234.

Chance, B. and G. Hollunger (1963). Inhibition of electron and energy transfer in mitochondria: IV. Inhibition of energy-linked diphosphopyridine nucleotide reduction by uncoupling agents. J. Biol. Biochem. 238:445-458.

Chance, B. (1982). Oxygen transport and oxygen reduction? In: Electron Transport and Oxygen Utilization. Ed. C. Ho, Elsevier, North Holland, p. 255-266.

Chance, B., J.S. Leigh, B.J. Clark, J. Maris, J. Kent, S. Nioka and D. Smith (1985a). Control of oxidative metabolism and oxygen delivery in human skeletal muscle: A steady-state analysis of the work/energy cost transfer function. Proc. Natl. Acad. Sci. USA 82:8384-8388.

Chance, B., J.S. Leigh and S. Nioka (1985b). Microheterogenity - The "Achilles heel" of NMR spectroscopy and imaging: Some calculations for brain ischemia and muscle exercise. News Metab. Res. 2:26-31.

Danek, S.J., J.P. Lynch, J.G. Weg and D.R. Dantzker (1980). The dependence of oxygen uptake on oxygen delivery in the adult respiratory distress syndrome. Am. Rev. Respir. Dis. 122:387-395.

Dantzker, D.R. (1987). Interpretation of data in the hypoxic patient. In: New Horizons. Oxygen Transport and Utilization. Vol. 2. Bryan-Brown, C.W. and Ayres, S.M. (Eds). Fullerton, CA. Society of Critical Care Medicine, pp. 93-108.

Desjardins, C. and B.R. Duling (1987). Microvessel hematocrit: measurement and implications for capillary oxygen transport. Am. J. Physiol. 252:H494-H503.

Duling, B.R. and B. Klitzman (1980). Local control in microvascular function: Role in tissue oxygen supply. Ann. Rev. Physiol. 42:373-382.

Federspiel, W.J. and I.H. Sarelius (1984). An examination of the contribution of red cell spacing to the uniformity of oxygen flux at the capillary wall. Microvas. Res. 27:273-285.

Gadian, D.G. (1982). Nuclear Magnetic Resonance and its Application to Living Systems. Oxford, Oxford University Press.

Gayeski, T.E. and C.R. Honig (1986). O_2 gradients from sarcolemma to cell interior in red muscle at maximal $\dot{V}O_2$. Am. J. Physiol. 251:H789-H799.

Gevers, W. (1977). Generation of protons by metabolic processes in heart cells. J. Mol. Cell. Card. 9:867-874.

Grisham, M.B. and J.M. McCord (1986). Chemistry and cytotoxicity of reactive oxygen metabolites. In: Physiology of Oxygen Radicals, Eds. A.E. Taylor, S. Matalon and P. Ward, Bethesda, American Physiological Society, pp 1-18.

Gutierrez, G. (1986). The rate of oxygen release and its effect on capillary O_2 tension: a mathematical analysis. Resp. Physiol. 63:79-96.

Gutierrez, G. and J.M. Andry. (1987). NMR measurements - Clinical Applications. Crit. Care Med. (in press)

Gutierrez, G. and R.J. Pohil (1986). Oxygen consumption is linearly related to O_2 supply in critically ill patients. J. Crit. Care 1; 45-53.

Gutierrez, G., R.J. Pohil, R. Strong and P. Narayana. Bioenergetics of rabbit skeletal muscle during hypoxemia and ischemia. J. Appl. Physiol. (in press)

Gutierrez, G., A.R. Warley and D.R. Dantzker (1986). Oxygen delivery and utilization in hypothermic dogs. J. Appl. Physiol. 60:751-757.

Gyulai L., Z. Roth, J.S. Leigh, Jr. and B. Chance (1985). Bioenergetic studies of mitochondrial oxidative phosphorylation using 31-Phosphorus NMR. J. Biol. Chem. 260:3947-3954.

Hochachka, P.W. (1985). Fuel and pathways as designed systems for support of muscle work. J. Exp. Biol. 115:149-164.

Hochachka, P.W. (1986). Defense strategies against hypoxia and hypothermia. Science 231:234-241.

Honig, C.R., C.L. Odoroff, and J.L. Frierson (1980). Capillary recruitment in exercise: rate, extent, uniformity, and relation to blood flow. Am. J. Physiol. 238:H31-H42.

Honig, C.R. and C.L. Odoroff (1984). Calculated dispersion of capillary transit times: Significance for oxygen exchange. Am. J. PHysiol. 27:H199-H208.

Idstrom, J.P., V.H. Subramanian, B. Chance, T. Schersten and A.C. Bylund-Fellenius (1985). Oxygen dependence of energy metabolism in contracting and recovering rat skeletal muscle. Am. J. Physiol. 248:H40-H48.

Kariman, K. and S.R. Burns (1985). Regulation of tissue oxygen extraction is disturbed in adult respiratory distress syndrome. Am. Rev. Respir. Dis. 132:109-114.

Klocke, R.A. (1986). Oxygen transfer from the red cell to the mitochondrion. In: New Horizons. Oxygen Transport and Utilization. Vol. 2. Eds. C.W. Bryan-Brown and S.M. Ayres, Fullerton, CA. Society of Critical Care Medicine, pp. 239-270.

Kontos, A.K. (1986). Regulation of the cerebral microcirculation in hypoxia and ischemia. In: New Horizons. Oxygen Transport and Utilization. Vol. 2. Eds. C.W. Bryan-Brown and S.M. Ayres, Fullerton, CA Society of Critical Care Medicine, pp 311-316.

Kreuzer, F. and S.M. Cain (1985). Regulation of peripheral vasculature and tissue oxygenation in health and disease. Crit. Care Clin. 1:453-470.

Krogh, A. (1919). The number and distribution of capillaries in muscles with calculations of the oxygen pressure head necessary for supplying the tissue. J. Physiol. (London) 52:409-415.

Kushmerick, M.J. and R.A. Meyer (1985). Chemical changes in rat leg muscle by phosphorus nuclear magnetic resonance. Am. J. Physiol. 248:C542-C549.

Lehninger, A. (1982). Principles of Biochemistry. New York, Worth.

Ljunggren, B., K. Norberg and B.K. Siesjo (1974). Influence of tissue acidosis upon restitution of brain energy metabolism following total ischemia. Brain Res. 77:173-186.

Mohsenifar, Z., P. Goldbach, D.P. Tashkin, and D.J. Campisi (1983). Relationship between oxygen delivery and oxygen consumption in the adult respiratory distress syndrome. Chest 84:266-271.

Moon, R.G. and J.H. Richards (1973). Determination of intracellular pH by 31P magnetic resonance. J. Biol. Chem. 248:7276-7278.

Myers, R.E. (1979). A unitary theory of causation of anoxic and hypoxic brain pathology. In: Advances in Neurology, Vol. 26, Eds. S. Fahn, J.N. Davis and L.P. Rowland. Raven Press, New York, 195-213.

Nishimura, N. (1984). Oxygen conformers in critically ill patients. Resuscitation 12:53-58.

Nuutinen, E.M., K. Nishiki, M. Erecinska and D.F. Wilson (1982). Role of mitochondrial oxidative phosphorylation in regulation of coronary blood flow. Am. J. Physiol. 243:H159-H169.

Nuutinen, E.M., D. Nelson, D.F. Wilson and M. Erecinska (1983). Regulation of coronary blood flow; effects of 2,4-dinitrophenol and theophylline. Am. J. Physiol. 244:H396-H405.

Pepe, P.E., and C.H. Culver (1985). Independently measured oxygen consumption during reduction of oxygen delivery by positive end-expiratory pressure. Am. Rev. Respir. Dis. 132:788-792.

Pittman, R.N. (1986). Determinants of oxygen exchange in the microcirculation In: New Horizons. Oxygen Transport and Utilization. Vol. 2. Eds. C.W. Bryan-Brown, S.M. Ayres, Fullerton, CA. Society of Critical Care Medicine, pp 271-292.

Pittman, R.N. (1987). Oxygen delivery and transport in the microcirculation. In: Microvascular Perfusion and Transport in Health and Disease. Ed. D. McDonagh Karger, Basel, 60-79.

Radda, G.K. (1986). The use of NMR spectroscopy for the understanding of disease. Science 233:640-645.

Rehncrona, S., L. Mela and B.K. Siesjo. (1979). Recovery of brain mitochondrial function in the rat after complete and incomplete ischemia. Stroke 10:437-446.

Sapega, A.A., R.B. Heppenstall and B. Chance (1985). Optimizing tourniquet application and release times in extremity surgery. J. Bone Joint Surg. 67-A:303-314.

Shibutani, K., T. Komatsu, K. Kubal, V. Sanchala, V. Kumar and D.V. Bizarri (1983). Critical level of oxygen delivery in anesthetized man. Crit. Care Med. 11:640-643.

Siegel, G., A. Walter, M. Thiel and B.J. Ebeling (1984). Local regulation of blood flow. Adv. Exp. Med. Biol. 169:515-540.

Siesjo, B.K. (1984). Cerebral circulation and metabolism. J. Neurosurg. 60:883-908.

Sinha, A.K. (1983). Oxygen uptake and release by red cells through plasma layer and capillary wall. In: Oxygen Transport to the Tissues. Fourth Edition. Eds. H. Bicher and D. Bruley New York, Plenum, 525-537.

Stainsby, W.N. and A.B. Otis. (1964). Blood flow, oxygen tension and oxygen transport in skeletal muscle. Am. J. Physiol. 206:858-866.

Sugano, T., N. Oshino and B. Chance (1974). Mitochondrial functions under hypoxic conditions. The steady states of cytochrome c reduction and of energy metabolism. Biochim. Biophys. Acta 347:340-358.

Taegtmeyer, H. (1986). Myocardial metabolism. In: Positron Emission Tomography: Principles and Applications for the Brain and Heart. M. Phelps, J. Mazziotta and H. Schelbert eds. Raven Press, New York, 149-195.

Vandegriff, K.D. and J.S. Olson (1984). Morphological and physiological factors affecting oxygen uptake and release by red blood cells. J. Biol. Chem. 259:12619-12627.

Warley, A.R. and G. Gutierrez. Chronic administration of sodium cyanate in dogs decreases the oxygen extraction ratio. J. Appl. Physiol. (in press)

Wilson, D.F., M. Erecinska, C. Drown and I.A. Silver (1977). Effect of oxygen tension on cellular bioenergetics. Am. J. Physiol. 233:C135-C140.

Wilson, D.F., M. Erecinska, C. Drown and I.A. Silver (1979). The oxygen dependency of cellular energy metabolism. Biochem. Biophys. 195:485-493.

SKELETAL MUSCLE MITOCHONDRIA: THE AEROBIC GATE?

Stan L. Lindstedt and Dominic J. Wells

Department of Zoology and Physiology
University of Wyoming
Laramie, WY 82071

ABSTRACT

At an animal's maximum aerobic capacity ($\dot{V}O_2$max), the O_2 flowing through the respiratory system is consumed by a functionally exclusive sink, skeletal muscle mitochondria. Thus, O_2 consumption will never exceed the muscles O_2 demand. If the system is ideally designed, structures upstream to the skeletal muscle O_2 sink must be built to insure adequate O_2 delivery to the working muscle. There are a number of structure-function solutions available to supply the demanded O_2 to the muscle; these have been found to vary, often ontogenetically, with hypoxia, training, etc. But there is one relationship that is invariant: Total O_2 uptake can be predicted by the total (active) skeletal muscle mitochondrial volume. In aerobic and sedentary animals, across a range of body sizes, maximum (in vivo) mitochondrial O_2 consumption is constant among mammals (at approximately 2000 O_2 molecules per square micron of inner mitochondrial membrane per second). Because the volume of mitochondria is one of the most plastic of all respiratory structures, we interpret this relationship as suggesting that skeletal muscle mitochondria alone sets the demand for O_2 and, thus, the volume of skeletal muscle mitochondria dictates an animal's maximum aeorbic capacity.

INTRODUCTION

An animal's peak aerobic performance is best understood operationally as the maximum rate of oxygen consumption ($\dot{V}O_2$max). It is identifiable in exercising animals as once reached, additional work output fails to elicit any further increase in oxygen consumption ($\dot{V}O_2$). This operational definition can be applied reproducibly to a variety of species, exercises, etc. to define an individual's maximum aerobic performance. Maximum aerobic performance is, thus, quite clearly defined; however, the limitations to O_2 uptake are far from clearly understood.

Oxygen travels through the respiratory system through two convective steps, pulmonary ventilation and cardiovascular circulation, and several diffusive steps through which oxygen moves down a gradient of partial pressure (e.g. alveolar-capillary membranes, erythrocytes, tissue capillary membranes, cell cytoplasm, etc.). Thus, a number of respiratory system structures share the task of transporting O_2 from the air to the tissue mitochondria; any or all of these could act to limit $\dot{V}O_2$max. In this chapter, we will focus on the final step in the O_2 cascade. Specifically, we will examine the structure of skeletal muscle and the insights it may provide in the broader picture of limitation to O_2 flow through the mammalian respiratory system.

Skeletal muscle is phenotypically very plastic in both its metabolic and contractile properties. If muscle is repeatedly called on to work aerobically, it will respond with shifts in its aerobic capacity, as documented by increases in oxidative enzymes. This is apparently accomplished via increased packing of mitochondria within the muscle cells (Taylor, 1987). Thus, the volume of mitochondria within a given muscle and the total volume within the animal, is apparently set by the muscle's need for aerobic ATP production. It is muscle energetics alone that determines the necessary mitochondrial volume and, hence, the required O_2 flow through the respiratory system. We believe that the volume of active mitochondria is sufficient information to predict maximum O_2 uptake ($\dot{V}O_2$max); i.e., the skeletal muscle mitochondria act as the aerobic gate through which O_2 flow is throttled. Upstream to the mitochondria, the design of the respiratory system must accommodate the required O_2 flow.

The Mitochondrial Oxygen Sink

A technique of systematic random sampling (developed in collaboration with Hans Hoppeler, University of Berne), allows us to calculate the averge mitochondrial density throughout an animal's entire skeletal musculature (Hoppeler et al., 1984). Random numbers are assigned to muscle segments in direct proportion to the relative mass of each segment. Selecting three random numbers then provides specific information of exact sampling locations for EM. By quantitative ultrastructural analysis, we can determine the absolute magnitude of an animal's mitochondrial sink.

In collaboration with J. Jones, C.R. Taylor, H. Hoppeler and E.R. Weibel, we have applied this modeling in two fundamentally different animal models: 1) Allometry. Across a size range, O_2 uptake varies as a regular function of body mass. 2) Sedentary vs. active animals. Within a size range, there are highly aerobic "athletes" and sedentary animals with limited aerobic scope. These may be directly compared.

To disprove our hypothesis, we need to examine the relationship between mitochondrial volume and $\dot{V}O_2$max. Is this relationship constant among species, independent of aerobic capacity and invariant with training? We also must examine similar relations between $\dot{V}O_2$max and other single steps, e.g., cardiovascular O_2 delivery.

Weight-specific maximum oxygen consumption ($\dot{V}O_2$max/Mb) varies greatly among animals of differing body size (woodmouse to steer) and within a size as a function of breeding (or selection) for aerobic performance (horses vs. steers). However, in all animals we have examined, there is a constant and linear relation between O_2 uptake and mitochondria volume (Fig. 1). Because this relation is linear, it allows us to calculate maximal mitochondrial $\dot{V}O_2$ across species. In all cases, skeletal muscle mitochondria consume O_2 at a maximum in vivo rate of 5 ml/(cm^3 mito min) (see Hoppeler and Lindstedt, 1985). Furthermore, because the inner surface area of skeletal muscle mitochondria is roughly constant among all mammals, we estimate that O_2 uptake per unit surface area of inner mitochondrial membrane must be constant and equal to about 2000 O_2 molecules per μm^2 of inner mitochondrial membrane per second. The volume of mitochondria is quantitatively coupled to the (aerobic) ATP demand of the working muscles; $\dot{V}O_2$max is in all cases a constant and an invariant multiple of mitochondrial volume. The mitochondria apparently do not vary qualitatively among mammalian species, only quantitatively.

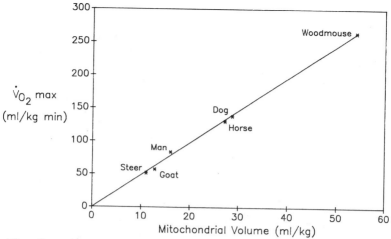

Fig. 1. When maximum O_2 uptake ($\dot{V}O_2$max) is plotted against mean mitochondrial volume density, there is a strikingly linear relationship between $\dot{V}O_2$max and mitochondrial volume density among animals spanning over 4 orders of magnitude in body mass and nearly 10-fold in their aerobic capacities. On average, the skeletal muscle mitochondria consume a maximum of about 2000 O_2 molecules per square micron of innermitochondrial membrane per second. This relationship, when extrapolated, intersects zero; from the least to the most aerobic animals this relation is invariant. (The single point for man has been calculated from Saltin, 1985 and Hoppeler et al., 1985. The remaining values have all been measured, see Hoppeler and Lindstedt, 1985.)

By examination of these interspecific data, we might conclude that mitochondrial volume alone sets the limit to O_2 uptake in mammals. We would like to evaluate this concept and integrate the apparent constancy of mito-chondrial volume into a broader picture of respiratory system design and limitation.

"Economically-Tuned Resistors"

In contrast to the fixed maximum mitochondrial O_2 consumption, the relationship with other (upstream) respiratory system structures is more variable. Nonetheless, single-step limtation has been suggested to occur, most often a limitation of O_2 delivery (Saltin and Gollnick, 1983; Ekblom, 1986). As early as 1912, Verzar had speculated that O_2 availability to the muscle likely limits O_2 uptake. Recently, many studies have indicated a partial (di Prampero, 1985) or complete (Buick et al., 1980; Saltin, 1985; Spriet et al., 1986) limitation to $\dot{V}O_2$max imposed by cardiovascular O_2 delivery. However, if we examine the evidence for such a limitation, convincing evidence within a single study becomes equivocal when examined across studies. For instance, in humans, hemoglobin concentration has been suggested to limit $\dot{V}O_2$max (Fig. 2). However, we believe that the data, taken together, strongly suggest that there is an upper limit of $\dot{V}O_2$max at any level of hemoglobin concentration, i.e., that hemoglobin concentration must be sufficient, but that it doesn't usually limit $\dot{V}O_2$max.

A stronger correlation is found when $\dot{V}O_2$max is plotted as a function of O_2 delivery (or, more appropriately, O_2 availability) in man, again across several studies (Fig. 3). In this figure, lines connect the data tracking the results of O_2 delivery manipulations within a single study. Usually, as

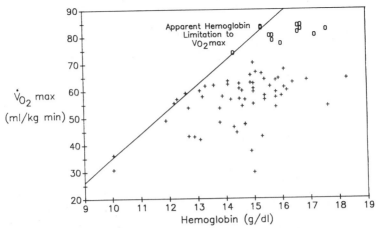

Fig. 2. $\dot{V}O_2$max in man is plotted against hemoglobin concentration combining a number of recent studies. The relationship between these two parameters, often suggested to be causal within a given study, is quantitatively highly variable between studies. Whether examined in healthy, but non-athletic individuals (+) or elite endurance athletes (o), any given level of $\dot{V}O_2$max may be associated with a broad range of hemoglobin concentration. The left-most points at each level of $\dot{V}O_2$max suggest that there may indeed be a hemoglobin-limitation to $\dot{V}O_2$max. (Data taken from: Anderson and Saltin, 1985; Buick et al., 1980; Celsing et al., 1986; Celsing et al., 1987; Coyle et al., 1986; Ekblom et al., 1972; Ekblom et al., 1975; Ekblom et al., 1976; Farrell et al., 1986; Hagberg et al., 1985; Holmgren and Astrand, 1966; Kanstrup and Ekblom, 1982; Kanstrup and Ekblom, 1984; Oelz et al., 1986; Robertson et al., 1982; Robertson et al., 1984; Spriet et al., 1986; Svedenhag et al., 1986; Thomson et al., 1982; Withers et al., 1981. Figure taken from Lindstedt et al., 1988.)

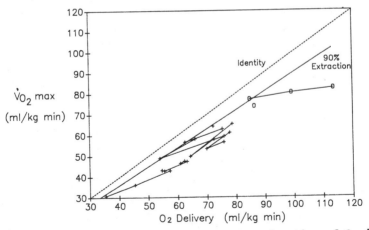

Fig. 3. $\dot{V}O_2$max is examined in man as a dependent function of O_2 delivery. Because extraction is not complete, more O_2 is delivered than can be utilized. The "best" extraction among the non-athletes (+) and the athletes (o) is very close to 90%. In examining these linked studies (i.e., manipulations of delivery), in nearly every case the highest extraction occurs with the lowest delivery. While the system may be driven at much higher than useful O_2 availability, increasing O_2 delivery would seem to result in ever decreasing returns. (Data taken from: Celsing et al., 1986; Ekblom and Hermansen, 1968; Ekblom et al., 1975; Ekblom et al., 1976; Robertson et al., 1984; Spriet et al., 1986; Thomson et al., 1982; Woodson et al., 1978. Figure taken from Lindstedt et al., 1988.)

O_2 delivery increases (e.g., with blood doping), any increase in $\dot{V}O_2$max is disproportionately less. With increasing O_2 availability, the muscle is simply not able to profit from all the O_2 made available to it. Hence, there is a decrease in O_2 extraction. There is an apparent loss of "structural efficiency". Interestingly, once the blood doping is completed, both $\dot{V}O_2$max and hemoglobin return to pre-transfusion levels.

We interpret these findings as evidence that the respiratory system resistors are "economically-tuned" to, rather than optimally designed for, the muscles' demand for O_2. Thus, we would modify the concept of symmorphosis (Taylor and Weibel, 1981), to include a consideration of cost-benefit analysis (Lindstedt and Jones, 1987). By increasing the capacitance of any single step in the cascade, $\dot{V}O_2$max may increase slightly as the remaining structures are pushed (e.g., by unusually high driving forces) beyond the limits for which they were evolved. As long as this apparent increase in $\dot{V}O_2$max is accomplished at a cost of reduced structural efficiency, we would contend that the resistors are no longer tuned; that single step limitation has not been demonstrated. Thus far, the data are almost exclusively from mammals (and most from a single mammal, man). Tipton (1986) has reviewed the available animal literature and finds a lack of data to make this comparison across species. If the observed patterns are constraints of design, we expect that they should hold across taxonomic lines. We are currently investigating training effect in sedentary and highly aerobic mammals and structure/function relations in other vertebrate classes (reptiles, fish and birds).

CONCLUSION

The O_2 consumption of an animal will never exceed the requirements for O_2 at the mitochondrial sink. How often, if ever, does the demand exceed the supply? We feel that the demand of the muscle mitochondria rarely, if ever, exceeds their (potential) supply, except under conditions of experimental manipulation of a single "rate-limiting" step. Some of the structural resistors in the respiratory system cascade have limited phenotypic plasticity while others respond rapidly to chronic changes in demand. Together these structures must be "economically-tuned" to the O_2 demand of the muscles. It would appear that the resistance steps with the least phenotypic plasticity will appear to be built excessively in all but the most aerobic individuals. Those with the most plasticity will be present in just sufficient quantity. When at their economically-tuned optimum, the cost to benefit ratio is maximized. Adding additional structure may result in an increase in $\dot{V}O_2$max, but it does so with a loss in structural efficiency (i.e., a decrease in O_2 consumption per unit of structure).

There may be multiple "solutions" available to insure adequate delivery; none of these is necessarily the best or only solution. Necessary O_2 delivery may be supplied by increasing any or all of the following: O_2 partial pressure, cardiac output, hemoglobin concentration, O_2 extraction, etc. The only structure/function relationship that may be invariant is the association between the volume of O_2-consuming mitochondria and $\dot{V}O_2$max. Ultimately, O_2 demand is best explained by the energetic demand of aerobically-working muscle (Lindstedt et al., 1985a,b).

Because the plasticity of skeletal muscle is great, we do not view the muscle mitochondria as limiting O_2 consumption. Rather, the aerobic capacity of the muscle is the gate or throttle that ultimately regulates the flow of O_2 through the respiratory system.

ACKNOWLEDGMENTS

This research was supported by NSF RII86-10680.

REFERENCES

Anderson, P. and B. Saltin (1985) Maximal perfusion of skeletal muscle in man. J. Physiol. (London) 366:233-249.

Buick, F.J., N. Gledhill, A.B. Froese, L. Spriet and E.C. Meyes (1980). Effect of induced erythrocythemia on aerobic work capacity. J. Appl. Physiol. 48:636-642.

Celsing, F., J. Nystrom, P. Pihlstedt, B. Werner and B. Ekblom (1986). Effect of long-term anemia and retransfusion on central circulation during exercise. J. Appl. Physiol. 4:1358-1362.

Celsing, F., J. Svedenhag, P. Pihlstedt and B. Ekblom (1987). Effects of anemia and stepwise-induced polycythaemia on maximal aerobic power in individuals with high and low haemoglobin concentrations. Acta Physiol. Scand. 129:47-54.

Coyle, E.F., M.K. Hemmert and A.R. Coggan (1986). Effects of detraining on cardiovascular responses to exercise: role of blood volume. J. Appl. Physiol. 60:95-99.

di Prampero, P.E. (1985) Metabolic and circulatory limitations to V_{O_2}max at the whole animal level. J. Exp. Biol. 115:319-331.

Ekblom, B. (1986) Factors determining maximal aerobic power. Acta Physiol. Scand. 128:15-19.

Ekblom, B., A.N. Goldbarg and B. Gullbring (1972). Response to exercise after blood loss and reinfusion. J. Appl. Physiol. 33:175-180.

Ekblom, B. and L. Hermansen (1968). Cardiac output in athletes. J. Appl. Physiol. 25:619-625.

Ekblom, B., R. Hout, E.M. Stein and A.T. Thorstensson (1975). Effect of changes in arterial oxygen content on circulation and physical performance. J. Appl. Physiol. 39:71-75.

Ekblom, B., G. Wilson and P.-O. Astrand (1976). Central circulation during exercise after venesection and reinfusion of red blood cells. J. Appl. Physiol. 40:379-383.

Farrell, P.A., A.B. Gustafson, T.L. Garthwaite, R.K. Kalkhoff, A.W. Cowley, Jr. and W.P. Morgan (1986). Influence of endogenous opioids on the response of selected hormones to exercise in humans. J. Appl. Physiol. 61:1051-1057.

Hagberg, J.M., W.K. Allen, D.R. Seals, B.F. Hurley, A.E. Ali and J.O. Holloszy (1985). A hemodynamic comparison of young and older endurance athletes during exercise. J. Appl. Physiol. 58:2041-2046.

Holmgren, A. and P.-O. Astrand (1966). DL and the dimensions and functional capacities of the O_2 transport system in humans. J. Appl. Physiol. 21:1463-1470.

Hoppeler, H., H. Howald, K.E. Conley, S.L. Lindstedt, H. Classsen, P. Vock and E.R. Weibel (1985) Endurance training in humans: Aerobic capacity and structure of skeletal muscle. J. Appl. Physiol. 59:320-327.

Hoppeler, H., S.L. Lindstedt, E. Uhlmann, A. Niesel, L.M. Cruz-Orive and E.R. Weibel (1984). Oxygen consumption and the composition of skeletal muscle tissue after training and inactivation in the European woodmouse (Apodemus sylvaticus). J. Comp. Physiol. B. 155:51-61.

Hoppeler, H. and S.L. Lindstedt (1985) Malleability of skeletal muscle tissue in overcoming limitations: Structural elements. J. Exp. Biol. 115:355-364.

Kanstrup, I.-L. and B. Ekblom (1982). Acute hypervolemia, cardiac performance and aerobic power during exercise. J. Appl. Physiol. 52:1186-1192.

Kanstrup, I.-L. and B. Ekblom (1984). Blood volume and hemoglobin concentration as determinants of maximal aerobic power. Med. Sci. Sport Exercise 16:256-263.

212

Lindstedt, S.L., H. Hoppeler, K.M. Bard and H.A. Thronson, Jr. (1985a) Estimate of muscle shortening rate during locomotion. Am. J. Physiol. 249:R699-R703.

Lindstedt, S.L., J.H. Jones, H. Hoppeler and H.A. Thronson, Jr. (1985b). Determinants of structure/function relations in the respiratory system: sufficiency vs. limitation. Physiologist 28:342 (abstract).

Lindstedt, S.L. and J.H. Jones (1987). Symmorphosis: the concept of optimal designed. In, M. Feder, A.F. Bennett, W. Burrgren and R. Huey (eds.), New Directions in Physiological Ecology. Cambridge University Press. (In press).

Lindstedt, S.L., D.J. Wells, J.H. Jones, H. Hoppeler and H.A. Thronson, Jr. (1988). Limitations to aerobic performance in mammals: interaction of structure and demand. Int. J. Sports Med. In press.

Olez, O., H. Howald, P.E. diPrampero, H. Hoppeler, H. Claassen, R. Jerri, A. Buehlmann, G. Ferretti, J-C. Brickner, A. Veicsteinas, M. Gussoni and P. Cerretelli (1986). Physiological profile of world-class high-altitude climbers. J. Appl. Physiol. 60:1734-1742.

Robertson, R.J., R. Gilcher, R.F. Metz, G.S. Skrinar, T.G. Allison, H.T. Bahnson, R.A. Abbott, R. Becker and J.E. Fanel (1982). Effect of induced erythrocythemia on hypoxia tolerane during physical exercise. J. Appl. Physiol. 53:490-495.

Robertson, R.J., R. Gilcher, K.F. Metz, C.J. Caspersen, T.G. Allison, R.A. Abbott, G.S. Skrinar, J.R. Krause and P.A. Nixon (1984). Hemoglobin concentration and aerobic work capacity in women following induced erythrocytemia. J. Appl. Physiol. 568-575.

Saltin, B. (1985) Hemodynamic adaptations to exercise. Am. J. Physiol. 55:42D-47D.

Saltin, B. and P.D. Gollnick (1983) Skeletal muscle adaptability: significance for metabolism and performance. Handbook of Physiology Skeletal muscle. L.D. Peachy, R.H. Adrian and S.R. Geiger (eds.), Williams & Wilkins, Baltimore, pp. 555-631.

Spriet, L.L., N. Gledhill, A.B. Froese and D.L. Wilkes (1986) Effect of graded erythrocytemia on cardiovascular and metabolic responses to exercise. J. Appl. Physiol. 61:1942-1948.

Svedenhag, J., A. Martinsson, B. Ekblom and P. Hjemdahl (1986). Altered cardiovascular responsiveness to adrenaline in endurance-trained subjects. Acta. Physiol. Scand. 126:539-550.

Taylor, C.R. (1987). Structural and Functional limits to oxidative metabolism: Insight from scaling. Annu. Rev. Physiol. 49:135-146.

Taylor, C.R. and E.R. Weibel (1981) Design of the mammalian respiratory system. I. Problem and strategy. Respir. Physiol. 44:1-10.

Thomson, J.M., J.A. Stone, A.D. Ginsburg and P. Hamilton (1982). Oxygen transport during exercise following blood reinfusion. J. Appl. Physiol. 1213-1219.

Tipton, C.M. (1986) Determinants of $\dot{V}O_2$max: Insights gained from non-human species. Acta Physiol. Scand. 128:33-43.

Withers, R.T., W.M. Sherman, J.M. Miler and D.L. Costill (1981) Specificity of the anaerobic threshold in endurance trained cyclists and runners. Eur. J. Appl. Physiol. 47:93-104.

Woodson, R.D., R.E. Wills and C. Lenfant (1978). Effect of acute and established anemia on O_2 transport at rest, submaximal and maximal workload. J. Appl. Physiol. 44:36-43.

Verzar, F. (1912) The gaseous metabolism of striated muscle in warm-blooded animals. J. Physiol. (Lond.) 44:243-258.

MODELS OF STEADY-STATE CONTROL OF SKELETAL MUSCLE $\dot{V}O_2$

EVALUATION USING TISSUE DATA

Richard J. Connett

University of Rochester
School of Medicine and Dentistry
Rochester, NY 14642

Oxygen delivered to the skeletal muscle ultimately serves as a substrate for mitochondrial oxidative phosphorylation. All models dealing with oxygen delivery to the mitochondria must include some quantitative description of oxygen consumption ($\dot{V}O_2$). Although a number of proposals for the regulation of mitochondrial $\dot{V}O_2$ have been made and tested with isolated mitochondria, there is still considerable uncertainty as to the best way of relating changes in $\dot{V}O_2$ in vivo to measurable changes in tissue metabolites. In a recent report (Connett, 1987) we demonstrated that during a rest-work transition in dog skeletal muscle, neither adenosine diphosphate concentration [ADP] nor inorganic phosphate [Pi] alone appeared to be rate-limiting and, therefore, regulating substrates. It was found that the best description of the changes in $\dot{V}O_2$ with time was obtained by relating $\dot{V}O_2$ to the phosphorylation potential with the charge balance of adenosine triphosphate [ATP] hydrolysis taken into account. This report deals with a preliminary evaluation of the same models using data from various rates of steady-state energy turnover.

Data

The data used in the analysis are all derived from vascularly-isolated, in situ, dog gracilis muscle sampled during steady-state isometric contractions at various frequencies. The detailed methods for both the muscle preparation and the tissue sampling and analysis have been reported previously (Connett et al., 1986; Gayeski et al., 1985). The data obtained for each muscle include:

1. Directly measured rates of oxygen consumption ($\dot{V}O_2$) using the Fick principle. Contents of metabolic intermediates measured on extracts of quickly frozen muscle: phosphocreatine (PCr), creatine (Cr), Pi, ATP, as well as a number of glycolytic intermediates. The distribution of intracellular oxygen contents measured by the microspectrophotometric method of Gayeski (1981).

2. Values computed directly from the measured variables using well defined enzyme equilibria. The calculated values include: cytosolic pH and free cytosolic [ADP] (Veech et al., 1979; Connett, 1985). Corrections for the binding of Mg^{2+}, K^+, and H^+ have been taken into account in the calculations. It was assumed that [K^+] and [Mg^{2+}] were constant at 0.1 M and 1 mM, respectively. All the calculation methods have been reported previously (Connett, 1985).

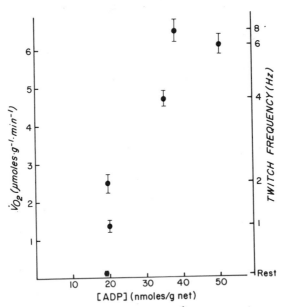

Fig. 1. Oxygen consumption rate ($\dot{V}O_2$) as a function of ADP concentration [ADP] at various frequencies of stimulation. [ADP] is calculated from the measured phosphocreatine, creatine and ATP values as indicated in the methods. Each point is plotted at the mean from at least 5 muscles. The error bars indicate 1 SEM.

Evaluation of Models

1. Regulation dominated by [ADP]. This model of $\dot{V}O_2$ regulation is derived from studies on isolated mitochondria and has been discussed in detail in a number of reports (e.g., Chance et al., 1986). The simplest formulation is that the dependence of $\dot{V}O_2$ on [ADP] shows a hyperbolic relationship, which fits a Michaelis-Menton equation. Because of the existence of significant bound pools of ADP in tissue, and hence the requirement for indirect estimation of free [ADP], this model is difficult to test with <u>in vivo</u> data. The large concentration of creatine and PCr in skeletal muscle and the observation that under almost all physiological conditions creatine kinase is at equilibrium, make data from this tissue some of the best available for testing this model. The [ATP]/[ADP] ratio can be calculated from measured [PCr]/[Cr] and the equilibrium constant of creatine kinase. [ADP] is then estimated by use of the measured ATP concentrations. The observed relationship between [ADP] and $\dot{V}O_2$ is shown in Fig. 1. Clearly the relationship is not the predicted hyperbolic one. Between rest and stimulation at 2 Hz while $\dot{V}O_2$ changes 30 fold, [ADP] shows no change. As $\dot{V}O_2$ approaches the maximum (at 8 Hz) there is an 2-fold increase in [ADP]. Thus, while changes in [ADP] may be important to achieve the maximum recruitment of $\dot{V}O_2$, this factor is clearly not the dominant regulator of $\dot{V}O_2$ in these muscles.

2. Thermodynamic Approach. This form of analysis relates the rate ($\dot{V}O_2$) to a linear combination of the driving forces (phosphorylation potential and redox potential). The redox term includes both [O_2] and mitochondrial NADH reduction state. The tissue oxygen concentration in these muscles has been measured using the cryogenic microspectrophotometric method of Gayeski (1981). The cellular oxygen concentrations at these levels of work (Connett, 1986) are greater than those limiting to $\dot{V}O_2$ (Gayeski et al., 1987). Previous studies on mitochon-

drial redox changes have indicated that at steady-state work at 70% $\dot{V}O_2$ max these are essentially unchanged from resting values (Olgin et al., 1986). Thus, the changes in the redox potential are probably negligible under the aerobic conditions used. Hence, the $\dot{V}O_2$ should be a linear function of the phosphorylation potential alone, i.e. $\dot{V}O_2 = a_o + a_1 \cdot \log([ATP]/[ADP] \cdot [Pi])$. When applying this model the charge balance of the ATP hydrolysis reaction must be taken into account as shown below:

$$ATP^{4-} + H_2O = ADP^{3-} + H_2PO_4^{-} \text{ (or } HPO_4^{2-} + H^+) \tag{1}$$

From this equation we can define the phosphorylation potential as:

$$\log(R_p^{-}) = \log([ATP^{4-}]/[ADP^{3-}][H_2PO_4^{-}]).$$

TABLE I

$$[ATP^{4-}] = [ATP]/b_t \tag{2.1}$$

$$[ADP^{3-}] = [ADP]/b_d \tag{2.2}$$

$$[HPO_4^{2-}] = [Pi]/b_p \tag{2.3}$$

where: [ATP], [ADP] and [Pi] are the total free concentrations of ATP, ADP and inorganic phosphate respectively.

and b_x are the correction terms for K^+, Mg^{2+} and H^+ binding to the compounds and take the form shown in equation 2.4 below:

$$b_x = 1 + K_x^{Mg}[Mg^{2+}] + K_x^{K}[K^+] + K_x^{H}[H^+](1 + K_x^{HMg}[Mg^{2+}]) \tag{2.4}$$

where: $K_x^{g} = [XY]/[X][Y]$

The phosphorylation potential can then be stated in terms of measured values:

$$\log(R_p)^{-} = \log(R_p) + pH + \log(b_d \cdot b_p/b_t) - \log(K_p^{H}) \tag{2.5}$$

where $R_p = [ATP]/[ADP][Pi]$ using the measured values.

If the reaction is with HPO_4^{2-} rather than $H_2PO_4^{-}$ then the term $-\log(K_p^{H})$ is eliminated from the equation.

From the creatine kinase equilibrium the equivalent equation is:

$$\log(R_p)^{-} = \log([PCr]/[Cr] \cdot [Pi] + \log(b_p/b_c) - \log(K_p^{H}/K_{cpk}) \tag{2.6}$$

where: b_c = correction term for binding to PCr

$$K_{cpk} = [ATP^{4-}][Cr]/([ADP^{3}][PCr][H^+]) = 3.6 \times 108$$

and PCr, Cr are the measured values of total phosphocreatine and creatine, respectively.

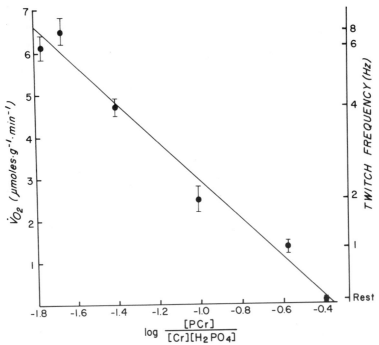

Fig. 2. $\dot{V}O_2$ vs log $(R_p)^-$. The graph shows the relationship
between the measured $\dot{V}O_2 \pm$ SEM and the phosphorylation potential calculated from measured concentrations as described in
Table I. The line is a least squares fit to the data and has
the equation:
$$\dot{V}O_2 = -1.65 - 4.55\log(R_p)^- \qquad R^2 = 0.98$$

Table I outlines the equations for estimating the phosphorylation potential
from the measured concentrations of ATP, ADP, Pi and pH or from phosphocreatine (PCr), free creatine (Cr) and Pi. If the linear thermodynamic
approach describes the control of $\dot{V}O_2$ then it should be a linear function of
$\log([PCr]/[Cr]\cdot[Pi])$. The results of such a fit are shown in Fig. 2. An
excellent linear fit is observed.

DISCUSSION

With the availability of a set of data which includes the rate of
oxygen consumption and concentrations of a number of phosphorylated intermediates as well as distribution of cellular PO_2 in the same muscles, it
becomes possible to test various models for describing the regulation of
cellular $\dot{V}O_2$. A dominating role for [ADP] either through adenine nucleotide
translocase or by any other mechanism is not supported. Application of the
linear thermodynamic approach using the charge balanced phosphorylation
potential gives an excellent fit to the data. The result is consistent with
the suggestion that mitochondrial redox potential is not a significant
variable in the regulation of $\dot{V}O_2$ (Connett, 1987). These preliminary
conclusions from data obtained at various levels of steady-state work are
identical to those obtained using data obtained from studies on the rest-
work transition (Connett, 1987). We conclude that the mitochondria are not
regulated by changes in one substrate but that the primary regulation is by
the cytosolic phosphorylation potential independent of whether the potential is changed via changes in [ATP], [ADP], or [Pi].

REFERENCES

Chance, B., J.S. Leigh, Jr., J. Kent and K. McCully (1986). Metabolic control principles and ^{31}P NMR. Fed. Proc. 45:2915-2920.

Connett, R.J. (1985). In vivo glycolytic equilibria in dog gracilis muscle. J. Biol. Chem. 260:3314-3320.

Connett, R.J. (1987). In vivo recruitment of mitochondrial $\dot{V}O_2$. Test of current models using tissue data. Adv. Exp. Med. Biol. 215:141-151.

Connett, R.J., T.E.J. Gayeski and C.R. Honig (1986). Lactate efflux is unrelated to intracellular PO_2 in a working red muscle in situ. J. Appl. Physiol. 61:402-408.

Gayeski, T.E.J. (1981). "Cryogenic microspectrophotometric method for measuring myoglobin saturation in subcellular volume; application to resting dog gracilis muscle (Ph.D. thesis). Univ. of Rochester, Rochester, NY (University Microfilms, No. DA8224720, Ann Arbor, MI, 1981).

Gayeski, T.E.J., R.J. Connett, and C.R. Honig (1985). Oxygen transport in rest-work transition illustrates new functions for myoglobin. Am. J. Physiol. 248:H914-H921.

Gayeski, T.E.J., R.J. Connett, and C.R. Honig (1987). Minimum intracellular PO_2 for maximum cytochrome turnover in red muscle in situ. Am. J. Physiol. 252 (Heart Circ. Physiol.) H906-915.

Olgin, J., R.J. Connett and B. Chance (1986). Mitochondrial redox changes during rest-work transition in dog gracilis muscle. Adv. Exp. Med. Biol. 200:545-554.

Veech, R.L., J.W.R. Lawson, N.W. Cornell and H.A. Krebs. (1979). Cytosolic phosphorylation potential. J. Biol. Chem. 254:6538-6547.

CRITICAL $P\bar{v}O_2$ VS CRITICAL OXYGEN TRANSPORT WITH

ACUTE HYPOXIA IN ANESTHETIZED ANIMALS

Esther P. Hill, David C. Willford and Francis C. White

Department of Medicine
University of California, San Diego
La Jolla, CA 92093

Oxygen delivery to peripheral tissues has often been described in terms of either the amount of oxygen delivered by the arterial blood or of that exiting the tissues on the venous side, because direct tissue oxygen measurements are often not feasible. Since the early studies of Verzar (1912) and Krogh (1919), physiologists have often assumed that mixed venous PO_2 ($P\bar{v}O_2$) reflects average tissue PO_2, but that assumption is certainly an oversimplification of a very complex situation (Tenney, 1974; Miller, 1982; Cain, 1983). Other attempts to quantitate oxygenation have concentrated instead on the arterial system by calculating the total oxygen transport (TOT) in ml/min as the product of blood flow rate ($\dot{Q}$, in dl/min) and the arterial oxygen concentration (CaO_2, in ml/dl):

$$TOT = \dot{Q} \cdot CaO_2$$

$$\cong 1.34\ \dot{Q}\ [Hb]\ SaO_2 \tag{1}$$

where [Hb] is the hemoglobin concentration in gm/dl and SaO_2 is the fractional oxyhemoglobin saturation in arterial blood. This product has also been termed "oxygen transport", "oxygen availability", and "oxygen delivery" by various authors.

TOT and $P\bar{v}O_2$ are related through the Fick equation and the oxygen dissociation curve. TOT and $P\bar{v}O_2$ both decrease during almost all forms of hypoxia, although not necessarily in direct proportion to each other. Our studies have attempted to compare and relate these variables under conditions of extreme hypoxia, both theoretically and experimentally. Our measurements were designed to determine "critical" TOT and "critical $P\bar{v}O_2$", which we define as the value of $P\bar{v}O_2$ or TOT below which the normal O_2 consumption rate cannot be maintained, and anerobic metabolism becomes observable. The concept of critical oxygen availability to tissues is an old concept which has been used extensively, but only recently explored in much detail (Stainsby and Otis, 1964; Cain, 1977; Cain and Bradley, 1983; Shibutani et al., 1983; Gutierrez et al., 1986). We sought to determine whether the tolerance to hypoxia (critical $P\bar{v}O_2$ or critical TOT) changes with hypothermia, and how the relationships change with different forms of hypoxia. Our experimental studies have dealt primarily with whole body measurements, so we usually report mixed venous PO_2 values.

EXPERIMENTAL METHODS

Immature domestic pigs were studied. As described previously
(Willford et al., 1986b) the animals were pretreated with ketamine and
atropine, anesthetized with thiamylal and halothane, ventilated, paralyzed
with succinylcholine, instrumented, and catheterized for obtention of
arterial and mixed venous blood gas and lactate samples. Following this the
animals were made progressively hypoxic by one of the following procedures
in order to determine critical TOT and critical $P\bar{v}O_2$:

Hypoxic hypoxia. The animals were made progressively hypoxic by
lowering inspired oxygen concentration (FIO_2) by adjusting nitrogen and
oxygen concentrations with a rotameter. Each reduction in FIO_2 was
estimated to lower arterial PO_2 by approximately 10% from its previous value
and was maintained for 10 minutes prior to collecting new steady state blood
samples. Cardiac output was measured with a thermal dilution computer
and/or with an electromagnetic flow meter, and oxygen consumption rates
determined by the Fick principle. Stepwise reductions in inspired oxygen
levels were continued until the animal died.

Anemic hypoxia. The procedure was similar to that above except that
FIO_2 was chosen to maintain arterial PO_2 in the range of 100 to 150 mm Hg,
and progressive anemia was produced by isovolemic hemodilution, using a

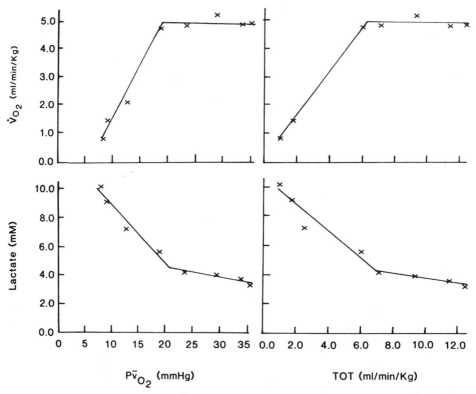

Fig. 1. Oxygen consumption and blood lactate concentration
plotted as a function of mixed venous PO_2 and total oxygen
transport. Data from one of the hypothermic animals. The
intersection of the two least-square line segments is defined
as the critical value of $P\bar{v}O_2$ and TOT. Notice the agreement
between critical values determined by $\dot{V}O_2$ vs lactate measurements.

Table I. Critical $P\bar{v}O_2$ and critical TOT values during hypoxic hypoxia, anemic hypoxia, and hypothermia.

	Hypoxic hypoxia	Anemic hypoxia	Hypothermia with hypoxic hypoxia
N	8	7	13
Temperature $^\circ$C	36.8±0.2	36.4±0.4	29.3±0.1
Critical $P\bar{v}O_2$ mm Hg	22.0±1.4	30.0±1.8	16.5±0.8
Critical TOT mL/min·kg	7.9±0.7	8.5±0.8	5.9±0.5

Values are mean ± SEM. N is number of experiments.

cell saver to remove red cells while returning plasma. Additional plasma from a donor animal was transfused to replace the volume of red cells removed.

Hypothermia and hypoxia. Some of the animals exposed to hypoxic hypoxia were studied at reduced body temperature (29°C). Hypothermia changes the relationship between arterial and venous oxygen levels and lowers metabolic rate and cardiac output. Body temperature was lowered by circulating blood from a jugular vein through a small condensing coil that was jacketed with ice water, and returning the blood to the animal. After the temperature reached 29°C, they were subjected to hypoxia using the protocol outlined above.

The data from each of the above procedures were analyzed assuming that the relationship between $\dot{V}O_2$ and TOT or $P\bar{v}O_2$ could be described by two line segments which intersect at the critical TOT or $P\bar{v}O_2$ value. Fig. 1 illustrates the method with data from one of the hypoxic hypothermic animals. Below the critical value, $\dot{V}O_2$ decreased approximately linearly with TOT or $P\bar{v}O_2$. The two line segments were fit to the data with a least-squares technique to minimize the perpendicular distance between the data and the line segments. Blood lactate data were also analyzed using the same least-squares technique. Table I summarizes results of critical TOT and critical $P\bar{v}O_2$ in all the animal groups.

THEORETICAL CALCULATIONS AND EXPERIMENTAL RESULTS

Changes in critical $P\bar{v}O_2$ and critical TOT during hypothermia. We expected that critical $P\bar{v}O_2$ would decrease with temperature because a smaller diffusion gradient would be required to supply the reduced oxygen consumption rate of the tissues. Using Fick's law of diffusion

$$\dot{V}O_2 = A\alpha O_2 \ D/T \ (P\bar{v}O_2 - Pmito) \tag{2}$$

where $\dot{V}O_2$ is oxygen consumption rate, $P\bar{v}O_2$ and Pmito are PO_2 values in the mixed venous blood and mitochondria, respectively, and A, T, αO_2 and D are effective area, diffusion distance, solubility and diffusivity of oxygen in the peripheral tissue. We assumed that $Q_{10} = 2.0$ (i.e. that $\dot{V}O_2$ decreases by a factor of 2 for a 10°C decrease in temperature), that αO_2 and D are functions of temperature as described by Grote (1967), that A/T does not change with temperature (we do not need to know absolute values of A, T, αO_2, or D; only the relative change with temperature), that Pmito = 2 mm Hg,

223

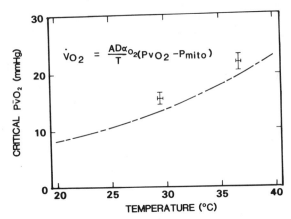

Fig. 2. Critical $P\bar{v}O_2$ as a function of temperature. The dashed line represents the theoretical relationship between $P\bar{v}O_2$ and temperature. The experimental measurements in anesthetized pigs (means ± SEM) are also represented. Calculations from Willford et al (1986a) and data from Willford et al (1986b).

and that the critical venous PO_2 = 20 mm Hg at $37^\circ C$. Fig. 2 shows the calculated decrease in critical venous PO_2 with a decrease in temperature (dashed line) and the actual data obtained in anesthetized pigs. The critical $P\bar{v}O_2$ actually measured at $37^\circ C$ was slightly higher than the arbitrarily assumed value of 20 mm Hg, but the change observed with temperature agrees well with predictions.

To calculate how critical TOT might change with temperature, we assumed that the relation between $\dot{V}O_2$ and TOT would remain unchanged below the critical TOT, and that the critical TOT would, therefore, decrease as temperature decreases, as suggested by Cain and Bradley (1983) (see insert on Fig. 3). We assumed a Q_{10} of 2.0, and a critical TOT at $37^\circ C$ of 9.0 mL/min·kg. The latter is the average of values previously reported by Cain (1977) of 9.8 mL/min·kg in dogs and by Shibutani (1983) of 8.2 mL/min·kg in humans. Our data appear to fit the prediction based on diffusion limitation ($P\bar{v}O_2$ vs. temperature) more closely than this prediction for TOT.

Relation between $P\bar{v}O_2$ and TOT. TOT and $P\bar{v}O_2$ are related through the Fick equation and the oxygen dissociation curve. By rearranging the Fick equation, and expressing the dissociation curve as a simple function of PO_2, we can obtain a useful expression which relates $P\bar{v}O_2$ and TOT:

$$\dot{V}O_2 = \dot{Q} \ (CaO_2 - C\bar{v}O_2)$$

$$= \dot{Q} \ CaO_2 \ (1 - C\bar{v}O_2/CaO_2)$$

$$= TOT \ (1 - C\bar{v}O_2/CaO_2)$$

$$\dot{V}O_2/TOT = 1 - C\bar{v}O_2/CaO_2$$

$$C\bar{v}O_2 = CaO_2 \ (1 - \dot{V}O_2/TOT)$$

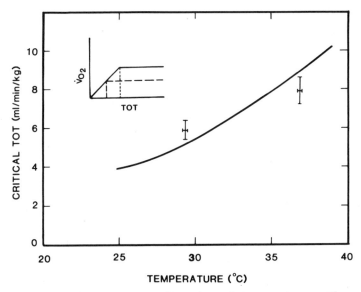

Fig. 3. Critical TOT as a function of temperature. Theoretical prediction of decreased critical total oxygen transport (TOT) with decreased temperature (solid line) and experimental data (means ± SEM) in pigs.

We now define β_a and $\beta_{\bar{v}}$ as the ratios of oxygen concentration to PO_2 (i.e. the slope of lines from the origin to the arterial and venous points on the oxygen dissociation curve):

$$\beta_a = CaO_2/PaO_2 \text{ and } \beta_{\bar{v}} = C\bar{v}O_2/P\bar{v}O_2$$

Then

$$P\bar{v}O_2 = (\beta_a/\beta_{\bar{v}})\ PaO_2\ (1 - \dot{V}O_2/TOT) \qquad (3)$$

Fig. 4 uses this equation to plot $P\bar{v}O_2$ as a function of $TOT/\dot{V}O_2$ (the reciprocal of the oxygen extraction ratio) under various conditions. Note that $\beta_{\bar{v}}$ is a function of $P\bar{v}O_2$, but $P\bar{v}O_2$ and $\beta_{\bar{v}}$ can be solved iteratively for a given dissociation curve, PaO_2 and $TOT/\dot{V}O_2$ ratio. Note from equation (1) that TOT is the product $\dot{Q}$ [Hb] 1.34 SaO_2. If TOT decreases by a decrease in $\dot{Q}$ or [Hb] at constant PaO_2 (and therefore constant SaO_2), then the effect on $P\bar{v}O_2$ and $\beta_{\bar{v}}$ should depend only on TOT and $\dot{V}O_2$ for any combination of [Hb] and $\dot{Q}$ values. This means that anemic hypoxia and stagnant hypoxia should show the same relationship between $P\bar{v}O_2$ and $TOT/\dot{V}O_2$, except for small effects due to dissolved oxygen or interactions with CO_2. Hypoxic hypoxia shows a more complicated relationship between $P\bar{v}O_2$ and $TOT/\dot{V}O_2$ because in this case both the TOT and PaO_2 decrease. TOT decreases as SaO_2 decreases, and the separate factor PaO_2 appearing in equation (3) also decreases. An additional complication is that for hypoxic hypoxia, the relationship of $P\bar{v}O_2$ to TOT depends on the relative magnitude of SaO_2 vs. $\dot{Q}$ [Hb] rather than on just the overall product $\dot{Q}$ [Hb] SaO_2.

These theoretical curves predict that for a given $TOT/\dot{V}O_2$ ratio, $P\bar{v}O_2$ will be lower during hypoxic than during anemic hypoxia, but do not predict at what values of $P\bar{v}O_2$ or TOT critical oxygen delivery will occur. Data from our measurements of critical TOT and critical $P\bar{v}O_2$ measurements are also plotted in Fig. 4. Critical $P\bar{v}O_2$ is significantly higher (P<0.005)

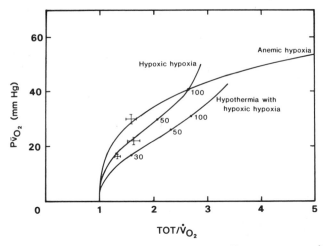

Fig. 4. Theoretical curves relating $\overline{Pv}O_2$ and $TOT/\dot{V}O_2$ for different forms of hypoxia. For anemic hypoxia (upper curve), we assumed PaO_2 = 100 mm Hg and P_{50} = 35 mm Hg (normal for pigs). This curve would also apply for stagnant hypoxia (decreased $\dot{Q}$). For hypoxic hypoxia (middle curve) we assumed [Hb] = 10 gm/dl (normal for pigs), $\dot{V}O_2/\dot{Q}$ = 0.05, and P_{50} = 35 mm Hg. The numerical values on the curves represent different PaO_2 values. For hypothermia (lower curve), we assumed the same values as for hypoxic hypoxia, except P_{50} = 26 mm Hg (Willford and Hill, 1986). Data for the anemic hypoxia and hypoxic hypoxia studies lie very close to predicted lines, but data for the hypothermic studies deviate significantly from the corresponding curve.

during anemic hypoxia than during hypoxic hypoxia, while the critical TOT does not differ between the two forms of hypoxia. The observation of higher critical $\overline{Pv}O_2$ in anemia agrees with that of Cain (1977) in dogs, although we see a smaller difference in critical $\overline{Pv}O_2$ between both forms of hypoxia.

DISCUSSION

Our data on hypothermic pigs showing that critical $\overline{Pv}O_2$ falls with temperature is in agreement with the predictions based on diffusion gradients limiting tissue oxygenation. It fits nicely with the concept of $\overline{Pv}O_2$ representing an average tissue PO_2 (Tenney, 1974) and tissue diffusion gradients limiting tissue oxygenation. Our approach has similarities with the graphical analysis introduced by Wagner et al. (this symposium) which relates mixed venous PO_2 or PO_2 at the venous end of muscle tissue and $\dot{V}O_2$ during maximal work. Our studies have analyzed critical $\overline{Pv}O_2$ and critical TOT in resting animals attempting to maintain constant $\dot{V}O_2$ levels and in animals with reduced $\dot{V}O_2$ during hypothermia, while Wagner's studies have emphasized oxygen limitation with elevated $\dot{V}O_2$ during maximal exercise.

Our theoretical calculations of $\overline{Pv}O_2$ vs $TOT/\dot{V}O_2$ present a concise way of analyzing the role of $\dot{Q}$, [Hb], SaO_2 and the dissociation curve in determining $\overline{Pv}O_2$ and give additional intuitive insights into these relationships. These calculations correctly predicted the difference in $\overline{Pv}O_2$ between anemic and hypoxic hypoxia that we observed at the same $TOT/\dot{V}O_2$ values in our animals. However these calculations were not able to predict when critical $\overline{Pv}O_2$ or critical TOT would occur. In addition, this analysis is disappointing in terms of its inability to explain observed changes in critical TOT during hypothermia.

The experimental measurements on anemic vs hypoxic hypoxia indicate that TOT is a better indicator of the limitation to tissue oxygenation than $P\bar{v}O_2$. This is in agreement with the notion that blood flow distribution during hypoxia may be regulated or at least influenced strongly by arterial oxygen levels (TOT). We are left with the conclusion that neither TOT nor $P\bar{v}O_2$ entirely describe the limit to tissue oxygenation. This is in agreement with observations of previous investigators, and should not be too surprising since these variables describe only inflow and outflow from the overall system, and neither addresses the details of oxygen exchange in the capillaries or tissues.

REFERENCES

Cain, S.M. (1977). Oxygen delivery and uptake in dogs during anemic and hypoxic and hypoxia. J. Appl. Physiol. 42:228-234.
Cain, S.M. (1983). Peripheral oxygen uptake and delivery in health and disease. Clinics in Chest Medicine 4:139-148.
Cain, S.M. and W.E. Bradley (1983). Critical O_2 transport values at lowered body temperatures in rats. J. Appl. Physiol. 55:1713-1717.
Grote, J. (1967) Die Sauerstoffdiffusionskonstanten im Lungengewebe und Wasser und ihre Temperaturabhangigkeit. Pflugers Arch. 295:245-254.
Gutierrez, G., A.R. Warley and D.R. Dantzker (1986). Oxygen delivery and utilization in hypothermic dogs. J. Appl. Physiol. 60:751-757.
Krogh, A. (1919). The supply of oxygen to the tissues and the regulation of the capillary circulation. J. Physiol. (London) 52:457-474.
Miller, M.J. (1982). Tissue oxygenation in clinical medicine: an historical review. Anesthesia and Analgesia 61:537-535.
Shibutani, K., T. Komatsu, K. Kubal, V. Sanchala, V. Kumar and D.V. Bizzarri (1983). Critical level of oxygen delivery in anesthetized man. Crit. Care Med. 11:640-643.
Stainsby, W.N. and A.B. Otis (1964). Blood flow, blood oxygen tension, oxygen uptake and oxygen transport in skeletal muscle. Am. J. Physiol. 206:858-866.
Tenney, S.M. (1974). A theoretical analysis of the relationship between venous blood and mean tissue oxygen pressures. Respir. Physiol. 20:283-296.
Verzar, F. (1912). The influence of lack of oxygen on tissue respiration. J. Physiol. (London) 45:39-52.
Wagner, P.D. (1987). An integrated view of the determinants of maximum oxygn uptake. (This symposium)
Willford, D.C. and E.P. Hill (1986). Modest effect of temperature on the procine oxygen dissociation curve. Respir. Physiol. 64:113-123.
Willford, D.C., E.P. Hill, W.Y. Moores (1986a). Theoretical analysis of oxygen transport during hypothermia. J. Clin. Monitor. 2:30-43.
Willford, D.C., E.P. Hill, F.C. White and W.Y. Moores (1986b). Decreased critical mixed venous oxygen tension and critical oxygen transport during induced hypothermia in pigs. J. Clin. Monitor. 2:155-168.

CAPILLARY CONFIGURATION IN CONTRACTED MUSCLES: COMPARATIVE ASPECTS

Odile Mathieu-Costello
Department of Medicine, M-023
University of California, San Diego
La Jolla, CA 92093

ABSTRACT

We compared the degree of orientation (anisotropy) of capillaries in skeletal muscles of animals with large differences in oxygen needs and/or tolerance to hypoxia (mammals of different size; reptiles; birds; mammals native to high altitude; diving mammals). In terrestrial mammals, we found a substantial increase in capillary tortuosity with fiber shortening, in muscles with large differences in capillary density (capillary counts/fiber mm^2 in transverse sections ranging 450-4350). There was no systematic difference in muscle capillary tortuosity with body size (mouse to pony), or with adaptation to high altitude (deer mice) or to prolonged periods of anoxia (Harbor seals), when account was taken of sarcomere length. A substantial increase in capillary tortuosity was also found in contracted skeletal muscles of the alligator with remarkably low capillary density (capillary counts/fiber mm^2 in transverse sections, 120-280). On the contrary, we found that in pigeon pectoralis, a highly aerobic muscle with large capillary density and a large number of capillary anastomoses running perpendicular to the muscle fiber axis, the decrease in capillary anisotropy with decreasing sarcomere length was smaller than in other muscles. Our results indicate that 1) sarcomere length at which samples are fixed needs to be taken into account when capillary counts in transverse sections are compared between muscles and/or after different experimental conditions, and 2) muscle capillary tortuosity is a consequence of fiber shortening, rather than an indicator of the O_2 requirements of the tissue.

INTRODUCTION

Capillary configuration is an important parameter for modelling blood-tissue exchange in muscles. Depending on capillary convolution among samples, capillary counts on transverse sections, a commonly used parameter to assess muscle capillarity, may not be directly comparable. The classical Krogh approach for modelling blood-tissue exchange assumes long straight capillaries oriented parallel to the muscle fiber axis (i.e., highly anisotropic capillaries). Based on their finding of a tight network of sinuous capillaries in contracted muscles using vascular casts, Groom et al., (1984) argued that it is not possible to use the Krogh cylinder geometry for analyzing oxygen transfer in shortened muscles. Although it had been recognized earlier (Ranvier, 1874; Krogh, 1919), the increased capillary tortuosity with muscle shortening in vivo has remained contro-

Table I: Ranges of capillary numerical densities per fiber sectional area of muscle fibers in transverse sections, QA(O), obtained in the muscles analyzed in each animal group.

Species	Muscles	QA(O) mm^{-2}
rat (340–750g)	soleus	880 – 1590
rat (90g)	gastrocnemius, soleus, gracilis	1670 – 3260
goat	gastrocnemius, semitendinosus	1310 – 1530
dog	gastrocnemius, vastus medialis, semitendinosus, sartorius	1130 – 2490
calf	gastrocnemius, vastus medialis	450 – 910
pony	gastrocnemius, vastus medialis, semitendinosus	590 – 1180
mice, sea level	calf, anterior thigh	2750 – 3820
mice, altitude	calf, anterior thigh	3210 – 4350
seal	gastrocnemius, vastus lateralis, vastus profundus, semitendinosus, latissiumus dorsi, diaphragm, subcutaneous	1030 – 1770
alligator	calf	120 – 280
pigeon	pectoralis	1760 – 5680
	anterior thigh	940 – 1760

versial (Gaehtgens, 1984; Lewis, 1984). It has been assumed that capillary tortuosity contributes to a small percentage (3-12%) of capillary length density in a variety of muscles fixed within physiological sarcomere lengths (Hoppeler, 1984; Piiper and Scheid, 1986). In contrast to this assumption, we found that in rat soleus muscle perfusion-fixed in situ at sarcomere lengths of 1.98 and 1.62 μm, capillary length per volume of muscle fiber was respectively 44 and 65% larger than revealed by capillary counts on transverse sections, i.e. as compared to straight capillaries parallel to the muscle fiber axis (Mathieu-Costello, 1987).

There have been only few comparative studies on capillary configuration in different muscles and/or after different experimental conditions. Increased capillary tortuosity and number of anastomoses have been described qualitatively in red as compared to white muscles (Ranvier, 1874; Romanul, 1965; Andersen and Kroeze, 1978), as well as in skeletal muscle of mice exposed and/or trained under conditions of high altitude (Appell, 1978), and in dog muscles during maturation (Aquin and Banchero, 1981). Highly oriented capillaries were reported in red and white muscle of lizards (Gleeson et al., 1984). In the above studies, however, sarcomere length, i.e., the degree of contraction or extension of the muscles compared was not taken into account.

This paper summarizes recent morphometric data on capillary anisotropy, as a function of sarcomere length, in skeletal muscles of animals with considerable differences in O_2 needs and/or tolerance to limited environmental O_2 availability.

METHODS

Various skeletal muscles (Table I) were perfusion-fixed in situ in:
a. terrestrial mammals: rats (Sprague-Dawley; 85-745 g), goat (Capra

hircus; 20 kg), dogs (Canis familiaris; 20-30 kg), calf (Bos taurus; 142 kg), pony (Equus caballus; 153 kg), and deer mice (Peromyscus maniculatus; 20-25g; born and raised either at 3800 m, UC Station at White Mountain, Barcroft, or at sea level)

b.   diving mammals: Harbor seals (Phoca vitulina; 18-41 kg)
c.   reptiles: alligators (Alligator mississipiensis; 1.5-3 kg)
d.   birds: pigeons (White King, White Carneau and Show Racer; 500-600 g).

Samples were obtained in muscles which were either passively extended or shortened by adequately positioning the knee or ankle joints or the wings (pigeon), before the perfusion was started. Tetanically contracted calf muscles were also perfused in rats and alligators.

All samples were fixed in 6.25% solution of glutaraldehyde in cacodylate buffer, postfixed in 1% solution of osmium tetroxide in buffer and embedded in plastic, as described previously (Mathieu-Costello, 1987). Four to eight tissue blocks were cut from each muscle, yielding four transverse and four longitudinal 1 μm thick sections. The angle of sectioning of each transverse and longitudinal section was carefully controlled, as described elsewhere in detail (Mathieu-Costello, 1987). Briefly, at least three sections were cut from each block, at three consecutive angles with respect to the muscle fiber axis (angles taken approximately 5 and 1 degrees apart for transverse and longitudinal sections, respectively), and sarcomere length was measured in each section. A section was considered longitudinal when changing the angle of sectioning by ± 1 degree with respect to the muscle fiber axis yielded fiber sections with larger sarcomere length. It was considered transverse when changing the angle of sectioning by ± 5 degrees relative to the fiber axis gave fiber sections with smaller sarcomere length. Each transverse and each longitudinal section was subsampled by as many systematic, non-overlapping micrographs as technically possible from the section area (typically, 20 to 60 micrographs per group of 4 transverse or longitudinal sections), using a Leitz Ortholux light microscope equipped with a Leica camera (magnification, 400x). The 35 mm films were projected on an A 144 square grid (see Weibel, 1979, Appendix 3), using a microfilm reader (final magnification, 2060 x). Capillary counts per fiber sectional area in transverse and longitudinal sections were estimated by point counting, using an Apple II+ computer.

The method used to estimate the degree of anisotropy of muscle capillaries from capillary counts in two sets of histological sections, taken parallel and perpendicular to the muscle fiber axis, has been described elsewhere in detail (Mathieu et al., 1983). Briefly, capillary length per volume of muscle fiber, JV, is related to the number of capillary sections per sectional area of muscle fibers, QA(α), by the following equation:

$$JV = c(K,\alpha) \cdot QA(\alpha)$$

where the numerical coefficient $c(K,\alpha)$ depends on K, concentration parameter of capillary segment orientation, and α, angle of the histological section relative to the muscle fiber axis. When two sets of histological sections, transverse ($\alpha=0$) and longitudinal ($\alpha= \pi/2$) to the muscle fiber axis are used, then:

$$QA(0) / QA(\pi/2) = c(k, \pi/2) / c(K,0)$$

Therefore, the following procedure can be used. First, the ratio QA(0) /QA($\pi/2$), obtained from capillary counts in histological sections, gives the ratio $c(K, \pi/2) / c(K,0)$. Knowing the latter ratio, one can obtain the value of $c(k,0)$ via a table (or a graph) of known coefficients (see Mathieu et al., 1983).

231

The value of the coefficient c(k,O), varies from 1 to 2. For capillaries with perfect anisotropy (K=+∝; straight capillaries, all parallel to the axis of orientation; no branching), c(K,O) = 1, and therefore JV = QA(O). For capillaries with no preferential orientation (K=O), the value of c(K,O) is 2, and JV = 2 QA(O). In other words, the coefficient c(K,O) is an index of the percentage of capillary length density contributed by tortuosity and/or branching. For example, a value c(K,O) = 1.50 indicates that capillary length density is 50% greater than it would be if all capillaries were straight, unbranched, and parallel to the axis of orientation.

RESULTS AND DISCUSSION

Distended preparations of the capillary network (Figs. 1-4) were obtained in all muscles. As found previously in rats (Mathieu-Costello, 1987), we experienced no difficulty in perfusing tetanically contracted calf muscles of alligators. The effect of muscle shortening on capillary anisotropy is best analyzed in preparations perfusion-fixed in situ. Partial and/or non-uniform muscle shortening and fiber kinking is found in muscle tissue excised without preservation of length and immersion-fixed (Zumstein et al, 1983). Capillaries can also appear artifactually wavy after immersion fixation, due to the collapse of the capillary wall between blood cells. A close to isotropic arrangement of the capillary network, as revealed by a value of the coefficient c(K,O) of up to 1.80, was obtained in immersion-fixed muscle biopsies (Mathieu et al., 1983).

The skeletal muscles of mammals analyzed in this study were perfusion-fixed in situ at sarcomere lengths ranging from 1.52 to 3.36 μm. Capillary counts per fiber mm^2 ranged from 450 (calf gastrocnemius) to 4350 (deer mouse limb) in transverse sections, and from 280 to 1970 in longitudinal sections of the same muscles. In all samples from mammals, capillaries were substantially more tortuous in shortened muscle (Fig. 1), than in extended muscle (Fig. 2). In rat soleus muscle, the value of the coefficient c(K,O) ranged from 1.05 to 1.14 in extended preparations (sarcomere length, 2.80-2.90 μm), and from 1.55 to 1.65 in tetanically contracted muscles (sarcomere length, 1.60-1.80 μm; Mathieu-Costello, 1987). In larger mammals (dog, goat, pony and calf), we also found a substantial decrease in capillary anisotropy with decreasing sarcomere length. The value of the coefficient c(K,O) was 1.14 and 1.70, in muscles fixed at sarcomere lengths of 2.34 (goat semitendinosus) and 1.57 μm (dog gastrocnemius), respectively. No systematic relationship was found between capillary anisotropy and either body size, animal athletic ability, or muscle capillary density (Mathieu-Costello et al., 1987a). A similar degree of anisotropy of capillaries was also found in deer mice limb muscles, as compared to other mammals (rats to

Fig. 1. Longitudinal section of pony vastus medialis, perfusion-fixed in a shortened position (sarcomere length, 1.9 μm). Magnification: x 335.

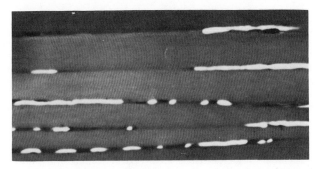

Fig. 2. Longitudinal section of dog sartorius muscle perfusion-fixed in an extended position (sarcomere length, 2.6 μm). Magnification: x 335.

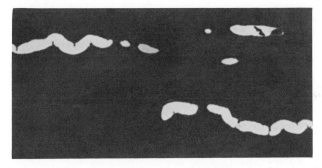

Fig. 3. Longitudinal section of alligator muscle, perfusion-fixed in tetanus (sarcomere length, 1.6 μm). Magnification: x 335.

Fig. 4. Longitudinal section of pigeon pectoralis muscle, perfusion-fixed in a shortened position (sarcomere length, 1.6 μm). Magnification: x 335.

pony), when account was taken of sarcomere length. In muscles of animals native to high altitude (deer mice), the degree of anisotropy of capillaries, at a given sarcomere length, was not different from that found in sea level animals (Mathieu-Costello, 1986). The data obtained in muscles of diving mammals (Bebout and Mathieu-Costello, 1987) also fit well into the comparison of terrestrial mammals at sea level, and high altitude mice. There is a curvilinear increase in the value of c(K,O) with decreasing sarcomere length in skeletal muscles of mammals (Mathieu-Costello et al., 1987a,b).

It has been a matter of discussion whether or not the dramatic increase in capillary tortuosity demonstrated by Potter and Groom (1983) in rat contracted muscles, using vascular corrosion casts, represented in vivo situations (Lewis, 1984). The possibility of contracture of the muscle fibers in material prepared for corrosion casts, as well as the effect of the casting material itself on capillary sinuosity, have been questioned. In a separate study, we combined the use of corrosion casting and stereological methods in the same muscle samples (rat gracilis, gastrocnemius and soleus muscles). We found a very good correlation between the degree of anisotropy of capillaries in material prepared for vascular casts and in muscles perfusion-fixed in situ. The dramatic increase in capillary tortuosity demonstrated by vascular corrosion casts in contracted muscles is the result of fiber shortening within physiological sarcomere lengths. It is not an artifact related to the casting procedure (Mathieu-Costello et al., 1987b).

A substantial degree of capillary tortuosity was also found in shortened hindlimb muscles of the alligator (Fig. 3). In tetanically contracted muscles (sarcomere length 1.42-1.57 µm), the value of the coefficient c(K,O) ranged from 1.41 to 1.62, which compares with the data obtained in fully shortened skeletal muscles of mammals (mouse to pony). It has been suggested that capillary tortuosity patterns are not related to the degree of contraction of the muscles, but are rather related to fiber types, and that capillaries are predominantly tortuous in red muscles, and straight in white muscles (Appell, 1984). Our results indicate that capillary tortuosity is a consequence of fiber shortening, rather than an indicator of the O_2 requirement of the tissue. A dramatic increase in the degree of sinuosity of capillaries is found in shortened muscles of mammals (mouse to pony), and also in those of reptiles, i.e. in muscles with considerable differences in capillary density.

Furthermore, a lesser degree of capillary tortuosity, and a systematically different arrangement of the capillary network, was found in pigeon flight muscle (aerobic capacity comparable to that of rat and pigeon heart), as compared to hindlimb muscles of mammals, reptiles and pigeons. In pigeon pectoralis muscle, capillary counts per fiber mm^2 ranged from 1490 to 5680 in transverse sections, and from 570 to 1380 in longitudinal sections. A large number of anastomoses were found perpendicular to the fiber axis, forming a tight capillary mesh around the muscle fibers (Fig. 4). With fiber shortening, capillary segments between anastomoses bow, but do not become tortuous. The decrease in capillary anisotropy with decreasing sarcomere length was smaller in pigeon pectoralis as compared to hindlimb muscles of pigeon, mammals and reptiles. At sarcomere length 1.6 µm, the value of c(K,O) was 1.21. For comparison, the value of c(K,O) was 1.47 in shortened thigh muscle of pigeon, fixed at the same sarcomere length (1.6 µm).

The tight network of capillaries found in pigeon M. Pectoralis suggests a more uniform blood supply around the fibers, and consequently the use of the Hill model (Hill, 1928; Groom et al., 1984) rather than the Krogh cylinder geometry, to model blood tissue transfer. The substantial convolution of capillaries with muscle shortening also suggests the use of the Hill approach to model blood tissue exchange in muscles of mammals during muscular work. Whether or not a uniform blood supply around the fibers can be assumed in a given muscle evidently depends not only on capillary tortuosity, but also on capillary-to-fiber ratio, sharing factor, and density of anastomoses. In this respect, a parameter to consider is the percent of the muscle fiber perimeter in contact with capillaries (Sullivan and Pittman, 1987). Our data indicate that sarcomere length at which samples are fixed needs to be taken into account when capillarity in transverse sections, and the degree of anisotropy of capillaries, are compared between muscles and/or in response to different experimental conditions.

ACKNOWLEDGEMENTS

This work was supported by Grants HL-17731 and HL-01534 from the National Institutes of Health.

REFERENCES

Andersen, P. and A.J. Kroeze (1978). Capillary supply in Soleus and Gastro-cnemius muscles of man. Pfluegers Arch. 375:245-249.

Appell, H-J. (1978). Capillary density and patterns in skeletal muscles. III. Changes of the capillary pattern after hypoxia. Pfluegers Arch. 377:R-53.

Appell, H-J. (1984). Variability in microvascular pattern dependent upon muscle fiber composition. Prog. appl. Microcirc. 5:15-29.

Aquin, L., and N. Banchero (1981). The cytoarchitecture and capillary supply in the skeletal muscle of growing dogs. J. Anat. 132:341-356.

Bebout, D.E. and O. Mathieu-Costello (1987). Capillary anisotropy in skeletal muscles of harbor seals. Fed. Proc. 46:352.

Gaehtgens, P. (1984). Summary of discussion. Prog. appl. Microcirc. 5:62-63.

Gleeson, T.T., C.J.M. Nicol, and I.A. Johnston (1984). Capillarization, mitochondrial densities, oxygen diffusion distances and innervation of red and white muscle of the lizard Dipsosaurus dorsalis. Cell Tis. Res. 237:253-258.

Groom, A.C., C.G. Ellis, and R.F. Potter (1984). Microvascular architec-ture and red cell perfusion in skeletal muscle. Prog. appl. Microcirc. 5:64-83.

Hill, A.V. (1928). The diffusion of oxygen and lactic acid through tissues. Proc. Roy. Soc. B. 104:39-96.

Hoppeler, H. (1984). Morphometry of skeletal muscle capillaries. Prog. appl. Microcirc. 5:33-43.

Krogh, A. (1919). The number and distribution of capillaries in muscles with calculations of the oxygen pressure head necessary for supplying the tissue. J. Physiol. (London) 52:409-415.

Lewis, D.H. (1984). Summary of discussion. Prog. appl. Microcirc. 5:109-110.

Mathieu-Costello, O. (1986). Capillary anisotropy in skeletal muscle of mice native to high altitude. Proc. Internat. Union Physiol. Siences 16:526.

Mathieu-Costello, O. (1987). Capillary tortuosity and degree of contrac-tion or extension of skeletal muscle. Microvas. Res. 33:98-117.

Mathieu, O., L.M. Cruz-Orive, H. Hoppeler, and E.R. Weibel (1983). Estimat-ing length density and quantifying anisotropy in skeletal muscle capillaries. J. Micr. 131:131-146.

Mathieu-Costello, O., H. Hoppeler, C.R. Taylor and E.R. Weibel (1987a). Capillary anisotropy in skeletal muscles of mammals of different size. Fed. Proc. 46:352.

Mathieu-Costello, O., R.F. Potter, C.G. Ellis and A.C. Groom (1987b). Capillary configuration and fiber shortening in muscle: correlation between corrosion casts and stereological measurements. Fed. Proc. 46:1534.

Piiper, J., and P. Scheid (1986). Cross-sectional PO_2 distributions in Krogh cylinder and solid cylinder models. Resp. Physiol. 64:241-251.

Potter, R.F. and A.C. Groom (1983). Capillary diameter and geometry in cardiac and skeletal muscle studied by means of corrosion casts. Microvasc. Res. 25:68-84.

Ranvier, M. (1874). Note sur les vaisseaux sanguins et la circulation dans les muscles rouges. Arch. Physiol. Ser. 2,1:446-450.

Romanul, F.C.A. (1965). Capillary supply and metabolism of muscle fibers. Arch. Neurol. 12:497-509.

Sullivan, S.M. and R.N. Pittman (1987). Relationship between mitochondrial volume density and capillarity in hamster muscles. Amer. J. Physiol. 252:H149-H155.

Weibel, E.R. (1979). _Stereological Methods, Vol. 1: Practical Methods for Biological Morphometry_. London/New York/Toronto, Academic Press.

Zumstein, A., O. Mathieu, H. Howald and H. Hoppeler (1983). Morphometric Analysis of the Capillary Supply in Skeletal Muscles of Trained and Untrained Subjects - Its Limitations in Muscle Biopsies. Pfluegers Arch. 397:277-283.

THE TOTAL IONIC STATUS OF MUSCLE DURING INTENSE EXERCISE

G.J.F. Heigenhauser and M.I. Lindinger

Department of Medicine
McMaster University Medical Centre
Hamilton, Ontario
Canada L8N 3Z5

INTRODUCTION

During intense exercise, the high rate of glycolysis results in a large accumulation of intracellular lactate (La^-) and increased hydrogen ion concentration ($[H^+]$) (Spriet et al., 1985). High intracellular $[H^+]$ during heavy exercise has often been implicated as a cause of muscle fatigue. A number of loci for fatigue have been suggested: excitation-contraction coupling (Fabiato and Fabiato, 1978), control of glycolytic flux at the level of phosphorylase (Chasiotis et al., 1983) and phosphofructokinase (Trivedi and Danforth, 1966) and impairment of ionic pumps and exchanges on the sarcoplasmic reticulum and sarcolemma (Nakamura and Schwartz, 1972).

To understand the factors influencing the $[H^+]$ in the intracellular fluid, we need to identify changes in the independent variables associated with the ionic systems in this compartment (Stewart, 1981; 1983) in terms of physicochemical systems involving both dependent and independent variables and obeying two fundamental laws of physical chemistry - Electrical Neutrality and Conservation of Mass. The concentrations of the dependent variables such as bicarbonate ($[HCO_3^-]$), hydroxyl ion ($[OH^-]$), hydrogen ion ($[H^+]$), the ionized ($[A^-]$) and unionized ($[HA]$) weak electrolytes are dependent on three independent variables: the PCO_2, the total concentration of weak electrolytes ($[Atot]$), and the strong ion difference ($[SID]$). The $[SID]$ is a term which describes the difference between the sum of the concentrations of strong basic cations ($[Na^+]+[K^+]+[Mg^{++}]$) and the sum of the strong acidic anions ($[Cl^-]+[La^-]$). The regulation of intracellular $[H^+]$ of muscle during intense exercise depends on changes in the total ionic composition of both the intracellular and extracellular fluid. The purpose of the present paper is to describe the ionic changes that occur within the intracellular fluid of fast twitch, high glycolytic muscle during exhaustive exercise and the implications of ionic regulation on muscle fatigue.

METHODS

Twenty-one male Sprague-Dawley rats were used in the study. The animals were randomly assigned to two groups: resting non-exercise control group and an experimental exercise group. Three hours prior to the study,

TABLE I. Physico-chemical constants used in the calculation of the dependent ionic variables in muscle.

	value
K'_w	4.4×10^{-14} Eq.1^{-1}
K_c	2.34×10^{-11} Eq.$1^{-1} \cdot$mmHg^{-1}*
K_3	6.0×10^{-11} Eq.1^{-1}
K_A (rest)	5.5×10^{-7} Eq.1^{-1}
K_A (exercise)	4.0×10^{-7} Eq.1^{-1}

*Calculated from $K_c = K \times S$, where K, the apparent dissociation constant for $CO_2 = 7.41 \times 10^{-7}$ Eq.1^{-1}, the CO_2 solubility coefficient $= 0.0351$ Eq.$1^{-1} \cdot$mmHg^{-1} at $37^\circ C$ and intracellular ionic strength.

the animals were injected via the tail vein with ^{3}H mannitol and ^{14}C-DMO for measurement of extracellular fluid volume and intracellular pH. The resting control group was killed by cervical dislocation. The exercise rats swam with an attached tail weight (5% of body mass) until they were unable to surface for 15s. The animals were removed from the water and killed by cervical dislocation. In both groups of animals, the abdomen was opened and 2 to 3 ml of blood were obtained in a heparinized syringe from the abdominal aorta. A sample of muscle (0.5g) was taken from the white gastrocnemius (WG). The muscle was immediately frozen and stored in liquid nitrogen until analyzed. The blood was analyzed for pH, PCO_2 and PO_2, ions (Na$^+$ K$^+$ Cl$^-$) and La$^-$ as previously described (Lindinger et al., 1986). Muscle total tissue water (TTW), intracellular and extracellular fluid volumes (ICFV and ECFV) were calculated from resting and exercised muscle (Lindinger and Heigenhauser, 1987). Intracellular ion concentrations were calculated from measurements of muscle ion contents measured by instrumental neutron activation analysis and muscle fluid volumes (Lindinger and Heigenhauser, 1987). Muscle pH was measured by a DMO technique. [Atot] and its dissociation constant, KA, were calculated from pH, [SID] and [HCO_3^-] obtained by titrating muscle homogenate samples with either CO_2 or NaOH and using the quantitative physico-chemical relationship described by Stewart (1981; 1983) (see Lindinger et al., 1987 for full description of methods). The dependent variables [H$^+$], [OH$^-$], [HCO_3^-], [A$^-$] and [HA] for the whole muscle were calculated from the independent variables ([SID], [Atot] and PCO_2) measured from whole muscle and calculated using the equation 7A.1.2 from Stewart (1983):

$$[H^+]^4 + \left[K_A + [SID]\right][H^+]^3 + \left[K_A \left([SID] - [A_{tot}]\right) - (K_c \times PCO_2 + K'w)\right][H^+]^2 - \left[K_A\right.$$
$$\left.(K_c \times PCO_2 + K'w) + K_3 \times K_c \times PCO_2\right][H^+] - K_A \times K_3 \times K_c \times PCO_2 = 0 \quad (1)$$

The constants used in equation 1 are listed in Table I.

RESULTS

The duration of swimming of exercised rats was 4.4 ± 0.5 min. Compared to rest, at the end of exercise there was a 2.5% increase in TTW associated

Table II. Intramuscular fluid volumes and intracellular ion concentrations of white gastrocnemius muscle at rest and at the end of exhaustive exercise.

	TTW	ECFV	ICFV	$[Na^+]$	$[K^+]$	$[Mg^{++}]$	$[Cl^-]$	$[La^-]$
Rest	758 ± 3	64 ±10	701 ±13	5.9 ±0.9	143 ± 6	31 ±2	8.7 ±1.1	7.2 ±1.1
Exercise	777 ± 3	78 ±18	714 ±14	14.6* ±2.6	128* ± 6	26* ±3	11.2 ±7.0	43.3* ±3.7

All values are means ± SEM. *Exercise values are significantly different from resting values (p<0.05).
TTW = total tissue water; ECFV = extracellular fluid volume; ICFV = intracellular fluid volume. Units: fluid volumes (ml·kg^{-1} wet wt.); strong ions (mEq·l^{-1} intracellular fluid).

Table III. Intracellular physico-chemical characteristics of the white gastrocnemius muscle at rest and at the end of exercise.

	Independent Variables		
Condition	[SID] mEq	[Atot] mEq	PCO_2 torr
Rest	161±3	184±4	50
Exercise	109±4*	182±7	82±12*

	Dependent Variables				
Condition	$[A^-]$ mEq	[HA] mEq	$[HCO_3^-]$ mEq	$[OH^-]$ nEq	(H^+) nEq
Rest	151±3	32±2	9.4±0.8	390±23	118±7
Exercise	101±4*	80±6*	7.2±0.4*	146±13*	320±33*

All values are means ± SEM. *Exercise values are significantly different from rest values (p<0.05).
All concentrations are for L of intracellular fluid. [SID] was calculated as the sum of the strong cations minus the strong anions: $[SID] = [Na^+] + [K^+] + [Mg^{++}]/2 - [Cl^-] - [La^-]$, where the $[Mg^{++}]$ is halved because about 50% of the intracellular Mg^{++} is nondiffusible. KA at rest = 5.5×10^{-7}; KA with exercise = 4.0×10^{-7} (Table I).

with an increase of 22% in ECFV and 2% increase in ICFV (Table II). The quantitatively important changes in intracellular strong ions with exercise were increases in $[La^-]$ and $[Na^+]$ and reduction in $[K^+]$ (Table II).

The intracellular independent physico-chemical variables in the WG at rest and in fatigue are shown in Table III. The [SID] decreased by 52 mEq.l^{-1} from rest to the end of exercise. The [Atot] calculated from the PCO_2 and NaOH titrations was not significantly different for resting and exercised muscle; however, KA decreased from 5.5×10^{-7} to 4.0×10^{-7} Eq. l^{-1}. The intramuscular PCO_2 was assumed to be approximately equal to the PCO_2 of 50 mmHg at rest and 82 mmHg after exercise. In resting and exercised muscle, the intracellular dependent physico-chemical variables

TABLE IV. Intramuscular pH measured by DMO and homogenate methods and calculated from physico-chemical equation.

Condition	pH–DMO	pH–physico-chem	pH–homog
Rest	6.98±.03	6.94±.02	6.91±.02
Exercise	6.64*±.03	6.51*±.04	6.47*±.06

Values are means ± SEM. *Significantly different from rest (P<0.05).

were calculated from the measured independent variables using equation 1 and the constants listed in Table I. Compared to rest, at exhaustion $[A^-]$ decreased, whereas [HA] increased by 50 mEq.l^{-1}, $[HCO_3^-]$ and $[OH^-]$ decreased by 2.2 mEq.l^{-1} and 2.44 nEq.l^{-1}, respectively, and $[H^+]$ increased by 202 nEq.l^{-1}.

The theoretical effects of changing the independent variables [SID], $[Atot]$, KA, and PCO_2 on $[H^+]$ were calculated from equation 1. Of the 202 nEq.l^{-1} increase in $[H^+]$ which occurred during exercise, 17% and 83% of the increase could be accounted for by an increase in PCO_2 from 50 to 82 mmHg and a decrease in [SID] from 161 to 109 mEq.l^{-1}. If KA did not decrease from 5.5×10^{-7} to 4.0×10^{-7} Eq.l^{-1}, a further 25% decrease in $[H^+]$ would have occurred.

The intracellular pH measured by DMO distribution and homogenate technique as well as the intracellular pH calculated by the physico-chemical equation were compared; no significant differences were observed among the methods in either resting or exercised muscle (Table IV). Compared to resting muscle, the pH of the exercised muscle was significantly lower.

DISCUSSION

The present study has provided quantitative evidence that the total ionic status of the intracellular fluid of muscle is determined by the independent physico-chemical variables [SID], [Atot], and PCO_2 in the intracellular fluids. Changes in the total ionic status cannot occur without a concomitant change in at least one of the three independent physico-chemical variables, which in turn will cause changes in all the dependent variables. In the WG, the $[H^+]$ increased from 118 nEq.l^{-1} at rest to 320 nEq.l^{-1} at exhaustion. In resting and exercised muscle, the values of pH calculated from equation 1 and the independent variables measured in the whole muscle are similar to those obtained by the distribution of DMO and by the homogenate technique.

An important factor in calculating pH by the physico-chemical equation is KA, the dissociation constant of weak electrolytes. The decrease in KA from 5.5×10^{-7} Eq.l^{-1} to 4.0×10^{-7} Eq.l^{-1} from rest to exhaustion has the effect of protecting the intracellular fluid against large increases in $[H^+]$. At rest, the KA is the histidine residues of intracellular proteins (Ka = 1.78×10^{-7} Eq.l^{-1}) and inorganic phosphate (Ka = 1.66×10^{-7} Eq. l^{-1}). During exercise, the concentration of proteins remains unchanged, but creatine phosphate decreases by 75% and inorganic phosphate increases by 4 fold. Thus, the contribution of creatine phosphate to the KA is reduced while those of creatine (Ka = 1.58×10^{-7} Eq.l^{-1}) and inorganic phosphate are increased. These combined effects reduce the KA of [Atot].

The decrease in [SID] was responsible for 87% of the increase in $[H^+]$ in the intracellular fluid of the exercised WG. The major contributors to the fall of [SID] were the increase in [La^-] and the decrease in [K^+]. The increase in [La^-] contributed to 67% of the decrease in [SID] and was the most important factor responsible for the ionic disturbances. During exercise, a reduction in the rate of glycolysis (Spriet et al., 1985) is an important mechanism for ion regulation together with removal of La^- from the intracellular fluid into the extracellular fluid and increased rates of La^- oxidation within the exercising muscle (Astrand, et al., 1986). Restoration of intracellular [K^+] by an increase in Na^+/K^+ ATPase activity will also contribute to a correction of the intracellular ionic status.

Fatigue of muscle may result when large increases in inorganic phosphate from the breakdown of creatine phosphate exert a direct inhibitory effect on actomyosin force generation (Wilkie, 1986). The increased [H^+] is also associated with relatively larger increases in the monobasic rather than dibasic forms of inorganic phosphate. The monobasic form of inorganic phosphate species exerts the inhibitory effect (Wilkie, 1986). This mechanism may be especially important in decreased force development during intense exercise.

An important factor in the regulation of glycolysis during high intensity exercise may be the decrease in [SID]. As shown in Table III, the concentrations of the ionized [A^-] and unionized [HA] weak acids associated with intracellular proteins changed with changes in [SID]. Thus, the changes in the two dependent variables [HA] and [A^-] with changes in [SID] may alter the catalytic activity of non-equilibrium glycolytic enzymes, which have classically been considered as "pH sensitive". Such "pH effects" have been proposed for the regulation of glycolysis at the level of phosphorylase (Chasiotis et al., 1983) and phosphofructokinase (Trivedi and Danforth, 1966). Also, there is evidence that the decrease in [K^+] found in the WG in the present study may also modify the activity of pyruvate kinase (Kachmar and Boyer, 1953). The mechanism by which K^+ exerts its allosteric regulatory effect appears to be mediated through conformational changes in the active site (Mildvan, 1974).

We conclude that the changes which occur in the three independent variables ([SID], PCO_2, [Atot]) result in changes in the dependent variables ([H^+], [OH^-], [HCO_3^-], [A^-]. The increased [La^-] in muscle is the prime contributor to changes in the dependent variables. Increases in the monobasic form of inorganic phosphate may contribute to a decrease in actomyosin ATPase activity. Changes that occur in the unionized and ionized state of the proteins (i.e., enzymes) may influence glycolytic flux during intense exercise. Therefore, during intense exercise in which there is a large accumulation of La^- in the intracellular fluid of muscle, it is necessary to regulate the total ionic status in the intracellular fluid to maintain muscle function.

REFERENCES

Astrand, P.-O., E. Hultman, A. Juhlin-Dannfelt and G. Reynolds (1986). Disposal of lactate during and after strenuous exercise in humans. J. Appl. Physiol. 61:338-343.
Chasiotis, D., E. Hultman and K. Sahlin (1983). Acidotic depression of cyclic AMP accumulation and phosphorylase b to a transformation in skeletal muscle of man. J. Physiol. (Lond.) 335:197-204.
Fabiato, A. and F. Fabiato (1978). Effects of pH on the myofilaments and the sarcoplasmic reticulum of skinned cells from cardiac and skeletal muscles. J. Physiol. (London).
Kachmar, J.F. and P.D. Boyer (1953). Kinetic analysis of enzyme reactions. II. The potassium activation and calcium inhibition of pyruvic phosphoferase. J. Biol. Chem. 200:669-682.

Lindinger, M.I., M. Ganagarajah and G.J.F. Heigenhauser (1987). Determinants of intramuscular H^+ concentration. Med. Sci. Sports Exerc. 19:S27.

Lindinger, M.I. and G.J.F. Heigenhauser (1987). Intracellular ion content of skeletal muscle measured by instrumental neutron activation analysis. J. Appl. Physiol. 63:426-433.

Lindinger, M.I., G.J.F. Heigenhauser and N.L. Jones (1986). Acid-base and respiratory properties of a buffered bovine erythrocyte perfusion medium. Can. J. Physiol. Pharmacol. 64:550-555.

Mildvan, A.S. (1974). Mechanism of enzyme action. Ann. Rev. Biochem. 43:357-399.

Nakamura, Y. and A Schwartz (1972). The influence of hydrogen ion concentration on calcium binding and release by skeletal muscle sarcoplasmic reticulum. J. Gen. Physiol. 59:22-32.

Spriet, L.L., C.G. Matsos, S.J. Peters, G.J.F. Heigenhauser and N.L. Jones (1985). Muscle metabolism and performance in perfused rat hindquarter during heavy exercise. Am. J. Physiol. 248:C109-C118.

Stewart, P.A. (1981). How to Understand Acid-Base: A Quantitative Primer for Biology and Medicine. New York: Elsevier North Holland.

Stewart, P.A. (1983). Modern quantitative acid-base chemistry. Can. J. Physiol. Pharmacol. 61:1444-1461.

Trivedi, B. and W.H. Danforth (1966). Effect of pH on the kinetics of frog muscle phosphofructokinase. J. Biol. Chem. 241:4110-4112.

Wilkie, D.R (1986). Muscular fatigue: effects of hydrogen ions and inorganic phosphate. Fed. Proc. 45:2921-2923.

LIMITATIONS TO OXYGEN TRANSFER

AN INTEGRATED VIEW OF THE DETERMINANTS OF MAXIMUM OXYGEN UPTAKE

Peter D. Wagner

Department of Medicine
University of California, San Diego
La Jolla, CA 92093

When one considers the pathway for transfer of O_2 from the atmosphere to the mitochondria, several distinct and sequential steps are classically identified: 1) convective gas flow of O_2 to the alveoli by ventilation. 2) diffusive gas mixing and transfer across the blood gas barrier into the capillary blood, 3) convective transport in the blood to the peripheral tissues, and 4) diffusive movement out of the tissue capillary to the mitochondria where O_2 is utilized.

Probably because these individual steps can be so readily separated, investigators have long sought to identify which step in this pathway is rate-limiting to O_2 utilization. Thus, the question of what determines maximum O_2 utilization ($\dot{V}O_2$max) is a very old one, yet today there remains considerable uncertainty about the answer. This is a different question from what determines fatigue (Bigland-Ritchie and Woods, 1984).

Most workers believe that cardiovascular transport is the limiting factor setting $\dot{V}O_2$max, and that the operative concept is the composite variable, oxygen delivery (Saltin, 1985). O_2 delivery, also called total oxygen transport, is the product of cardiac output, arterial O_2 saturation and hemoglobin content (neglecting physically dissolved O_2). Thus, it is well-known that $\dot{V}O_2$max falls in normal subjects during hypoxic exposures such as ascent to altitude (Dempsey et al., 1982; Pugh et al., 1964; West, 1962), and most workers have been content to accept the mechanism as a fall in O_2 delivery due to the reduced arterial saturation caused by hypoxia. $\dot{V}O_2$max is increased in hyperoxia (Kaijser, 1970; Welch, 1982) and after blood transfusion (Buick et al., 1980; Gledhill, 1982; Thomson et al., 1982; Williams et al., 1981) further supporting these notions.

The elegant experiments of Rowell et al., (1986) and Andersen and Saltin (1985) clearly demonstrate that when specific muscle blood flow is increased by restricting exercise to isolated muscle groups in man, $\dot{V}O_2$max is also increased well beyond values seen for the same muscle group in whole body exercise. Such data add to the weight of evidence that $\dot{V}O_2$max and O_2 delivery are closely linked, and it is tempting to ascribe cause and effect to this relationship.

Experiments such as those just mentioned, as well as others scattered in the literature (Pirnay et al., 1972; Hartley et al., 1973) however, provide an important clue suggesting that O_2 delivery, while undeniably a major factor, interacts with another process to actually set $\dot{V}O_2$max under any set of circumstances. In other words, O_2 delivery as "the" limiting factor to $\dot{V}O_2$max becomes an unsatisfactory concept. The clue, usually glossed over in the papers referred to, is that effluent muscle venous blood is far from depleted of its O_2 at $\dot{V}O_2$max. In fact, if the muscles' boundary conditions (O_2 delivery in particular) are systematically altered experimentally, there is evidence that when $\dot{V}O_2$max is increased (e.g., by increasing O_2 delivery), <u>so too is effluent muscle venous PO_2</u>. This is at first a paradoxical result since we usually associate increased $\dot{V}O_2$ with <u>greater</u> O_2 extraction, not less, and thus would expect a lower effluent PO_2.

This paper explores the significance of data currently available pertaining to venous PO_2 at $\dot{V}O_2$max and builds, from these observations, an integrative view of what determines $\dot{V}O_2$max. What clearly emerges from the analysis is: a) the critical role played by diffusion of O_2 from the muscle capillary to the mitochondria as a determinant of $\dot{V}O_2$max, and b) one should not seek "the" limiting factor to $\dot{V}O_2$max. Rather, as will be presented, $\dot{V}O_2$max is set by the integrative interplay between each and every step involved in O_2 transfer from the atmosphere to the mitochondria: interference at <u>any</u> point in the chain will act to reduce $\dot{V}O_2$max in a quantitatively predictable fashion.

In the analysis to follow, the position is taken that $\dot{V}O_2$max is never limited by the biochemical constraints of the oxidative enzymes (i.e., the muscles are able to use all the O_2 that can be extracted from the blood within the currently known spectrum of feasible studies). Eventually one could imagine an experimental setting where massive O_2 delivery exceeded metabolic capabilities to use O_2, but here we assume with good logic (Rowell et al., 1986) that this condition of biochemical limits to $\dot{V}O_2$max <u>in vivo</u> is never reached.

ANALYSIS

The basic concept of this paper is that there is interaction between the two components of delivery of O_2 within the muscle (convective delivery by capillary blood flow and diffusion of O_2 from the Hb molecule to the mitochondria). This interaction determines $\dot{V}O_2$max. Intuitively, convective delivery largely determines capillary PO_2, while subsequent diffusive transport out of the capillary is the product of PO_2, and what might be called the tissue diffusing capacity for O_2, in accordance with Fick's First Law of Diffusion. More analytically, we can express these concepts by two different equations:

$$\dot{V}O_2 = \dot{Q} \, [CaO_2 - CvO_2] \tag{1}$$

$$\dot{V}O_2 = DO_2 \, [PvO_2 - PmitO_2] \tag{2}$$

Equation (1), the well-known mass balance (or Fick equation), simply expresses $\dot{V}O_2$ as the product of blood flow and arteriovenous O_2 content difference. For simplicity, assume that the outflow of O_2 from the muscle capillary occurs at a single point, at which there is a stepdown in PO_2 from arterial to venous values. By Fick's Law of Diffusion (equation (2)), $\dot{V}O_2$ will then also be expressed by the product of DO_2 (the tissue O_2 diffusing capacity) and the driving partial pressure gradient: capillary PO_2 (PvO_2) minus mitochondrial PO_2 ($PmitO_2$). The convenience of considering O_2 efflux at a single point sufficiently illustrates the concept without having the mathematical complexity of integration of these two equations at all points along the capillary.

246

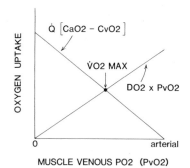

As generally believed, mitochondrial PO_2 at $\dot{V}O_2$max is so close to zero
that for the purposes of the present analysis it may be neglected. Equation
(2) then simplifies to:

$$\dot{V}O_2 = DO_2 \cdot PvO_2 \tag{3}$$

It is easiest to continue the analysis graphically. The constructs of
equations (1) and (3) suggest a plot where the ordinate is the variable $\dot{V}O_2$
and the abscissa the variable PvO_2. For some given, constant values for $\dot{Q}$,
CaO_2 and DO_2, equation (1), when plotted has an intercept (where $PvO_2 = 0$)
equal to O_2 delivery ($\dot{Q} \cdot CaO_2$) and a negative slope equal numerically to the
product of blood flow ($\dot{Q}$) and the slope of the oxyhemoglobin dissociation
curve. Since $\dot{Q}$ is a constant and the slope of the dissociation curve is
relatively constant in the normal range of venous PO_2 values, equation (1)
can be approximated (in the physiological range) as a negatively sloped,
straight line as shown in Fig. 1 and is indicated by the term $\dot{Q}[CaO_2 - CvO_2]$.

On the other hand, equation (3) is simply a straight line of positive
slope (given by the value of DO_2, the tissues O_2 diffusing capacity) passing
through the origin. This line is indicated by $DO_2 \times PvO_2$ in Fig. 1.

The point of intersection of these two lines gives $\dot{V}O_2$max for the given
values of $\dot{Q}$, CaO_2 and DO_2. This statement is the major point of this paper
and is argued as follows: the positively sloped "Fick Law" line expresses,
at any given capillary PO_2, the maximum $\dot{V}O_2$ that can be delivered by
diffusion from the red cell to the mitochondria. However, $\dot{V}O_2$ must also be
given by the Fick principle simply to preserve mass conservation. Thus,
each and every point on the "Fick Principle" line obeys mass conservation,
but all values above and to the left of the point of intersection of the
lines in Fig. 1 are infeasible from the standpoint of diffusive O_2 transport: the capillary PO_2 would be too low to accomplish such high $\dot{V}O_2$
values. Hence, the highest $\dot{V}O_2$ that both satisfies mass conservation and
the limits imposed by a finite tissue diffusing capacity is found at the
intersection of the two lines and is $\dot{V}O_2$max.

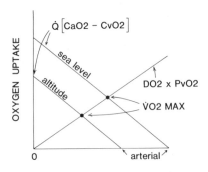

Fig. 2. Extension of the concepts of Fig. 1 to include the effects of altitude. Altitude reduces arterial oxygen content and thereby displaces downward the line for the Fick principle relationship as shown. The lower point of intersection with the positively sloped line indicates that altitude produces not only a reduced oxygen uptake, but necessarily a corresponding reduction in muscle venous PO_2. In particular, the relationship between maximum oxygen uptake and muscle venous PO_2 as the subject moves from sea level to altitude is seen to form a straight line through the origin whose slope characterizes the tissue diffusing capacity.

The implications of this simple graphical analysis are many, and they allow an explanation of many experimental observations. Perhaps the first extension is to examine the effect of acute exposure to high altitude. In this setting, maximum cardiac output is essentially not different from that at sea level (Gale et al., 1985), yet $\dot{V}O_2$max is clearly reduced, together with arterial O_2 content, due to hypoxia.

Fig. 2 adds to the two lines of Fig. 1 a third line reflecting high altitude where arterial PO_2 is lower. This third line represents equation (1) with the same blood flow but lower arterial O_2 content, and is thus similar in shape and slope to the sea level line, although at lower ordinate values because the intercept, $\dot{Q}$ CaO_2, is less. This altitude "Fick Principle" line intersects the "Fick Law" line at a lower point and shows that not only must $\dot{V}O_2$max be reduced at altitude <u>but so too must muscle venous PO_2</u>, assuming the same value for DO_2. At least with acute hypoxia the assumption of a constant DO_2 at $\dot{V}O_2$max is reasonable.

Not only does Fig. 2 predict a lower venous PO_2 along with a lower $\dot{V}O_2$max, but it also predicts that <u>all</u> alterations in O_2 delivery ($\dot{Q}$ CaO_2) will produce points on the graph of Fig. 2 that must lie on the "Fick Law" straight line of positive slope and passing through the origin.

Fig. 2 also predicts that any maneuver that results in an <u>increased O_2</u> delivery by the circulation can produce a higher $\dot{V}O_2$max but that this must be accompanied by an increased muscle venous PO_2 (again, at constant DO_2). There are considerable data showing that O_2 delivery can be increased by blood transfusion or breathing elevated concentrations of O_2, and although such changes in O_2 delivery are small, there is an associated acute increase in $\dot{V}O_2$max. Whether this is actually accompanied by a corresponding increase in venous PO_2 remains to be determined.

Before discussing other implications of Figs. 1 and 2, it is worth examining what data are available in the literature to support or refute the hypothesis that convective and diffusive processes interact to determine $\dot{V}O_2$max.

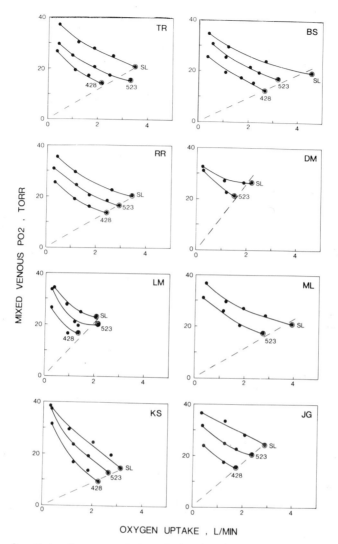

Fig. 3. Data from an acute altitude simulation study (Wagner et al., 1986) done at Duke University showing pulmonary arterial PO_2 at near maximum exercise as a function of oxygen uptake for 8 human subjects. Three altitudes are shown: sea level, 523 torr barometric pressure and 428 torr barometric pressure. Notice that at the maximal values shown the points lie close to a straight line through the origin in accordance with the prediction of Fig. 2.

Relatively few relevant experiments have been performed. What is needed are measurements of effluent muscle PO_2 at $\dot{V}O_2$max and under at least two different levels of O_2 delivery in order to place the $\dot{V}O_2$max-$P\bar{v}O_2$ points on Figs. 1 and 2. Four papers in the literature have useful data in this regard. However, in none were the data collected with the present ideas in mind, and so they are not all optimally designed to evaluate these concepts. Well-designed prospective experiments will be needed to properly support or refute the concepts of this paper. Moreover, while the concepts are illustrated most easily using venous PO_2 as the X axis, the better variable for the abscissa of Figs. 1 and 2 would be mean capillary PO_2. This cannot of course be measured, but it could be estimated by a numerical integration procedure if one assumes that <u>at $\dot{V}O_2$max</u> the muscle functions as a homogene-

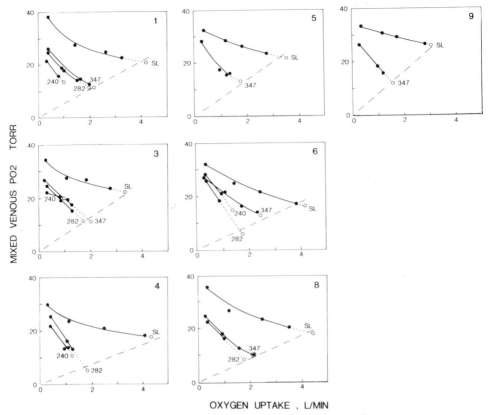

Fig. 4. Corresponding data to those shown in Fig. 3 but obtained in a chronic altitude simulation study referred to as Operation Everest II (OE II) (Sutton et al., 1986). In this study, data are shown for 7 subjects at sea level, and at a variety of barometric pressures as low as the equivalent summit of Mount Everest (chamber pressure 240 torr). Except at this extreme altitude, the data appear to agree with the concepts of Fig. 2. Solid lines join measured points, while dashed lines are extrapolations of mixed venous PO_2 to independently measured maximum oxygen uptake values.

ous structure in terms of the ratio of $\dot{V}O_2$ to perfusion. For the present, we shall assume that effluent venous PO_2 and mean capillary PO_2 are directly proportional. It is not necessary to assume the two values are equal, just that they are related by a constant factor (at $\dot{V}O_2$max, as O_2 delivery is experimentally altered).

In each of the four papers reporting $\dot{V}O_2$ and PvO_2 values at $\dot{V}O_2$max under conditions of at least two levels of O_2 delivery in man, it was also observed that as O_2 delivery was reduced, so were $\dot{V}O_2$max and venous PO_2. Thus, Pirnay et al., (1972) found with β-adrenergic blockade in normal subjects that $\dot{V}O_2$max was reduced by 12% and femoral vein PO_2 at $\dot{V}O_2$max was reduced by 21%. Hartley et al., (1973) found that reducing PIO_2 to levels equivalent to an altitude of 15,000 feet in a hypobaric chamber (P_B = 464 torr) produced almost identical (25%) reductions in $\dot{V}O_2$max and femoral venous PO_2. More recently we made similar, incidental observations on the relationship between mixed venous PO_2 and $\dot{V}O_2$max in two groups of normal subjects, one exposed to acute (Wagner et al., 1986) and the other to chronic (Sutton et al., 1986) hypobaric hypoxia. These data are summarized in Figs. 3 and 4.

250

To date, no data have been found that refute the concept of Figs. 1 and 2: whenever measured, there is an essentially linear relationship between effluent muscle PO_2 and $\dot{V}O_2$max as convective O_2 delivery is altered experimentally.

DISCUSSION

1. The basic tenet of this paper is that the central factor determining $\dot{V}O_2$max is the capability of tissue diffusional transport of O_2 from Hb inside the red cell to the mitochondria. Convective O_2 delivery by the circulation plays a conceptually secondary role in that it determines PO_2 levels in the capillary, and thus diffusion gradients for O_2 transport through the tissues. Together, the two processes set $\dot{V}O_2$max as described.

Perhaps the most important theory to emerge is that this analysis explicitly demands that each and every step in the O_2 transport pathway will play a direct role in setting the value of $\dot{V}O_2$max. O_2 delivery is determined by PIO_2, lung function, Hb concentration, and cardiac output (or muscle blood flow). A reduction in any of these components must, by the current analysis, reduce $\dot{V}O_2$max. For any combination of these variables, the tissue diffusing capacity ("DO_2" of equation (3)) will determine actual $\dot{V}O_2$max.

Thus, each and every component of the pathway directly affects $\dot{V}O_2$max and the concept of "one rate limiting step" is not perceived as conceptually satisfactory.

2. It is easy to see from the concepts and data of this paper, how well-correlated are convective O_2 delivery and $\dot{V}O_2$max. For example, as one increases in altitude in Fig. 2, O_2 delivery and $\dot{V}O_2$max are, indeed, seen to fall together. There is thus nothing in the present analysis that does not account for such correlative observations. They are in fact expected, and this analysis, if correct, should place these observations in their proper perspective. Convective O_2 delivery determines $\dot{V}O_2$max only insofar as it sets capillary PO_2 values; for any such capillary PO_2 values, actual $\dot{V}O_2$max is independently determined by the level of tissue diffusing capacity for O_2. Thus, O_2 delivery is an essential component of, but far from, the sole determinant of $\dot{V}O_2$max.

3. In examining existing data in the light of the concepts of this paper, considerable support is observed, and not one paper could be found that refutes the hypothesis. However, the assumptions currently made should be kept in mind: 1) At $\dot{V}O_2$max, muscle is functionally homogeneous in terms of distribution of $\dot{V}O_2$ to $\dot{Q}$ ratios. 2) Mitochondrial PO_2 is essentially zero (a value as high as 1 or 2 torr would not make much difference to the analysis, however, and there is evidence that at high work levels, mitochondrial PO_2 is certainly no higher than this (Honig et al., 1984)). 3) Tissue diffusing capacity at $\dot{V}O_2$max is constant as O_2 delivery is altered acutely by, for example, reducing arterial content by ascent to altitude. 4) Finally, observable venous PO_2 values (femoral vein or pulmonary arterial) are proportional (not necessarily equal) to mean muscle capillary PO_2 in any one individual. All are reasonable assumptions for normal subjects at $\dot{V}O_2$max, but eventually it would be beneficial if they could be confirmed experimentally.

4. The analysis of Figs. 1 and 2 can be extended to provide a number of interesting hypotheses and testable questions that go far beyond the basic concepts of this paper and have quite practical endpoints. These are now briefly discussed.

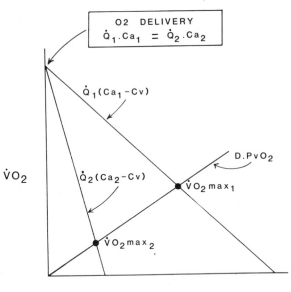

Fig. 5. Explanation of how two different modes of achieving
the same oxygen delivery could, by the current theory, produce
different values of maximum oxygen uptake. The two negatively
sloped lines again reflect the Fick principle for two
different experiments having the same oxygen delivery. The
line producing $\dot{V}O_2max_1$ achieves an oxygen delivery by means of
low cardiac output coupled to a high arterial oxygen content,
while the line producing $\dot{V}O_2max_2$ achieves the same oxygen
delivery by means of a high cardiac output and low arterial
oxygen content. The lines have different slopes because
cardiac output is a major determinant of the slope of the line.
This analysis shows that at the same oxygen delivery it is
possible to achieve different oxygen uptakes for the same
tissue diffusing capacity.

A. However logical the main analysis of this paper may appear, it is true
that both O_2 delivery and effluent muscle venous PO_2 will closely correlate
with $\overline{V}O_2max$, as evident from the preceding discussion. One might ask if
anything in the present context can be used to dissociate the roles of O_2
delivery and tissue O_2 diffusion from one another in setting $\dot{V}O_2max$. Based
on the relationships in Fig. 1, there is indeed an extension of the concepts
that could form the basis of a feasible experiment to dissociate O_2 delivery
from diffusion. This is illustrated in Fig. 5. Here, using the same
coordinate system as in Fig. 1, three lines are drawn. One depicts Fick's
Law of Diffusion as before and has zero intercept plus a positive slope.
The other two have negative slopes, originate from the same intercept, and
both reflect the Fick mass balance principle. Because the intercepts are
the same, O_2 delivery ($\dot{Q} \times CaO_2$) is identical for both of these lines.
However, line 1 achieves O_2 delivery by means of a lower blood flow coupled
to a higher arterial O_2 content, while line 2 achieves the same O_2 delivery,
but by combining a higher cardiac output with a lower arterial O_2 content.
Based on equation (1), line 1 must have a less negative slope (because $\dot{Q}$ is
lower) than line 2. Accordingly, lines 1 and 2 intersect the Fick law line
at different points.

The consequence of this is that, comparing two situations in which O_2
delivery is identical, the present theory predicts that when O_2 delivery is
achieved by a lower flow plus higher arterial O_2 content, $\dot{V}O_2max$ will be

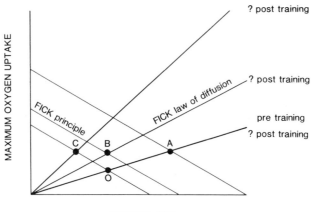

Fig. 6. Illustration of hypothetical effects of exercise training on the relationship between maximum oxygen uptake and effluent muscle venous PO_2. Point 0 is the starting point prior to exercise training and following a period of training, maximum oxygen uptake is increased. Points A, B and C indicate three of many hypothetical locations of post-training values. All three are drawn at the same vertical point, reflecting the same $\dot{V}O_2$max, but at quite different venous PO_2's. A achieves maximum oxygen uptake along the same diffusing capacity line as 0 and therefore reflects only an augmentation of oxygen delivery (upward displacement of the Fick principle line). C reflects the precise converse, namely, no change in oxygen delivery compared to 0, but a substantial increase in tissue diffusing capacity associated with an increased slope of the Fick diffusion line. B represents a compromise in which both components are elevated. By means of such an analysis, it should be possible to quantitatively separate the roles of central and peripheral factors in determining how $\dot{V}O_2$max is augmented by training.

higher along with effluent venous PO_2, compared to a setting with the same O_2 delivery generated by a high cardiac output and low arterial O_2 content.

Thus, a variety of $\dot{V}O_2$max levels may result from a given level of O_2 delivery, depending on how that delivery is achieved. Preliminary data from an isolated, perfused dog gastrocnemius preparation is in line with this prediction (Hogan et al., 1987).

B. The plot of $\dot{V}O_2$ against effluent muscle venous PO_2 shown in Fig. 1 might become useful as a means of analyzing the response to muscle training. Thus, the partitioning of the increase in $\dot{V}O_2$max (brought about by training) into fractions due to improved O_2 delivery and to improved tissue diffusing capacity will become possible as indicated in Fig. 6. In this manner, "central" convective and "peripheral" diffusive components of $\dot{V}O_2$max can be determined.

In an identical manner, the response to other stimuli such as prolonged exposure to altitude, could be studied. Correlations between structural changes in mitochondrial density, capillarity, and fiber type and size as well as tissue diffusing capacity (determined from paired measurement of $\dot{V}O_2$max and effluent muscle venous PO_2) before and after training or before and after altitude acclimatization may well throw light on which component of the diffusion pathway contributes the greatest resistance to O_2 transfer.

253

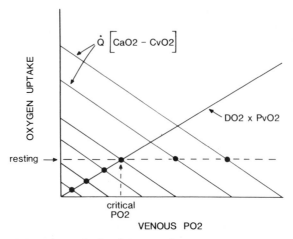

Fig. 7. Extension of the interaction between tissue diffusing capacity and oxygen delivery to explain the concept of a critical PO_2 below which oxygen uptake cannot be maintained. Solid points show actual oxygen uptake levels that would occur as oxygen delivery is progressively reduced. Above the critical PO_2, diffusing capacity exceeds the need for delivery of oxygen, but below the critical PO_2 oxygen uptake would be limited by tissue diffusing capacity, as shown. Such a two-phase relationship is commonly observed in this experimental situation.

We predict that exercise training will increase tissue O_2 diffusing capacity (permitting a lower venous PO_2 for a given $\dot{V}O_2$) while chronic altitude exposure will not alter diffusing capacity. These predictions are based on the hypothesis that the major resistance to O_2 diffusion is between the Hb molecule and the muscle cell wall. Training increases this contact surface area while altitude seems only to cause muscle atrophy and not to increase the number of capillaries (Banchero, 1975; Tenney and Ou, 1970). These predictions remain to be tested.

C. Finally, the entire construct of the present analysis has been directed towards $\dot{V}O_2$max. However, as shown in Fig. 7, the same relationships can be used to explain the observations of a critical venous PO_2 and O_2 delivery below which resting $\dot{V}O_2$ cannot be maintained. $\dot{V}O_2$ will fall linearly toward zero below that $Pv\dot{O}_2$ value which, for the given diffusing capacity, can no longer supply resting $\dot{V}O_2$ by diffusion. Willford et al., (1986) also suggest that diffusion limitation may explain the existence of a critical PO_2.

SUMMARY AND CONCLUSIONS

This paper develops an analysis of the interaction between circulatory convective and tissue diffusive O_2 transport that explains how $\dot{V}O_2$max is set by the integrated response to each and every step in the O_2 transport chain from atmosphere to mitochondria. We suggest that it is not useful to search for "the" rate limiting step determining $\dot{V}O_2$max; rather, all steps contribute. Beyond these basic concepts, the theory accounts for observations of $\dot{V}O_2$max at altitude, permits experimentally testable predictions of $\dot{V}O_2$max under a variety of conditions, explains the concept of critical O_2 delivery below which $\dot{V}O_2$ cannot be maintained, and suggests an analysis to separate the contributions of central convective from peripheral diffusive adaptations to stimuli such as exercise training and chronic altitude exposure.

ACKNOWLEDGEMENTS

This work was supported by NIH grant HL 17731 and the California Lung Association Established Investigator Award. I also wish to thank Tania Davisson for preparation of this paper.

REFERENCES

Anderson, P. and B. Saltin (1985). Maximal perfusion of skeletal muscle in man. J. Physiol. 366:233-249.

Banchero, N. (1975). Capillary density of skeletal muscle in dogs exposed to simulated altitude. Proc. Soc. Exp. Biol. N.Y. 148:435-439.

Bigland-Ritchie, B. and J.J. Woods (1984). Changes in muscle contractile properties and neural control during human muscular fatigue. Muscle Nerve 7:691-699.

Buick, F.J., N. Gledhill, A.B. Froese, L. Spriet and E.C. Myers (1980). Effect of induced erythrocythemia on aerobic work capacity. J. Appl. Physiol. 48:636-642.

Cain, S.M. (1977). Oxygen delivery and uptake in dogs during anemic and hypoxic hypoxia. J. Appl. Physiol. 42:228-234.

Dempsey, J., P. Hanson, D. Pegelow and A. Claremont (1982). Limitations to exercise capacity and endurance: pulmonary system. Can. J. Appl. Sport Sci. 7:4-13.

Gale, G.E., J. Torre-Bueno, R.E. Moon, H.A. Saltzman and P.D. Wagner (1985). Ventilation-perfusion inequality in normal humans during exercise at sea level and simulated altitude. J. Appl. Physiol. 58(3):978-988.

Gledhill, N. (1982). Blood doping and related issues: a brief review. Med. Sci. Sports Exercise 14(3):183-189.

Hartley, L.H., J.A. Vogel and M. Landowne (1973). Central, femoral and brachial circulation during exercise in hypoxia. J. Appl. Physiol. 34(1):87-90.

Hogan, M.C., J. Roca, P.D. Wagner and J.B. West (1987). Dissociation of VO_2max from O_2 delivery in isolated, in situ canine gastrocnemius. Submitted to American Physiological Society Annual Fall Meeting, October 11-15.

Honig, C.R., T.E.J. Gayeski, W. Federspiel, A. Clark, Jr. and P. Clark (1984). Muscle O_2 gradients from hemoglobin to cytochrome: new concepts, new complexities. Adv. Exp. Med. Biol. 169:23-38.

Kaijser, L. (1970). Limiting factors for aerobic muscle performance. The influence of varying oxygen pressure and temperature. Acta Physiol. Scand. Suppl. 3246:1-96.

Pirnay, F., M. Lamy, J. Dujardin, R. Deroanne and J.M. Petit (1972). Analysis of femoral venous blood during maximum muscular exercise. J. Appl. Physiol. 33:289-292.

Pugh, L.G.C.E., M.B. Gill, S. Lahiri, J.S. Milledge, M.P. Ward and J.B. West (1964). Muscular exercise at great altitudes. J. Appl. Physiol. 19:431-440.

Rowell, L.B., B. Saltin, B. Kiens and N.J. Christensen (1986). Is peak quadriceps blood flow in humans even higher during exercise with hypoxemia? Am. J. Physiol. 251:H1038-H1044.

Saltin, B. (1985). Hemodynamic adaptations to exercise. Am. J. Cardiol. 55:42D-47D.

Sutton, J.R., J.T. Reeves, P.D. Wagner, B.M. Groves, A. Cymerman, P. Young, M.K. Malconian and C.S. Houston (1986). Oxygen uptake during exercise at extreme simulated altitude is maintained by marked reduction in mixed venous oxygen tension - Operation Everest II. Fed. Proc. 45:4231.

Tenney, S.M. and L.C. Ou (1970). Physiological evidence for increased tissue capillarity in rats acclimatized to high altitude. Respir. Physiol. 8:137-150.

Thomson, J.M., J.A. Stone, A.D. Ginsburg and P. Hamilton (1982). O_2 transport during exercise following blood reinfusion. J. Appl. Physiol. 53:1213-1219.

Wagner, P.D., J.T. Reeves, J.R. Sutton, A. Cymerman, B.M] Groves, M.K. Malconian and P.M. Young (1986). Possible limitation of maximal O_2 uptake by peripheral tissue diffusion. Am. Rev. Respir. Dis. 133: A202.

Wagner, P.D., G.E. Gale, R.E. Moon, J.R. Torre-Bueno, B.W. Stolp and H.A. Saltzman (1986). Pulmonary gas exchange in humans exercising at sea level and simulated exercise. J. Appl. Physiol. 61:260-270.

Welch, H.G. (1982). Hyperoxia and human performance: a brief review. Med. Sci. Sports Exercise 14:253-262.

West, J.B. (1962). Arterial oxygen saturation during exercise at high altitude. J. Appl. Physiol. 17:617-621.

Willford, D.C., E.P. Hill and W.Y. Moores (1986). Theoretical analysis of oxygen transport during hypothermia. J. Clin. Monit. 2:30-43.

Williams, M.H., S. Wesseldine, T. Somma and R. Schuster (1981). The effect of induced erythrocytemia upon 5-mile treadmill run time. Med. Sci. Sports Exercise 13:169-175.

BREATHING DURING EXERCISE: DEMANDS, REGULATION, LIMITATIONS

H.V. Forster and L.G. Pan

Department of Physiology
Medical College of Wisconsin
Milwaukee, WI 53226

ABSTRACT

In humans alveolar ventilation ($\dot{V}A$) is adjusted almost perfectly to the metabolic demands of mild and moderate exercise. For example, in exercise transitions and in the steady state, $PaCO_2$ rarely deviates by more than 1 to 3 mmHg from the value at rest. This near-homeostasis contrasts to most other mammaliam species; equines for example, demonstrate a progressive hypocapnia and alkalosis as exercise intensity is increased to moderate levels. In equines, the control systems seem programmed for a specific hyperventilation that contributes to maintenance of PaO_2 homeostasis. Generally, during heavy exercise all species hyperventilate creating hypocapnia, increased PAO_2, widened A-a O_2 gradient, and PaO_2 homeostasis.

The origin of the metabolic ventilatory stimulus remains controversial. Evidence exists for: a) "neural" mediation, either central command or peripheral afferent in nature; and b) "humoral" mediation with an intrathoracic metabolite receptor being a possibility. The mechanism of the species differences in hyperventilation during exercise does not appear to be due to species variation in chemoreceptor "fine tuning". Contrary to traditional thinking, recent findings suggest that the hyperventilation during heavy exercise might not be mediated by lactacidosis stimulation of chemoreceptors.

The increase in $\dot{V}A$ during exercise is achieved efficiently in that airway diameter is modulated and the pattern of breathing and the recruitment of respiratory muscles are set to minimize the O_2 cost of breathing. It has been postulated that mechanoreceptors in airways, lung parenchyma and the chest wall are important to efficient breathing. Their role and contribution to the exercise hyperpnea has been shown by reductions in respiratory neural output within breath when respiratory impedance is reduced via helium breathing. Hilar nerve afferents do not appear to be critical to this response. However, carotid chemoreceptors appear essential for "fine tuning" of $\dot{V}A$ when respiratory impedance is reduced.

In most healthy exercising mammals, the efficiency component of the exercise stimulus does not compromise $\dot{V}A$. There are two known major exceptions. One is the extremely fit human athlete during very high workloads when atypically there is minimal or no hyperventilation resulting in

Table I. Major demands imposed on the pulmonary system during exercise.

	Humans		
	Rest	Moderate Work	Heavy Work
$\dot{V}O_2 \sim \dot{V}CO_2$, $1.min^{-1}$	0.3	1.0	3.0
$(a-\bar{v})O_2$ Content, Vol%	5.0	9.0	14.0
$P\bar{v}CO_2$, mmHg	46.0	52.0	65.0
Qc, $1.min^{-1}$	5.0	11.0	20.0
	Ponies		
$\dot{V}O_2 \sim \dot{V}CO_2$, $1.min^{-1}$	0.5	3.0	9.0
$(a-\bar{v})O_2$ Content, Vol%	5.0	12.0	17.0
$P\bar{v}CO_2$, mmHg	52.0	57.0	80.0
Qc, $1.min^{-1}$	11.0	33.0	>50.0

arterial hypoxemia. That indeed the high O_2 cost of breathing compromises $\dot{V}A$ is indicated by hyperventilation and alleviation of hypoxemia with resistance unloading through helium breathing. A second example of a compromise of $\dot{V}A$ is that of a galloping racehorse at very high workloads. In this instance locomotor and breathing efficiency require "entrainment" that is a one for one relationship between breathing and foot plant frequency. The achievable $\dot{V}A$ does not meet the metabolic needs resulting in arterial hypercapnia and hypoxemia. Contrary to traditional thinking, the pulmonary response may contribute to exercise limitations even in healthy mammals.

INTRODUCTION

As exercise intensity increases, $P\bar{v}CO_2$ and the arterial-mixed venous O_2 difference increase progressively (Table I). Accordingly, with the increase in cardiac output, there is a progressive increase in the demands on the pulmonary system for replenishment of O_2 and elimination of CO_2. One of the major pulmonary responses to this demand is an increase in alveolar ventilation ($\dot{V}A$). This response is achieved efficiently in that airway diameter is modulated and the pattern of breathing and recruitment of respiratory muscles are set to minimize the O_2 consumption of the respiratory muscles. Thus, the exercise breathing stimulus consists of a component regulating $\dot{V}A$ and a component minimizing the O_2 cost of breathing. The present review is limited to these aspects of the pulmonary response to the demands imposed by muscular exercise.

Some Problems in Studying the Breathing Responses to Exercise

Insights into the adequacy and mechanisms of the breathing response to exercise can be gained by defining the "true" temporal pattern of alveolar and arterial blood gases during exercise. This objective is not easily achieved, and, thus, until recently, very little data existed. The difficulty lies in circumventing the effects of multiple confounding factors that can influence breathing, alveolar gas tensions, and arterial blood gases. The instrumentation required to obtain alveolar gases can stimulate or inhibit breathing (Askanazi et al., 1980; Bisgard et al., 1978; Pan et al., 1983). The catheterization of an artery required for obtaining multiple samples of blood tends to cause hyperventilation, particularly at rest. Changing from the supine to the sitting to the standing postures at rest causes a progressive hyperventilation (Forster et al., 1986; Matalon and Fahri, 1979). Finally, many ill-defined psychogenic stimuli cause hyperventilation at rest. Each of these factors then potentially prevents obtaining "true" resting or baseline data. Thus, it is essential to have a relaxed subject familiar with the test protocol, unencumbered by any apparatus, and isolated from extraneous visual, auditory and psychogenic stimuli. The influence of many of these factors may be eliminated or diminished by exercise; thus the temporal pattern of $PaCO_2$ between two levels of steady-state exercise probably provides the best opportunity to evaluate the breathing responses to increased demands. Accordingly, we have performed these types of studies in humans and ponies to gain insights into the adequacy and mechanisms of the breathing responses to exercise. We sampled arterial blood at rest and at short, prescribed intervals throughout 8 minutes of exercise with the work rate either increased or decreased after 4–5 minutes.

Blood gases during exercise

Homeostasis of alveolar and arterial blood gases throughout exercise would imply adequacy of breathing and other pulmonary responses. Indeed, it is generally believed that alveolar and arterial blood gas homeostasis in humans is maintained throughout mild and moderate exercise (Asmussen, 1965;

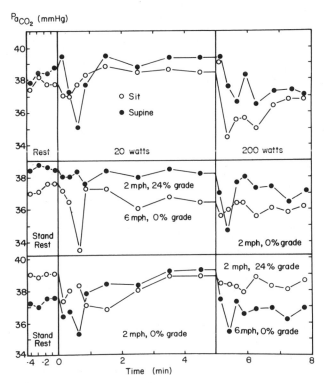

Fig. 1. The temporal pattern of $PaCO_2$ between rest and steady-state exercise and the temporal pattern between two levels of steady-state exercise for one human subject during 4 different exercise tasks on the treadmill (bottom 2 panels) and 2 different tasks on a bicycle. Note trend toward hypocapnia during first minute after metabolic rate was changed.

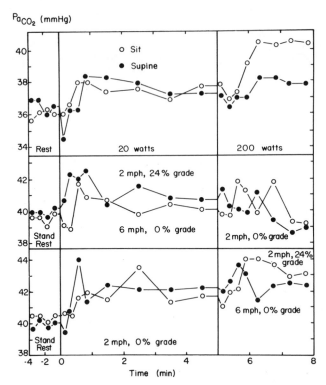

Fig. 2. The temporal pattern of $PaCO_2$ between rest and steady-state exercise and the temporal pattern between two levels of steady-state exercise for one human subject during 4 different exercise tasks on the treadmill (bottom 2 panels) and 2 different tasks on a bicycle. Note, in contrast to the hypocapnia for the subject depicted in Fig. 1, this subject consistently demonstrated hypercapnia in rest to work and work to work transitions.

Dejours, 1965; Whipp, 1981). However, recent data indicate slight disruptions of homeostasis. For example, most of the subjects we recently studied demonstrated a transient hypocapnia during both rest-to-work and low-to-moderate workload transitions (Fig. 1; Forster et al., 1986). On the other hand, a few subjects consistently demonstrated a transient hypercapnia during work transitions (Fig. 2). Our data thus confirms the findings of others who found a disruption of $PETCO_2$ or $PaCO_2$ homeostasis in rest-to-work transition (Barr et al., 1964; Fordyce et al., 1982; Oldenburg et al., 1979; Young and Woolcock, 1978). Finally, we also found statistically significant 1 to 2 mmHg differences in $PaCO_2$ between different levels of mild and moderate treadmill and bicycle exercise (Fig. 3, Forster et al., 1986). Even though homeostasis was not strictly maintained, the demands for gas exchange were indeed most often met and exceeded, and it is remarkable how minimal the $PaCO_2$ disruptions were. However, during exhaustive exercise in humans, hyperventilation is generally observed; it is usually progressive in nature as the work intensity exceeds 60% of maximal capacity, and it can result in as much as a 10-15 mmHg hypocapnia (Asmussen, 1965; Dejours, 1965; Whipp, 1981).

Most non-human species hyperventilate during even mild and moderate exercise (Bisgard et al., 1982; Clifford et al., 1986; Flandrois et al., 1974; Forster et al., 1984; Fregosi and Dempsey, 1984; Kiley et al., 1980; Kuhlmann et al., 1985; Pan et al., 1983; Smith et al., 1983; Szlyk, 1981;

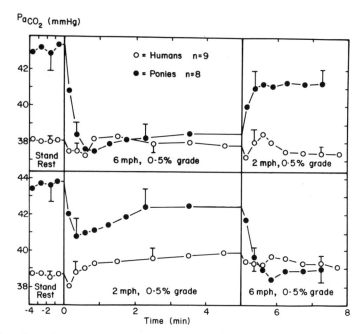

Fig. 3. The temporal pattern of PaCO$_2$ between rest and steady-state exercise and the temporal pattern between two levels of steady-state exercise for humans and ponies during 2 different treadmill exercise protocols. Note that the disruption of PaCO$_2$ homeostasis is greater in ponies than it is in humans.

Watkins et al., 1962). In ponies (Fig. 3) and dogs (Clifford et al., 1986), the hyperventilation is maximal during the first minute of a rest-to-work or work-to-work transition. After the first minute, PaCO$_2$ increases to a steady-state level of hypocapnia. In ponies (Forster et al., 1984), ducks (Kiley et al., 1980), and goats (Smith et al., 1983), the magnitude of the hypocapnia is a function of metabolic rate. In fact, in ponies the linear relationship between hypocapnia and metabolic rate extends from rest to maximal exercise, irrespective of anaerobiosis (Fig. 4; Pan et al., 1986). Furthermore, regardless of whether metabolic rate is increased by increasing treadmill grade or treadmill speed and, in ponies, regardless of whether exercise is on all four legs or only on two legs, PaCO$_2$ decreases linearly as $\dot{V}O_2$ and $\dot{V}CO_2$ increase (Forster et al., 1984; Smith et al., 1983). The mechanism of the species differences in exercise hyperventilation is discussed in a subsequent section.

"Benefits" of exercise hyperventilation

Hyperventilation contributes to sustained arterial and mixed venous alkalosis during mild and moderate exercise in ponies (Fig. 5). A second factor contributing to the alkalosis is the increase in buffer base resulting from the mobilization of splenic erythrocytes (Pan et al., 1984). Whether a mild alkalosis in extracellular fluid is "beneficial" for exercise performance remains to be determined.

Hyperventilation increases alveolar PO$_2$ which increases the driving force for diffusion of O$_2$ from the lungs to the blood; thus PaO$_2$ is also affected by hyperventilation. In ponies, the hyperventilation at the onset of exercise is associated with a PaO$_2$ increased above normal (Fig. 5). However, as exercise is continued, PaO$_2$ returns to levels at rest, irrespec-

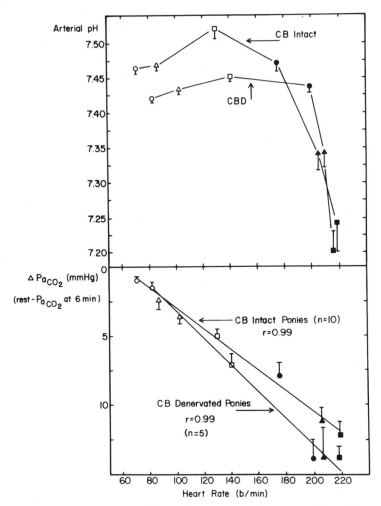

Fig. 4. The relationship of heart rate (as an index of work-rate) to exercise hypocapnia (as an index of hyperventilation) and to arterial pH in normal and carotid body denervated ponies. All data points were obtained between 4 and 6 minutes of treadmill exercise. $\Delta PaCO_2$ refers to the difference in $PaCO_2$ between rest and each workrate.

tive of the workload. Is, then, hyperventilation required in non-human species, even at moderate workloads to maintain PaO_2 homeostasis in the steady state? Conceivably, the alveolar-capillary adjustments to exercise might be less in the non-human species, rendering them more dependent on hyperventilation. The importance of hyperventilation to PaO_2 homeostasis is further demonstrated by examples provided in the last section of this review.

Mechanism of the exercise hyperpnea

Very few topics in physiology have been researched to the extent, yet remain as controversial, as the mechanism of the exercise hyperpnea. The controversy centers aroung the origin of the stimulus that provides for a rapid, rather precise adjustment in V̇A to meet the metabolic demands so that PaO_2 does not fall and $PaCO_2$ does not increase. Most of the several theories that have been advanced fit into one of two categories. One

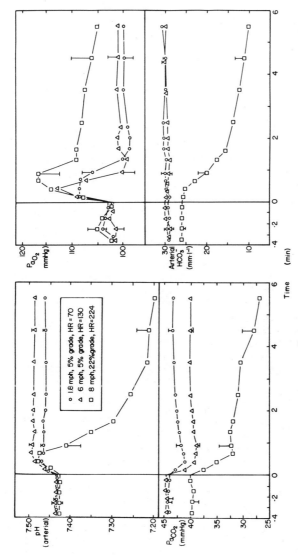

Fig. 5. PaCO$_2$, arterial pH, arterial HCO$_3^-$, and PaO$_2$ in groups of normal ponies at rest and during mild (1.8-5% grade), moderate (6 mph-5%), and exhaustive (8 mph-10%) treadmill exercise. Note particularly: 1) the sustained alkalosis during mild and moderate exercise; and 2) the homeostasis of PaO$_2$ over rest and each level of steady-state exercise.

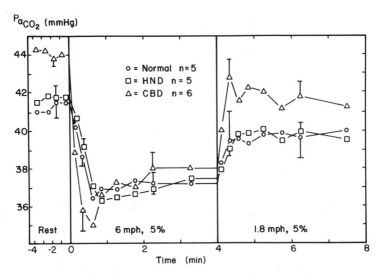

Fig. 6. The temporal pattern of $PaCO_2$ between rest and steady-state exercise and the temporal pattern between two levels of steady-state exercise in normal, hilar nerve denervated (HND), and carotid body denervated (CBD) ponies. Note that HND did not alter the $PaCO_2$ response to exercise, but CBD accentuated the disruption of homeostasis in exercise transitions.

popular notion is that the stimulus is "humoral", that is, exercise causes a change in some blood-borne agent which stimulates VA at a receptor located outside of the exercising limbs. CO_2 has been most often suggested as the likely agent (Wasserman et al., 1977; Whipp, 1981); it has been postulated to act at known carotid (Cross et al., 1982; Wasserman et al., 1979; Yamamoto and Edwards, 1960) and postulated pulmonary chemoreceptors (Green and Sheldon, 1983; Sheldon and Green, 1982; Green et al., 1986). The second popular notion is that the stimulus is "neural", meaning that the signal is not blood-borne; it originates in the brain (DiMarco et al., 1983; Eldridge et al., 1985; Krogh and Lindhard, 1913) or in the exercising muscles (Bennett, 1984; Mitchell et al., 1977; Kao, 1977; Dejours, 1965) and then it is transmitted to the medullary centers by either descending or ascending neural pathways. Most perplexing has been the fact that proponents of each theory provide data which seem to indicate that each can totally account for the hyperpnea. These apparently contradictory findings led us to pursue the topic in what we thought was a systematic manner.

One approach was to study fully awake ponies before and after surgical removal of potential receptor mechanisms. Since it has been postulated that the carotid body mechanism might be capable of mediating the exercise hyperpnea in spite of steady-state blood gas homeostasis in humans (Yamamoto et al., 1960; Wasserman et al., 1979), we studied ponies before and after carotid body denervation (CBD). After CBD, the ventilatory response to i.v. injections of NaCN was abolished, verifying the absence of arterial chemoreception (Pan et al., 1983). This deficit resulted in an accentuated rather than attenuated hyperventilation during exercise, particularly during transition conditions (Fig. 6; Pan et al., 1983). These data are inconsistent with mediation of the hyperpnea by these chemoreceptors. Others have proposed that CO_2 delivery to the lung is a critical factor, implying that a receptor mechanism within the lung might mediate the hyperpnea (Green and Sheldon, 1983; Sheldon and Green, 1982; Phillipson et al., 1981). As a result we vagally denervated the lung, which eliminated lung volume reflexes under anesthesia and altered the pattern of breathing

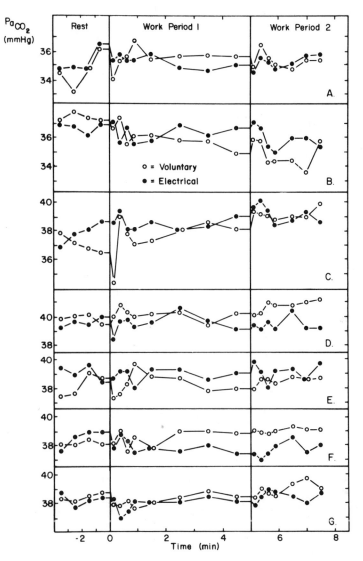

Fig. 7. The temporal pattern of $PaCO_2$ in normal human subjects during voluntary and electrically induced muscle contractions that increased the metabolic rate two- (work period 1) to fourfold (work period 2). Each panel represents one subject. Each subject was studied three times during each condition.

in awake ponies, as expected after vagotomy. However, arterial blood gases (Fig. 6), $\dot{V}A$ and $\dot{V}E$ at rest and during exercise did not differ from predenervation (Flynn et al., 1985). These data following CBD and pulmonary vagotomy are consistent with other studies on the role of the carotid bodies (Bisgard et al., 1982; Flandrois et al., 1974) and lung afferents during exercise (Clifford et al., 1986; Favier et al., 1982). Accordingly, there does not appear to be a critical dependence on carotid and pulmonary receptor mechanisms for mediation of the exercise hyperpnea.

We have also directly tested the neural mediation of the hyperpnea in ponies (Pan et al., 1987). Afferent information from the exercising hind-limbs to the brain was compromised through partial bilateral surgical

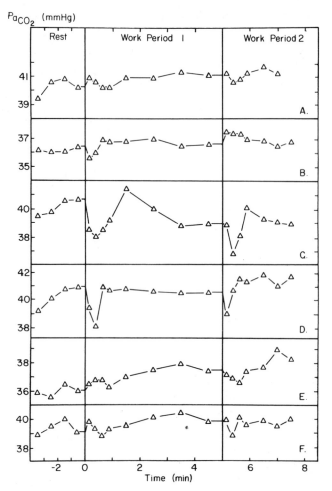

Fig. 8. The temporal pattern of $PaCO_2$ in paraplegic human subjects during electrically induced muscle contractions that increased metabolic rate two- (work period 1) to fourfold (work period 2). Each panel represents one subject who was studied three times.

lesions of the dorsal lateral columns at the L-1 vertebral level. One month and one year after lesioning, we found $PaCO_2$ during exercise slightly higher than pre-lesioning, while $\dot{V}E$ and f were slightly lower. These unique studies in fully awake ponies suggest that spinal afferents participate in the control of breathing and blood gases during exercise. The importance of such a role may be underestimated by these data due to only partial spinal lesion and/or intact innervation of forelimbs.

In contrast to the aforementioned pony studies, our studies on humans appear to support humoral rather than neural mediation of the exercise hyperpnea. For example, we found in normal humans that the $\dot{V}E$, f, VT and blood gas responses (Fig. 7) did not differ between voluntary and electrically induced muscle contractions (Pan et al., 1986b). Furthermore, we found that the responses of these parameters did not differ between normal and paraplegic humans when the muscles of the legs were electrically stimulated (Fig. 8; Brice et al., 1986). These findings confirm and extend earlier work by others (Asmussen et al., 1943; Adams et al., 1984). Accordingly, it appears that the ventilatory responses to the elevated metabolic needs of muscle contractions are not critically dependent on "central-command" or spinal afferent pathways.

Our data are thus representative of the current controversy regarding the exercise hyperpnea mechanism. Strong evidence exists for and against both neural and humoral mediation of the hyperpnea. We are assuming, of course, that our findings are not due merely to species differences. It seems that a hypothesis forwarded by Yamamoto deserves strong consideration and testing (Yamamoto et al., 1977). Yamamoto speculates that multiple mechanisms are capable of mediating the hyperpnea. As a result, eliminating a single "humoral" or "neural" mechanism may not affect the hyperpnea and, moreover, given the proper experimental protocol, a single mechanism may be capable of mediating the entire hyperpnea. This hypothesis provides a potential explanation for the vast amount of seemingly contradictory evidence in the literature.

Just as mediation of the exercise hyperpnea is controversial, so also is the hyperventilation of heavy exercise. Traditional theory links this hyperventilation in humans to the onset of anaerobic metabolism (Nielson and Asmussen, 1963; Wasserman et al., 1975; and Wasserman et al., 1977). Arterial acidosis supposedly stimulates carotid chemoreceptors resulting in hyperventilation and respiratory acid-base compensation. These conclusions are based on correlative data and on demonstrations that carotid body-resected asthmatics do not hyperventilate during heavy exercise (Wasserman et al., 1975). However, some recent studies have shown a dissociation of hyperventilation and lactacidosis in humans (Dempsey et al., 1984; Green et al., 1983; Hagberg et al., 1982; Heigenhauser et al., 1983; Hughes et al., 1982). To more directly test the traditional theory, we studied both normal and CBD ponies (Pan et al., 1986a). As already indicated for normals, we found in both groups a linear relationship between exercise hypocapnia and work rate (Fig. 4). In other words, there was no indication of an "added" ventilatory drive at the onset of anaerobiosis. Moreover, the CBD ponies hyperventilated more than the normal ponies. Clearly, then, our data do not provide support for the traditional theory on heavy exercise hyperventilation. For ponies, at least, the data indicate that during heavy exercise there is no stimulus "added" to the progressively increasing "exercise" stimulus.

Mechanism of species difference in $\dot{V}A/\dot{V}CO_2$ response to exercise

In ponies at least, the ventilatory control system seems programmed for a specific hyperventilation at every level of work. This conclusion is supported by unchanged hyperventilation during experimental interventions which would otherwise markedly alter $\dot{V}A$. For example, tracheostomy breathing decreases respiratory dead space and resistance; thus, if there were no adjustments in breathing, $\dot{V}A$ would change accordingly (Fig. 9; Forster, 1985). However, both tidal volume (VT) and breathing frequency (f) change such that $PaCO_2$ remains within one mmHg of normal nares breathing at the same metabolic rate. Another example suggesting a programmed hyperventilation are the effects of vagal denervation of the lungs (HND). In both ponies and dogs, $\dot{V}E$, $\dot{V}A$, and $PaCO_2$ at a given metabolic rate are not altered by HND despite a reduced f and increased VT after HND (Flynn et al., 1986; Clifford et al., 1986b). In this instance, VD must have changed with the altered pattern of breathing; otherwise $\dot{V}A$ and the resulting hypocapnia would have been greater after HND. Finally, if, as suggested earlier, the hyperventilation is required to maintain steady-state PaO_2 homeostasis at each workload, it seems logical that a specific hyperventilation is programmed for each workload. Accordingly, we hypothesize that in contrast to the nearly constant $\dot{V}A/\dot{V}CO_2$ ratio in humans during submaximal exercise, in most non-human species $\dot{V}A$ is programmed to increase proportionally more than $\dot{V}CO_2$ resulting in a progressive hypocapnia.

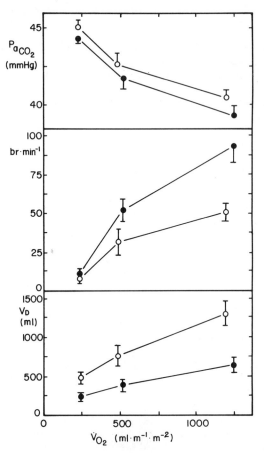

Fig. 9. Physiologic dead space (VD), breathing frequency
($br \cdot min^{-1}$), and $PaCO_2$ in ponies (n=6) at rest and during 2
levels of treadmill exercise while breathing through the nares
(open symbol) and while breathing through a chronic tracheostomy (closed symbol). Note that the $PaCO_2$ difference between
nares and tracheostomy breathing is independent of metabolic
rate ($\dot{V}O_2$) in spite of the VD and breathing frequency differences increasing as $\dot{V}O_2$ was increased.

Minimization of the O_2 Cost of Breathing

The O_2 cost of breathing during exercise can be estimated by measuring
the mechanical work of breathing. For example, Margaria et al., (1960)
found that the relationship between ventilation and ventilatory work was
curvilinear. At ventilations of 50, 100, and 125 $l \cdot min^{-1}$, work was approximately 15, 50, and 100 $cal \cdot min^{-1}$. Data of Thoden et al., (1969) provide
similar values for ventilatory work. Assuming an efficiency of the respiratory muscles of 0.25 (Milic-Emili and J.M. Petit, 1960), the respiratory
work of 100 $cal \cdot min^{-1}$ represents about 3% of the total energy uptake. These
values of ventilation and ventilatory work represent maximal exercise
values for most humans. However, elite human athletes achieve ventilations
exceeding 150 $l \cdot min^{-1}$. Because of the curvilinear relationship between
ventilatory work and ventilation, Margaria et al., (1960) predict that at
ventilations approximating 180-200 $l \cdot min^{-1}$, the work has reached a point
where all the additional O_2 is needed by the respiratory muscles. Accordingly, in elite athletes, ventilatory work might be limiting to exercise

performance. What, then, are the factors that minimize the work of breathing?

To achieve efficiently the $\dot{V}A$ required for gas exchange during exercise, precise control is required of f, VT, airway diameter, and respiratory muscle recruitment. For example, a pattern of breathing is selected which minimizes the total elastic and flow-resistive work of breathing (Mead, 1960; Otis et al., 1950). Accordingly, in humans at low workloads, the hyperpnea is accomplished primarily through an increase in tidal volume (Dempsey and Rankin, 1967). Both inspiratory and expiratory reserve volume are utilized so that the relationship between changes in transpulmonary pressure and lung volume remain on the linear pressure-volume relationship. However, a VT in excess of 65% of vital capacity encroaches on the alinear P-V relationship; thus, at higher workloads, f contributes progressively more to the hyperpnea (Dempsey, 1986). In a species such as equines with a stiffer chest wall than humans and, therefore, a greater amount of elastic work for a given VT, the hyperpnea is achieved primarily by increasing f at low workloads (Bisgard et al., 1978). As the hyperpnea increases further, VT then becomes increasingly more important in equines. In addition to breathing pattern, airway diameter must also be regulated because the increased $\dot{V}A$ during exercise requires increases in both inspiratory and expiratory flow rates in a shorter period of time. Airway modulation thus prevents major changes in airway resistance. The large increase in upper airway VD in exercising equines (Fig. 9) is evidence of airway dilation (Forster et al., 1985). A second change to reduce resistance is elimination of expiratory "braking" (Remmers and Bartlett, 1977). The expiratory braking at rest serves to minimize the O_2 cost of breathing (Remmers and Bartlett, 1977). However, with the flow rates demanded during exercise, braking would increase O_2 cost. Accordingly, this is but one example of where a change in the recruitment pattern of respiratory muscles during exercise is of importance in minimization of the O_2 cost of breathing. Not only is braking eliminated during exercise but, in contrast to quiescence at rest, in most species, expiratory muscles are activated during exercise to increase expiratory flow rate and to reduce functional residual capacity (Dempsey, 1986). Utilizing the expiratory reserve volume has three effects: 1) it increases VT with a minimum of elastic work; 2) it results in a "passive" component of inspiration, thus sparing the inspiratory muscles; and 3) it lengthens the diaphragm which permits generation of increased tension for the same amount of phrenic nerve activation (Grimby et al., 1976). Essentially, then, the O_2 cost of breathing during exercise is minimized through appropriate selection of the rate, magnitude, and pattern of activation of inspiratory, expiratory and airway muscles.

It is the general view that afferents from mechanoreceptors in the chest wall, lung and, possibly the diaphragm, provide the crucial information to the respiratory controller for efficient breathing (Mead, 1960; Otis et al., 1950). In support of this idea are findings that "wasteful" expiratory pressure generation is avoided during exercise (Dempsey, 1986). Below approximately 60% of vital capacity during a forced expiration at rest, increases in pleural pressure beyond a certain value do not cause additional increases in expiratory flow because of dynamic airway compression. The muscular effort to generate the extra pressure is wasteful. During exercise, high and near maximum expiratory flow rates are generated just as during a voluntary forced expiration at rest. However, during exercise, the extra, wasteful pressures are avoided. In some manner, wasteful pressure or effort by the respiratory muscles is sensed, and the aforementioned receptors would seem the most suitable to provide the necessary information. Also supportive of this concept are data obtained when the mechanical impedance of the lung is reduced by switching from a N_2-O_2 to a He-O_2 inspiratory gas mixture. This change in inspired gases decreases airway resistance by more than 50% (Hussain et al., 1985; Pan et al.,

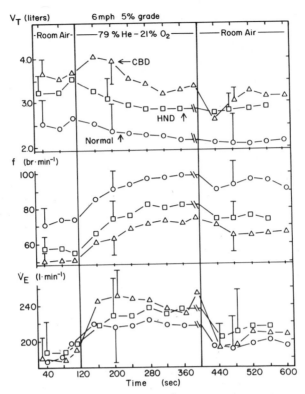

Fig. 10. Pulmonary ventilation V̇E), breathing frequency (f)
and tidal volume (VT) in normal, carotid body denervated
(CBD), and hilar nerve denervated (HND) ponies during exercise
at 6 mph-5% grade while breathing room air and while breathing
a 79% He-21% O_2 gas mixture.

1987b). Within nearly the first breath of reduced resistance, there is a
20-40% reduction in the amplitude and rate of rise of the diaphragmatic
electromyogram (EMG) (Hussain et al., 1985). Accordingly, during normal
air breathing, the higher EMG must be due to some reflex sensitive to the
resistive load or impedance presented by the lung to the respiratory musculature. The magnitude of this He-O_2 reduction in diaphragmatic activity is
workload-dependent. It follows, then, that a significant portion of the
exercise hyperpnea is due to mechanoreceptor-proprioceptive feedback from
the thorax.

Our recent study provides new insights into this proprioceptive feedback and load compensation during exercise (Pan et al., 1987b). We found
that the He-O_2 induced changes in breathing were virtually identical in
normal and hilar nerve denervated ponies (Fig. 10). These findings indicate
that the changes in breathing are not critically dependent upon lung
afferents. In the normal and HND ponies, the adjustment to the reduced
resistance was almost perfect in that there was only minimal hyperventilation. However, in CBD ponies, there was a workload-dependent hyperventilation during He-O_2 breathing which was greater than that seen in normal
ponies. The greater hyperventilation was due to an abnormally high VT. In
other words, neural output to the respiratory muscles was not as dampened as
in chemoreceptor-intact ponies. It follows, then, that in chemoreceptor-
intact ponies, a significant portion of the reduced neural output during He-
O_2 breathing must have been chemoreceptor-mediated. Mechanoreceptor feedback alone was insufficient to compensate breathing perfectly for the
reduced pulmonary impedance during He-O_2 breathing.

Compromises Between Metabolic and Efficiency Components of Exercise Hyperpnea

The traditional view is that the pulmonary system meets the metabolic demands of maximal exercise efficiently and, thus, this system does not contribute to limitations in exercise performance in untrained subjects (Dempsey, 1986). This conclusion is based largely on the common observation that PaO_2 is maintained at or slightly above resting levels, even during maximal exercise (Asmussen, 1960; Dempsey et al., 1971, 1984; Holmgren and Linderholm, 1958). However, even though the lung over-all is not limiting, the limits of alveolar-capillary gas exchange adjustments to exercise might be reached. This conclusion is based on findings of a widened alveolar-arterial PO_2 gradient during heavy exercise (Asmussen, 1960, 1965; Dempsey and Rankin, 1967). In other words, alveolar hyperventilation, and, thus, elevated PAO_2 appear necessary to compensate for absence of additional alveolar-capillary adjustments (Dempsey, 1986). Two recent findings demonstrate the importance of the hyperventilation to PaO_2 homeostasis. Moreover, these findings also demonstrate limitations in the breathing responses to maximal exercise.

In contrast to the hyperventilation commonly observed in untrained and less than world-class athletes, many world-class athletes hyperventilate minimally during heavy exercise (Dempsey et al., 1984; Rowell, 1964). The attenuated hyperventilation correlates with a reduced PaO_2 which, in some subjects, approaches 60 mmHg. In all subjects that became hypoxemic during heavy exercise while breathing room air, hyperventilation was enhanced and PaO_2 returned toward and, in some cases, to resting levels during heavy exercise while breathing an $He-O_2$ mixture (Dempsey et al., 1984). The potential causes of the minimal hyperventilation while breathing room air are complex. The $He-O_2$ data suggest, however, a contribution from lung-chest wall mechanoreceptors. It should be emphasized that due to the hyperbolic $\dot{V}A/PaCO_2$ relationship, an athlete exercising at a $\dot{V}CO_2$ exceeding 6 $l \cdot min^{-1}$ will require a $\dot{V}A$ nearly double that of an individual exercising maximally at a $\dot{V}CO_2$ of 3 $l \cdot min^{-1}$ (Dempsey, 1986). These extreme ventilatory demands may reach the mechanical limitation of the system. Mechanical impedance of the lung is not matched by feedback-related neural output. PaO_2 homeostasis is compromised simply because it costs too much in terms of O_2 consumption to maintain homeostasis (Margaria et al., 1960).

A galloping thoroughbred racehorse provides another example of an apparent compromise between blood gas homeostasis and breathing efficiency. Racehorses achieve $\dot{V}O_2$'s of 150 $ml \cdot kg^{-1} \cdot min^{-1}$; cardiac output exceeds 600 $l \cdot min^{-1}$; mixed venous $\dot{P}CO_2$ exceeds 100 mmHg, and $P\bar{v}O_2$ is reduced to below 15 mmHg (Snow et al., 1985; Thorston et al., 1983). The pulmonary system indeed faces a formidable task of gas exchange. Ideally accomplished, this task should not be at the expense of breathing and locomotor efficiency. For efficiency, breathing and stride frequency must be linked or entrained, simply because the thoracic muscles serve ventilatory, locomotor and stabilization needs (Bramble, 1984). Accordingly, expiration always occurs as the forelimbs are planted. Expiration is thereby facilitated by the transfer of ground forces to the rib cage via chest wall musculature and by shifting the abdominal contents forward against the diaphragm causing thoracic compression. Inspiration, coupled with forelimb extension, is facilitated by enlargement of the anterior rib cage which results from rotation of the shoulder girdle forward, protraction of the forelimbs, and extension of the head and neck, all a part of locomotion. Entrainment is, therefore, beneficial for overall efficiency. However, the compromise is a sacrifice of PaO_2 and $PaCO_2$ homeostasis during very heavy exercise (Bayle, 1983; Thorston, 1983). In this instance, the pulmonary limitation is in part ventilatory as $PaCO_2$ increases above 50 mmHg and PaO_2 reaches 60 mmHg (Bayle, 1983). The hypoventilation and CO_2 retention are a result of the constraints of entrainment on breathing. The very high breathing frequencies require high dead space ventilation and, because of the limited time available for airflow, VT only achieves about 30% of vital capacity (Hornicke et al., 1982). Alveolar ventilation cannot thus increase sufficiently to meet the metabolic demands.

The elite human and equine athletes both provide examples of where the control system compromises metabolic needs for efficiency during exercise. Underlying both instances of blood gas disruptions is the apparent avoidance of extremely large energy expenditures by the respiratory muscles. Undoubtedly, the hypoxemia and thus the lungs are contributing to the limitations of exercise performance. It is probably a reasonable assumption that performance would be compromised even more had breathing efficiency been compromised for blood gas homeostasis.

ACKNOWLEDGMENTS

Reviewed work of ours was supported by National Heart, Lung and Blood Institute Grant HL-25739 and by the Veterans Administration. Drs. G.E. Bisgard, A.G. Brice, C. Flynn, R.D. Wurster and R.P. Kaminski contributed to our studies. We thank Therese Gauthier for assistance in preparing this review.

REFERENCES

Adams, L., H. Frankel, J. Garlick, A. Guz, K. Murphy and S.J.G. Semple (1984). The role of spinal cord transmission in the ventilatory response to exercise in man. J. Physiol. (London) 355:85-97.

Askanazi, J., P.A. Silverberg, R.J. Foster, H.I. Hyman, J. Milic-Emili, and J.M. Kinney (1980). Effect of respiratory apparatus on breathing frequency. J. Appl. Physiol. 48:577-580, 1980.

Asmussen, E., M. Nielsen and G. Weith-Pedersen (1943). Cortical or reflex control of respiration during muscular work? Acta Physiol. Scand. 6:168-175.

Asmussen, E. and M. Nielsen (1960). Alveolar-arterial gas exchange at rest and during work at different O_2 tensions. Acta Physiol. Scand. 50:153-166.

Asmussen, E. (1965). Muscular exercise. In: Handbook of Physiology, Respiration. Washington, DC; Am. Physiol. Soc., Sec. 3., Vol. II, Chapt. 36, p. 939-978.

Barr, P.O., M. Beckman, H. Bjurstedt, J. Brismar, C.M. Hessler, and G. Matell (1964). Time course of blood gas changes provoked by light and moderate exercise in man. Acta Physiol. Scand. 60:1-17.

Bayle, W.M., B.D. Grant, R.G. Breeze and J.W. Kramer (1983). The effects of maximal exercise on acid-base balance and arterial blood gas tensions in thoroughbred horses. In: Equine Exercise Physiology, D.H. Snow, S.G. Persson, and R.J. Rose (Eds.) Cambridge: Granta Editions, pp. 400-404.

Bennett, F. (1984). A role for neural pathways in exercise hyperpnea. J. Appl. Physiol. 56:1559-1564.

Bisgard, G.E., H.V. Forster, B. Byrnes, K. Stanek, J. Klein and M. Manohar (1978). Cerebrospinal fluid acid-base balance during muscular exercise. J. Appl. Physiol. 45:94-101.

Bisgard, G., H.V. Forster, J. Messina and R.G. Sarazin (1982). Role of the carotid body in hypernea of moderate exercise in goats. J. Appl. Physiol. 52:1216-1222.

Bramble, D.M. (1984). Locomotive-respiratory integration in running mammals. In: Proceedings of the Association of Equine Sports Medicine 42-53.

Brice, G., H.V. Forster, L. Pan, A. Funahashi, M. Hoffman, T. Lowry and C. Murphy (1986). Is the hyperpnea during electrically induced exercise critically dependent on spinal afferents? Fed. Proc. 45:2117.

Clifford, P.S., J.T. Litzow and R.L. Coon (1986). Arterial hypocapnia during exercise in beagle dogs. J. Appl. Physiol. 61:599-603.

Clifford, P.S., J.T. Litzow, J.H. von Colditz and R.L. Coon (1986). Effect of chronic pulmonary denervation on ventilatory response to exercise. J. Appl. Physiol. 61:603-610.

Cross, B.A., A. Davey, A. Guz, P.G. Katona, M. Maclean, K. Murphy, S.J.C. Semple and R. Stidwell (1982). The pH oscillations in arterial blood during exercise: a potential signal for the ventilatory response in the dog. J. Physiol. (London) 329:57-73.

Dejours, P. (1965). Control of respiration in muscular exercise. In: Handbook of Physiology, Respiration. Washington, DC. Am Physiol. Soc., Sect 3, Vol. 1, Chapt. 25, pp. 631-648.

Dempsey, J.A. and J. Rankin (1967). Physiologic adaptations of gas transport systems to muscular work in health and disease. Am. J. Phys. Med. 46:582-647.

Dempsey, J.A., W.G. Reddan, J. Rankin, M.L. Birnbaum, H.V. Forster, J.S. Thoden and R.F. Grover (1971). Effects of acute through life-long hypoxic exposure on exercise pulmonary gas exchange. Respir. Physiol. 13:62-87.

Dempsey, J.A., P. Hanson and K. Henderson (1984). Exercise induced arterial hypoxemia in healthy humans at sea level. J. Physiol. (London) 355:161-175.

Dempsey, J.A. (1986). Is the lung built for exercise? Med-Sci in Sports and Exercise 18:143-155.

DiMarco, A.F., J.R. Romaniuk, C. von Euler and Y. Yamamoto (1983). Immediate changes in ventilation and respiratory pattern associated with onset and cessation of locomotion in the cat. J. Physiol. (London) 343:1-16.

Eldridge, F.L., D.E. Milhorn, J.P. Kiley and T.G. Waldrop (1985). Stimulation by central command of locomotion, respiration, and circulation during exercise. Respir. Physiol. 59:313-337.

Favier, R., G. Kepenekian, D. Desplanches and R. Flandrois (1982). Effects of chronic lung denervation on breathing pattern and respiratory gas exchange during hypoxia, hypercapnia, and exercise. Respir. Physiol. 47:107-119.

Flandrois, R., J.F. Lacour and J.F. Eclache (1974). Control of respiration in exercising dog: interaction of chemical and physical humoral stimuli. Respir. Physiol. 21:169-181.

Flynn, C., H.V. Forster, L.G. Pan and G.E. Bisgard (1985). Role of hilar nerve afferents in hypercapnia of exercise. J. Appl. Physiol. 59:798-806.

Fordyce, W.E., F.M. Bennett, S.K. Edelman and F.G. Grodins (1982). Evidence for a fast neural mechanism during the early phase of exercise hyperpnea. Respir. Physiol. 48:27-43.

Forster, H.V., L.G. Pan, G.E. Bisgard, C. Flynn, S.M. Dorsey and M.S. Britton (1984). Independence of exercise hypocapnia and limb movement frequency in ponies. J. Appl. Physiol. 57:1885-1893.

Forster, H.V., L.G. Pan, G.E. Bisgard, C. Flynn and R.E. Hoffer (1985). Changes in breathing when switching from nares to tracheostomy breathing in awake ponies. J. Appl. Physiol. 59:1214-1221.

Forster, H.V., L.G. Pan and A. Funahashi (1986). Temporal pattern of $PaCO_2$ during exercise in humans. J. Appl. Physiol. 60:653-660.

Fregosi, R.F. and J.A. Dempsey (1984). Arterial blood acid-base regulation during exercise in rats. J. Appl. Physiol. 57:396-402.

Green, H.J., R.L. Hughson, G.W. Orr and D.A. Ranney (1983). Anaerobic threshold, blood lactate, and muscle metabolites in progressive exercise. J. Appl. Physiol. 54:1032-1038.

Green, J.F. and M.I. Sheldon (1983). Ventilatory changes associated with changes in pulmonary blood flow in dogs. J. Appl. Physiol. 54:997-1002.

Green, J.F., E.R. Schertel, H.M. Coleridge and J.C.G. Coleridge (1986). Effect of pulmonary arterial PCO_2 on slowly adapting pulmonary stretch receptors. J. Appl. Physiol. 60:2048-2055.

Grimby, G., M. Goldman and J. Mead (1976). Respiratory muscle actions inferred from rib cage and abdominal V-P partitioning. J. Appl. Physiol. 41:739-751.

Hagberg, J.M., E.F. Coyle, J.E. Carroll, J.M. Miller, W.H. Martin and M.H. Brooke (1982). Exercise hyperventilation in patients with McArdles disease. J. Appl. Physiol. 52:991-994.

Heigenhauser, G.J.T., J.R. Sutton and N.L. Jones (1983). Effect of glycogen depletion on the ventilatory response to exercise. J. Appl. Physiol. 54:470-474.

Holmgren, A. and H. Linderholm (1958). Oxygen and carbon dioxide tension of arterial blood during heavy and exhaustive exercise. Acta Physiol. Scand. 44:203-215.

Hornicke, H., R. Meixner and U. Pollman (1982). Respiration in exercising horses. In: Equine Exercise Physiology., D.H. Snow, S.G. Persson and R.J. Rose (Eds.) Cambridge: Granta Editions, p. 7-16.

Hughes, E.F., S.C. Turner and G.A. Brooks (1982). Effect of glycogen depletion and pedaling speed on "anaerobic threshold". J. Appl. Physiol. 52:1598-1607.

Hussain, S.N.A., R.L. Pardy and J.A. Dempsey (1985). Mechanical impedance as determinant of inspiratory neural drive during exercise in humans. J. Appl. Physiol. 59:365-375.

Kao, F.F. (1977). The peripheral neurogenic drive: An experimental study. In: Muscular Exercise and the Lung, ed. by J.A. Dempsey and C.E. Reed, Madison, WI: Univ. of Wisconsin Press, pp. 71-85.

Kiley, J.P., W.D. Kuhlmann and M.R. Fedde (1980). Arterial and mixed venous blood gas tensions in exercising ducks. Poultry Science 59:914-917.

Krogh, A. and J. Lindhard (1913). The regulation of respiration and circulation during the initial stages of muscular work. J. Physiol. (London) 47:112-136.

Kuhlmann, W.D., D.S. Hodgson and M.R. Fedde (1985). Respiratory, cardiovascular, and metabolic adjustments to exercise in the Hereford calf. J. Appl. Physiol. 58:1273-1280.

Margaria, R., G. Milic-Emili, J.M. Petit and G. Cavagna (1960). Mechanical work of breathing during muscular exercise. J. Appl. Physiol. 15:354-358.

Matalon, S.V. and L.E. Farhi (1979). Cardiopulmonary readjustments to passive tilt. J. Appl. Physiol. 47:503-507.

Mead, J. (1960). Control of respiratory frequency. J. Appl. Physiol. 15:325-336.

Milic-Emili, G. and J.M. Petit (1960). Mechanical efficiency of breathing. J. Appl. Physiol. 15:359-362.

Mitchell, G.S., T.T. Gleason and A.F. Bennett (1981). Ventilation and acid-base balance during activity in lizards. Am. J. Physiol. 240:R29-R37.

Mitchell, J.H., W.C. Reardon and P.I. McCloskey (1977). Reflex effects on circulation and respiration from contracting skeletal muscle. Am. J. Physiol. 23:H374-H378.

Nielsen, M. and E. Asmussen (1963). Humoral and nervous control of breathing in exercise. In: The Regulation of Human Respiration, ed. by D.J.C. Cunningham and B.B. Lloyd. Philadelphia, PA; Davis, pp. 504-513.

Oldenburg, F.A., D.O. McCormack, J.L.C. Morse and N.L. Jones (1979). A comparison of exercise responses in stairclimbing and cycling. J. Appl. Physiol. 46:510-516.

Otis, A.B., W.O. Fenn and H. Rahn (1950). Mechanics of breathing in man. J. Appl. Physiol. 2:592-607.

Pan, L.G., H.V. Forster, G.E. Bisgard, R.P. Kaminski, S.M. Dorsey and M.A. Busch (1983). Hyperventilation in ponies at the onset of and during steady-state exercise. J. Appl. Physiol. 54:1394-1402.

Pan, L.G., H.V. Forster, G.E. Bisgard, S.M. Dorsey and M.A. Busch (1984). O_2 transport in ponies during treadmill exercise. J. Appl. Physiol. 57:744-751.

Pan, L.G., H.V. Forster, G.E. Bisgard, C.L. Murphy and T.F. Lowry (1986a). Independence of exercise hyperpnea and acidosis during high intensity exercise in ponies. J. Appl. Physiol. 60:1016-1024.

Pan, L.G., H.V. Forster, G. Brice, A. Funahashi, M. Hoffman, C. Murphy and T. Lowry (1986b). Ventilatory response to voluntary versus electrically induced exercise in normal, healthy humans. Fed. Proc. 45:2116.

Pan, L.G., H.V. Forster, R.D. Wurster, A.G. Brice, T.F. Lowry, C.L. Murphy and G.S. Bisgard (1987a). Effect of partial spinal ablation on the exercise hyperpnea in ponies. Fed. Proc. 46:1973.

Pan, L.G., H.V. Forster, G.E. Bisgard, T.F. Lowry and C.L. Murphy (1987b). Role of carotid chemoreceptors and pulmonary vagal afferents during helium:O_2 breathing in ponies. J. Appl. Physiol.:62:1020-1027, 1987.

Phillipson, E.A., J. Duffin, J.D. Cooper (1981). Critical dependence of respiratory rhythmicity on metabolic CO_2 load. J. Appl. Physiol. 50:45-54.

Remmers, J.E. and D. Bartlett (1977). Reflex control of expiratory airflow and duration. J. Appl. Physiol. 42:80-87.

Rowell, L.B., H.L. Taylor, Y. Wang and W.B. Carlson (1964). Saturation of arterial blood with oxygen during maximal exercise. J. Appl. Physiol. 19:284-286.

Sheldon, J.I. and J.F. Green (1982). Evidence for pulmonary CO_2 chemosensitivity: effects on ventilation. J. Appl. Physiol. 52:1192-1197.

Smith, C.A., G.S. Mitchell, L.C. Jameson, T.I. Musch and J.A. Dempsey (1983). Ventilatory response of goats to treadmill exercise: grade effects. Respir. Physiol. 54:331-341.

Snow, D.H., R.C. Harris and S.P. Gash (1985). Metabolic response of equine muscle to intermittent maximal exercise. J. Appl. Physiol. 58:1689-1697.

Szlyk, P.C., B.W. McDonald, B.W. Pendergast and J.H. Krasney (1981). Control of ventilation during graded exercise in the dog. Respir. Physiol. 46:345-365.

Thoden, J.S., J.A. Dempsey, W.G. Reddan, M.L. Birnbaum, H.V. Forster, R.F. Grover and J. Rankin (1969). Ventilatory work during steady-state response to exercise. Fed. Proc. 28:1316-1320.

Thorston, J., B. Essen-Gustavsson, A. Lindholm, D. McMicken and S. Persson (1983). Effects of training and detraining on oxygen uptake, cardiac output, blood gas tensions, pH, and lactate concentrations during and after exercise in the horse. In: Equine Exercise Physiology, D.H. Snow, S.G. Persson, and R.J. Rose (Eds.) Cambridge: Granta Editions, pp. 470-486.

Wasserman, K., B.J. Whipp, S.N. Koyal and M.G. Cleary (1975). Effect of carotid body resection on ventilatory and acid-base control during exercise. J. Appl. Physiol. 39:354-358.

Wasserman, K., B.J. Whipp, R. Casaburi, W.L. Beaver and H.V. Brown (1977). CO_2 flow to the lungs and ventilatory control. In: Muscular Exercise and the Lung, ed. by J.A. Dempsey and C.E. Reed, Madison, WI: University of Wisconsin Press, pp. 103-135.

Wasserman, K., B.J. Whipp, R. Casaburi, M. Golden and W.L. Beaver (1979). Ventilatory control during exercise in man. Bul. Europ. Physiopath. Resp. 15:27-47.

Watken, R.L., H.H. Rostosfer, S. Robinson, J.L. Newton and M.D. Baillie (1962). Changes in blood gases and acid-base balance in the exercising dog. J. Appl. Physiol. 17:656-660.

Whipp, B. (1981). Control of exercise hyperpnea. Ed. T.F. Hornbein, In: Regulation of Breathing, Part II. New York: Dekker, pp. 1069-1140.

Yamamato, I.H. and M.W. Edwards (1960). Homeostasis of CO_2 during intravenous infusion of CO_2. J. Appl. Physiol. 15:807-818.

Yamamoto, W.S. (1977). Looking at the regulation of ventilation as a signalling process. Eds. J.A. Dempsey and C.E. Reid, In: <u>Muscular Exercise and the Lung</u>. Madison: University of Wisconsin Press, p. 137-149.

Young, I.H. and A.J. Woolcock (1978). Changes in arterial blood gas tension during unsteady-state exercise. J. Appl. Physiol. 44:93-96.

PROBLEMS WITH THE HYPERVENTILATORY RESPONSE TO EXERCISE AND HYPOXIA

Jerome A. Dempsey

John Rankin Laboratory of Pulmonary Medicine
University of Wisconsin
Department of Preventive Medicine
504 N. Walnut Street
Madison, WI 53706

To achieve the optimal ventilatory response to heavy exercise, several criteria must be satisfied. First, alveolar hyperventilation must be adequate to meet not only the rising CO_2 flow to the lungs but also to compensate the progressive development of metabolic acidosis and to ensure a high overall $\dot{V}A/\dot{Q}c$ and diffusion gradient in the face of marked reductions in mixed venous PO_2 and shortened capillary transit time. Secondly, the mechanical limits of the lung and chest wall - in terms of respiratory muscle force, energy stores and rate of energy supply - must be sufficient to meet these ventilatory demands. Third, the regulation of respiratory muscle length and the timing of muscle contraction must be regulated so as to optimize chest wall function.

The neuro-mechanical characteristics of the chest wall, lung, and control system are almost ideally suited to meet most of these requirements - at least in the untrained healthy person exercising heavily at sea level (Dempsey et al., 1985; Dempsey, 1986). In these circumstances ($\dot{V}CO_2 < 3.5-4$ L/min), arterial oxygenation is maintained, substantial compensatory hypocapnia is achieved and the flow:volume limits of each breath are not exceeded. Further, as demonstrated in rats run to exhaustion, glycogen utilization and/or lactate production in respiratory muscles are relatively small relative to those of locomotor muscles of similar fiber type (Fregosi and Dempsey, 1986). Some imperfections in this response include the fact that respiratory muscles must consume at least some of the total energy available and sufficient alveolar hyperventilation is not achieved to completely compensate for metabolic acidosis engendered at maximum work. On the other hand, there are severe deficiencies and costs associated with this hyperventilatory response in the highly trained individual and in the sojourner at high altitude. We will concentrate here on these problems encountered in these two situations.

MAGNITUDE AND IMPORTANCE OF THE HYPERVENTILATORY RESPONSE

The diversity of ventilatory responses during very brief progressive exercise and during short-term sustained exercise in highly trained subjects is shown in Fig. 1A and B. Note in Fig. 1A that at maximal exercise

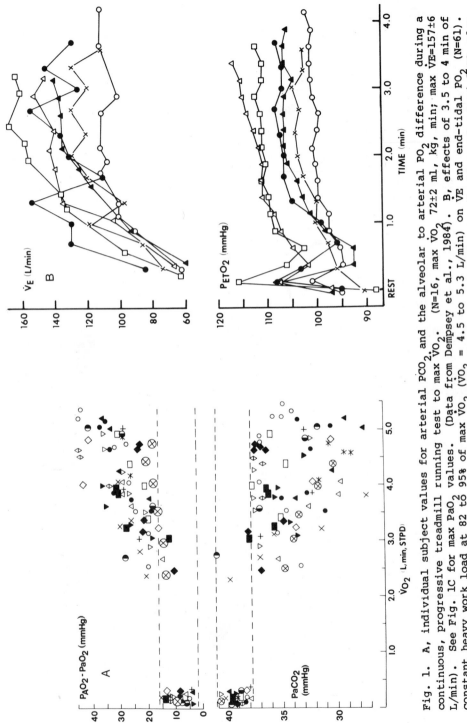

Fig. 1. A, individual subject values for arterial PCO_2 and the alveolar to arterial PO_2 difference during a continuous, progressive treadmill running test to max $\dot{V}O_2$. (N=16, max $\dot{V}O_2$ 72±2 ml, kg, min; max $\dot{V}E$=157±6 L/min). See Fig. 1C for max PaO_2 values. (Data from Dempsey et al., 1984). B, effects of 3.5 to 4 min of constant heavy work load at 82 to 95% of max $\dot{V}O_2$ ($\dot{V}O_2$ = 4.5 to 5.3 L/min) on $\dot{V}E$ and end-tidal PO_2 (N=61). Arterial pH fell progressively and averaged 7.26 (7.20 to 7.34) at exercise termination and arterial PO_2 also fell in 5 of 6 subjects and averaged 68 mmHg (55 to 77) at end-exercise.

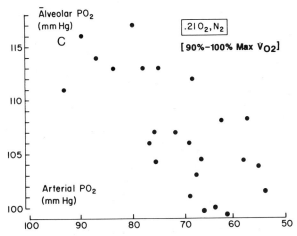

Fig. 1C. Correlation of alveolar PO_2 to arterial PO_2 at max exercise. (r=0.70, P<.01). A-aDO_2 averaged 35 mmHg (range 25 to 45 mmHg).

($\dot{V}CO_2$ 5-6 l/min) $PaCO_2$ varies from just a few mmHg to as much as 10-12 mmHg below resting levels; a similar diversity in ventilatory response exists over the time course of hyperventilation achieved during 3-5 minutes of heavy constant-load exercise at 80-95% of max $\dot{V}O_2$, as shown in Fig. 1B. Note here that all subjects showed an initial hyperventilation (acute rise in $PETO_2$) but thereafter alveolar PO_2 was reduced and between 3-4 minutes of exercise varied from 20 mmHg above to virtually no change from resting levels. This hyperventilatory response is critical to maintaining arterial PO_2 at the very heavy work loads because alveolar to arterial O_2 difference widens to >30-35 mmHg (Fig. 1A); thus PAO_2 must, of course, rise concomitantly. In those cases in which this did not occur ($PaCO_2$ 36-41 mmHg and PAO_2 <110 mmHg) substantial hypoxemia often developed with PaO_2 in the 55-70 mmHg range (see Fig. 1C). We also emphasize that in some cases the A-aDO_2 increase is so marked, i.e. 40-45 mmHg, that even a substantial hyperventilatory response could not prevent hypoxemia at these exercise levels (Fig. 1A) (Dempsey et al., 1984).

High altitude hypoxia (3100 and 4300 m) greatly accentuates the hyperventilatory response to exercise in all subjects, even in mild exercise (Fig. 2A-C). The level of hyperventilation becomes quite severe during heavy exercise with $PaCO_2$ reduced to slightly above 20 mmHg at 3100 m and in the 15-20 mmHg range at 4300 m. Heavy exercise of long duration causes a progressive time-dependent hyperventilation. The longer the exposure to hypoxia the greater the hyperventilation - both at rest and exercise at least over the initial two weeks of hypoxic exposure (Fig. 2A). This hyperventilatory response to exercise in hypoxia is essential to O_2 transport in the sojourner to high altitude. Indeed, during the first few days at high altitude the only significant accommodation to hypoxia made by the O_2 transport system is this hyperventilatory response. The ventilatory response by itself over time at high altitude causes a substantial increase in arterial blood oxygenation during heavy exercise amounting at 3100 m to 15-20 mmHg (PaO_2 58 vs. 40 mm Hg), a 12% increase in HbO_2 saturation and 2-3 volumes percent increase in CaO_2. In addition, the marked degree of hyperventilation enables PaO_2 during heavy exercise to remain within 5 mmHg of resting levels despite a marked widening of the A-aDO_2. Further, after one or two weeks of accommodation to hypoxia, increased red cell production, combined with the hyperventilation is sufficient to bring CaO_2 during heavy exercise back to normal sea level values. Unfortunately, these combined changes do not restore systemic O_2 transport to normal during exercise as

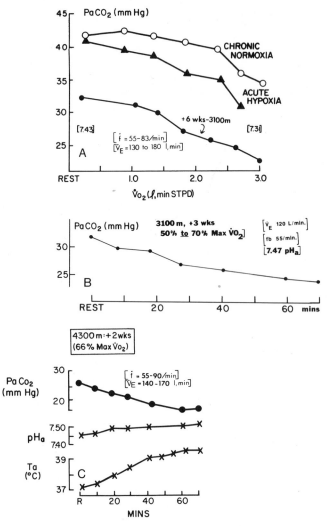

Fig. 2. A, effects of acute <u>vs</u> short-term hypoxia on the ventilatory response to progressive treadmill exercise in natives of sea-level (N=10). PaO_2 averaged 50-55 mmHg. B, long-term exercise effects on $PaCO_2$ and $\dot{V}E$ after acclimatization to 3100 m. C, long-term exercise effects on $PaCO_2$ after acclimatization to 4300 m.

total cardiac output is now reduced and therefore, at any given $\dot{V}O_2$, arterial to femoral venous O_2 difference widens (Fig. 3) and femoral venous PO_2 and O_2 content may be reduced to extremely low levels (Dempsey et al, 1975).

Finally, we note that the magnitude of the hyperventilatory response to hypoxia and the protection it affords PaO_2 might be attenuated at the very high work levels achieved in the highly trained athlete (4.5-5.5 l/min $\dot{V}O_2$). For example, upon acute exposure to even as moderate an altitude as 1600 m the hyperventilatory response was often less and arterial PO_2 fell considerably more (to 40-50 mmHg range) than in the untrained subject working at a lower max $\dot{V}O_2$. This effect of hypoxia tended to occur in highly trained persons whether or not they had shown substantial arterial HbO_2 desaturation under normoxic conditions.

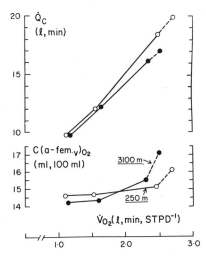

Fig. 3. Short-term moderate hypoxia (3100 m) causes a reduced cardiac output and therefore widened arterial to femoral venous O_2 content difference. Arterial O_2 content was equal to that at sea-level (19 ml/100 ml). Femoral venous O_2 contents were 0.5 to 1.3 ml O_2/100 ml.

PROBLEMS WITH THE HYPERVENTILATORY RESPONSE

Insufficient hyperventilation at high metabolic demand. Why is there less hyperventilation in many highly trained subjects working at ~6 l/min $\dot{V}CO_2$ vs the untrained working at 3-4 l/min $\dot{V}CO_2$? We think this "inadequate" response implies that the mechanical limits of the ventilatory systems are being approached at very high metabolic rates observed in the highly trained subjects. Three types of indirect evidence support this contention. First, the required level of ventilation which must be achieved and sustained in order to even partially compensate metabolic acidosis and to prevent hypoxemia is inordinately large. For example, to achieve a $PaCO_2$ of ~30 mmHg and PAO_2 ~115 mmHg, at $\dot{V}CO_2$ 3-4 l/min, the required $\dot{V}E$ is ~120 l/min; whereas at $\dot{V}CO_2$ 5-6 l/min, the required $\dot{V}E$ is 240 l/min (Dempsey et al., 1985). We think these ventilatory volumes could not be achieved and sustained by most subjects, although this hypothesis has not yet been adequately tested. Secondly, flow and volume during tidal breathing are often - but not always - at maximal limits of the expiratory flow-volume loop (Grimby et al., 1971). Thirdly, we have previously shown that within a subject exercising at a given work load, the level of ventilation does not adjust to the time-dependent increases in metabolic acidosis or to the developing hypoxemia (Dempsey et al., 1984; 1985). Similarly among subjects, the level of ventilation is poorly correlated with the magnitude of humoral stimuli during heavy and maximal exercise (see Fig. 4A). Furthermore, if hypoxemia or hyperoxia is superimposed during heavy exercise - even in the presence of substantial metabolic acidosis and high circulating levels of norepinephrine - very little hyper- or hypoventilatory response occurs (see Fig. 4B). It is as if the ventilatory response under these high exercise conditions becomes dissociated from the prevailing level of chemical stimuli. Interestingly, the largest ventilatory response during heavy exercise was not to additional humoral stimuli, but to reducing mechanical impedance by helium breathing (Dempsey et al., 1984); and normoxic helium breathing also significantly increased endurance performance time in highly trained subjects (Aaron et al., unpublished observations).

So just as the control system in the untrained refuses to grant suffi-

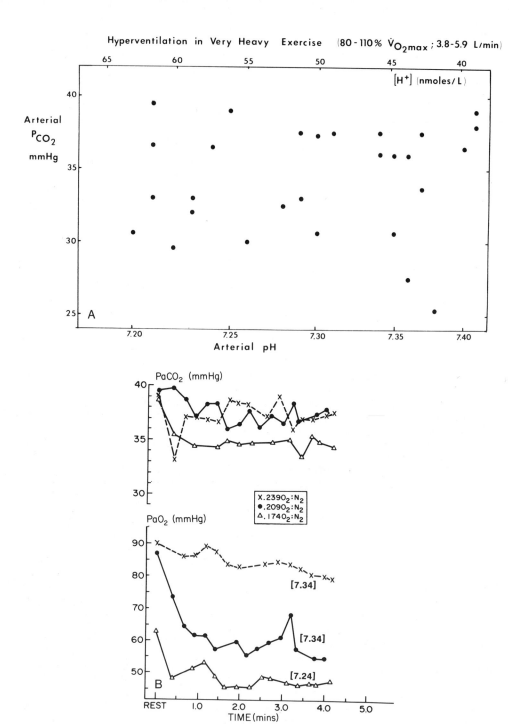

Fig. 4. A, top: lack of a significant relationship between the
hyperventilatory response to maximum exercise and the coinci-
dent arterial [H+] in highly trained subjects at sea-level.
(Also see Fig. 1A). B, bottom: effects of hyperoxia and mild
acute hypoxia (simulation of ~7000 ft altitude) at high work
load (VO_2=70 ml/kg·min, 97% max $\dot{V}O_2$). Note the relatively
severe exercise-induced hypoxemia and the relatively small
additional hyperventilation induced by this hypoxemia.

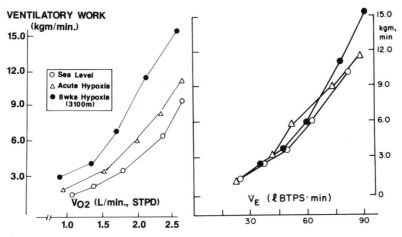

Fig. 5. Total ventilatory work during mild to heavy steady-state exercise at sea-level with acute moderate hypoxia (3100 m) and after short-term acclimatization to 3100 m. The hyperventilation experienced from normoxia to acute hypoxia to chronic hypoxia explains most of the effect of hypoxia on total ventilatory work.

cient alveolar ventilation to provide <u>complete</u> compensation for metabolic acidosis in heavy exercise, many trained persons working at much higher ventilatory demands will not provide the herculean ventilatory volumes to even ensure maintenance of arterial PO_2. Perhaps respiratory muscle fatigue occurs or demand exceeds maximum capability for pressure development under these conditions - but we think it more likely that feedback inhibition of inspiratory (and expiratory) muscle contractility may occur to <u>prevent</u> their fatigue.

<u>The inordinate cost of hyperventilation in hypoxia</u>. It is unfortunate that survival in hypoxic environments, at least in the lowlander sojourning at high altitude, is so critically dependent upon the hyperventilatory response, because this compensatory mechanism is so costly. At least five types of "cost" might be identified.

1. <u>Increased mechanical work</u>. First, the work load imposed on the respiratory muscles is a substantial one. It is also quite different than that seen at sea level because the hyperventilation, especially in heavy exercise in hypoxia, is achieved at high breathing frequencies (see Fig. 2). Note in Fig. 5 that at any given metabolic rate there is substantial increase in total ventilatory work performed by the respiratory muscles as one moves from normoxia to acute hypoxia to short-term hypoxia (Thoden et al., 1969). The reason for this augmented work is to be found almost exclusively in the magnitude of the hyperventilatory response as suggested by the similarity in response when normoxic and hypoxic conditions are compared at equal minute ventilations (right side Fig. 5). (Also note here that there was no apparent effect of decreased air density on ventilatory work at this altitude, but some effect in reducing total ventilatory work would be expected at altitudes above 4000 m.)

2. <u>Increased glycogenolysis</u>. Metabolically, respiratory muscles undergo greater changes during exercise in hypoxia than in normoxia (Fregosi and Dempsey, 1986). In rats, significant glycogenolysis occurs consistently in the diaphragm, and muscle lactate concentrations increase markedly following long term exhaustive exercise at PaO_2 50-60 mmHg and $PaCO_2$ 17-24 mmHg (see Fig. 6). These effects differ significantly from those obtained during

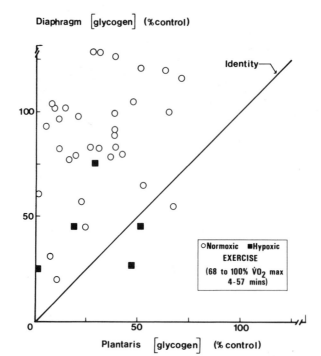

Fig. 6. Effects of short-term maximal exercise and long-term heavy exercise to exhaustion on diaphragm and plantaris muscle glycogen. Note that in normoxia, diaphragm glycogen usually was either unchanged by exercise or reduced <25% below control with some exceptions at 50% reduction. In hypoxic exhaustive exercise (PaO$_2$ 50-65 mmHg) in 4 of 5 animals, diaphragm glycogen was reduced to <50% of control.

exhaustive exercise in normoxia: in this case reductions in glycogen were relatively small and inconsistent and even though lactate concentration increased substantially, this was attributed to uptake from arterial perfusate rather than to a significant anerobic production of lactate by the respiratory muscles. These effects of hypoxia were attributed to a combination of increased ventilatory demand (55-70% increase in $\dot{V}A/\dot{V}CO_2$), and ~30% reduction in CaO$_2$ and possibly to presumed hypoxic effects on increasing circulating catecholamines. We emphasize that these changes in the diaphragm during hypoxic exercise were considerably smaller than those in locomotor muscles with similar fiber type and that the reduction in diaphragm glycogen still only amounted to 50-75% of total stores available.

3. Derecruitment of expiratory muscles. A third problem concerns the efficiency with which respiratory muscles generate a breath. In normoxic exercise of even moderate intensity, active expiration becomes very important to the preservation of inspiratory muscle function for the following reasons. First, as expiratory time shortens with hyperpnea, expiratory muscles are activated, expiratory flow is greatly increased and end-expiratory lung volume is reduced below its resting levels - as much as a liter or more during very heavy exercise (Henke et al., 1987). This is illustrated in Fig. 7 by the exercise-induced increase in end-expiratory intrathoracic and abdominal pressures during exercise. This means that the functional residual capacity (FRC) has fallen; thus, the diaphragm is at a longer length and is, therefore, able to generate more tension for a given phrenic motor activation on the subsequent inspiration. It also means that

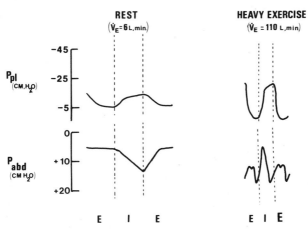

REST
($\dot{V}_E = 6$ L,min)

HEAVY EXERCISE
($\dot{V}_E = 110$ L,min)

P_{pl} (CM,H$_2$O)

P_{abd} (CM H$_2$O)

E   I   E   E I E

Fig. 7. Changes in esophageal pressure (Ppl) and gastric pressure (Pab) at rest and during heavy exercise. At rest, with no active expiration, the diaphragm descends throughout inspiration and Pab increases. During exercise, expiration becomes active (note increased end-expiratory Ppl) and as expiratory muscles cease contraction at the initiation of inspiration, Pab moves in a negative direction, thereby generating a negative pleural pressure and partially sparing the diaphragm.

energy is stored in expiratory muscles during expiration and upon its release at end-expiration it may actually assist the diaphragm so that at least a portion of inspiration is "passive". Indirect evidence for this is the negative deflection of abdominal pressure observed at the beginning of inspiration during exercise in humans (see Fig. 7, Grimby et al., 1976). More directly, we have recently observed in dogs that diaphragmatic EMG is significantly delayed following the initiation of inspiratory flow during states of hyperpnea when expiratory muscles are highly active (Smith et al., 1987). Hypoxia - or the accompanying hypocapnia - may interfere with this important process because most expiratory muscles are either not recruited or even derecruited with the hyperventilation accompanying hypoxia. The result is that end-expiratory lung volume is either not reduced or may even increase in hypoxia, thus the benefits available to the inspiratory muscles in terms of increased length and passive "assistance" during other types of hyperpnea are not forthcoming in hypoxia. This effect of hypoxia has been demonstrated in humans at rest (Garfinkel and Fitzgerald, 1978) and in the awake dog at rest and during mild hypoxic exercise (Smith et al., 1987). If this also applies to exercise in humans in hypoxia, then the inspiratory muscles will operate in an inefficient manner and at high metabolic cost.

Taken together, these types of evidence point strongly to the combination of increased pressure development, reduced energy supplies, augmented glycogenolysis and anerobic metabolism, and markedly reduced efficiency on the part of inspiratory muscles during exercise in hypoxia. We do not know if these factors result in respiratory muscle fatigue; this is not evidenced by overt CO_2 retention with progressive exercise. At the very least, these factors must cause increased energy cost by the respiratory muscles and contribute significantly to the exertional dyspnea experienced during exercise in hypoxia.

4.   Compromised cerebral blood flow. Hypoxic effects on vascular resistance may be either advantageous or disadvantageous, depending on the

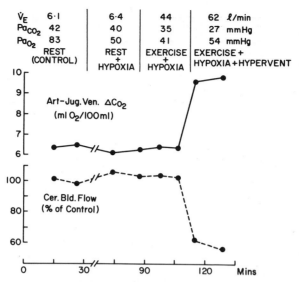

$\dot{V}_E$	6·1	6·4	44	62 ℓ/min
Pa_{CO_2}	42	40	35	27 mmHg
Pa_{O_2}	83	50	41	54 mmHg
	REST (CONTROL)	REST + HYPOXIA	EXERCISE + HYPOXIA	EXERCISE + HYPOXIA+HYPERVENT

Fig. 8. Vasoconstrictive effects of superimposed hypocapnia on cerebral blood flow during exercise in moderate hypoxia. During exercise $PaCO_2 < 30$ mmHg always caused reduced cerebral blood flow and these reductions were sustained for up to one hour of constant load exercise.

vascular bed in question. For example, global vasoconstriction of pulmonary arterioles in an hypoxic environment causes pulmonary hypertension, which is more severe when cardiac output is increased, as during exercise. Pulmonary vasoconstriction in these conditions no longer promotes a more uniform $\dot{V}A/\dot{Q}c$ distribution as would be expected with <u>local</u> alveolar hypoxia. On the other hand, hypoxemia exerts powerful vasodilatory effects in other vascular beds such as coronary skeletal muscle and cerebral arterioles. Unfortunately, hypocapnia exhibits powerful vasoconstrictor effects on the cerebral circulation, probably via reduced cerebral ECF [H+], and this effect may be dominant in some situations during hypoxic exercise. As shown in the example in Fig. 8, we tested this effect in normal humans during moderate exercise in acute hypoxia by measuring O_2 content and acid-base differences between arterial and jugular venous blood (Dempsey et al., 1975). Note that during exercise in hypoxia, cerebral blood flow and therefore the arterial-to-jugular venous O_2 content difference remained at control resting levels with the milder levels of hyperventilation. However, when $PaCO_2$ was voluntarily lowered to the 25-30 mmHg range as exercise continued, cerebral blood flow was then reduced 25-40% below control and venous O_2 content fell substantially despite the hyperventilatory effect on raising arterial (inflow) O_2 content. Thus, even a relatively mild hypocapnia superimposed on a relatively mild hypoxemia during exercise exerted a dominant vasoconstrictive effect on cerebral blood flow and this effect was sustained for up to an hour of prolonged exercise so long as $PaCO_2$ remained <30 mmHg.

5. <u>Disordered breathing in hypoxic sleep</u>. Perhaps the most profound negative effect of the hyperventilatory response to oxygen exposure is to be found not during heavy exercise but at the other end of the spectrum of metabolic demand, i.e., during quiet sleep. The hyperventilatory response to hypoxia is not state-dependent, so that just as in the resting awake state or during exercise, hyperventilation also appears during all sleep states and the magnitude of this hyperventilation increases with duration of hypoxic exposure (Berssenbrugge et al., 1984). On the other hand, sleep does cause a marked sensitization to the <u>inhibitory</u> effects of hypocapnia so

that significant apnea occurs when $PaCO_2$ is reduced only 3 or 4 mmHg, i.e. to the normal waking level of $PaCO_2$ (Dempsey and Skatrud, 1986). These sleep effects on control system gains cause 20 to 30 second periods of oscillation between marked excitation (hyperventilation) interspersed with marked inhibition (apnea), i.e., periodic breathing. This pattern is initiated within a few minutes of sleep initiation in hypoxia and persists so long as slow wave sleep occurs - although some modulation of this periodicity seems to occur over the initial weeks of acclimatization to hypoxia (Berssenbrugge et al., 1984). The consequences are substantial in terms of marked fluctuations in arterial O_2 saturation and pulmonary hypertension and in disruption of sleep state, leading to daytime somnolence and fatigue. Transient O_2 desaturation during sleep is of a much greater magnitude than that obtained even at the highest levels of exercise at the same altitude.

SUMMARY - ACCOMMODATION VS ADAPTATION

The hyperventilatory response to hypoxia is truly paradoxical. On the one hand, it is the critical - in fact, the only-compensatory mechanism of significance available in acute and short-term hypoxia to protect arterial O_2 content. Indeed, this response becomes even more important as stroke volume and cardiac output are reduced with longer durations of hypoxic exposure and is a critical acid-base regulator throughout the body during heavy and prolonged exercise in hypoxia. The carotid chemoreceptors responsible for these changes possess all of the salient characteristics of an excellent negative-feedback error detector: ideal anatomical location, fast response, and high gain. And yet this same feedback system also produces inordinate costs in terms of ventilatory work and the inefficient use of respiratory muscles; and the resultant hypocapnia is clearly inappropriate for many homeostatic needs, as evidenced by a compromised cerebral O_2 transport, inhibited expiratory muscle recruitment, and a markedly disrupted breathing pattern in hypoxic sleep.

Perhaps this chemoreceptor-induced hyperventilatory response to hypoxia in the human sojourner at high altitude may be viewed as an attempt at short-term "accommodation" until the subject can either return to his native normoxic environment or "survive" long enough until more efficient means of minimizing losses in oxygen transport are achieved. This idealized adaptation may be found in many of the "survivors" of chronic hypoxic exposure in the native or long-term resident of high altitude. Some of the key differences between the sojourner and resident are shown in Fig. 9. Note that the high altitude resident does hyperventilate at rest. The primary distinguishing feature is to be found during exercise in hypoxia where the resident shows an iso-capnic ventilatory response (which mimics that of the sea level native at sea level) and contrasts sharply with the hyperventilation of the sojourner in hypoxia. The resident does not experience the extreme tachypnea shown by the sojourner. Severe exercise-induced arterial hypoxemia is avoided in the native by maintaining a relatively narrowed alveolar-to-arterial O_2 difference. During sleep the native hyperventilates but avoids repeated apneas and periodic breathing by his blunted chemoreceptor responsiveness, which does not permit a hyperventilatory reaction to any added hypoxemia resulting from transient ventilatory inhibition (Lahiri et al., 1983).

The two key adaptations in chronic hypoxia both involve structural changes, in the carotid chemoreceptors rendering them less responsive to severe hypoxemia and in the lung's expanded gas exchange surface area, thereby, increasing alveolar-capillary diffusion capacity (Arias-Stella and Valcancel, 1973; Cerny et al., 1973). Apparently the sojourner's relatively brief exposure to hypoxia is insufficient to promote these truly adaptive changes and his only solution is an enhanced but costly ventilatory responsiveness.

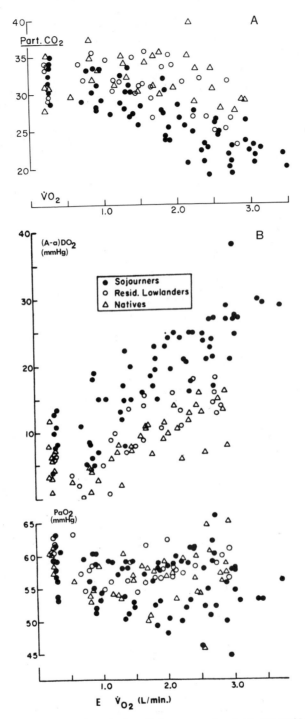

Fig. 9.   A, exercise-induced hypocapnic hyperventilation in sojourners to 3100 m and iso-capnic hyperpnea in long-term residents and natives of 3100 m.   Alveolar PO_2 averaged 60 to 72 mmHg in the residents and 67 to 83 in the sojourner.   B, narrowed A-aDO_2 and similar PaO_2 in the high altitude resident vs the sojourner during exercise at 3100 m.   Diffusion capacity for CO (DLCO) was consistently higher in the altitude resident at rest and exercise.   (See Dempsey et al., 1971).

IMPLICATIONS FOR EXERCISE LIMITATION

Does pulmonary gas exchange and/or alveolar ventilation provide significant limitations to systemic oxygen transport and to the achievement of maximum $\dot{V}O_2$?

The answer is clearly affirmative in those situations in which arterial O_2 content is reduced sufficiently, because we know that acutely reducing $S\dot{a}O_2$ a few percent by breathing even mildly hypoxic gas mixtures causes a significant reduction in $\dot{V}O_2$ max. For example, SaO_2 reduced from 92 to 85% causes a 7-10% reduction in max $\dot{V}O_2$. So, under what exercise conditions would a sufficient reduction in arterial O_2 saturation occur in health? a) In many highly trained persons during heavy work at sea level; b) in most athletes even at relatively mild elevations in altitude, especially upon <u>acute</u> hypoxic exposure; and c) even in the untrained at greater than 3000 m upon acute exposure and at greater than 3500 m in the acclimatized sojourner.

Exactly how might the reduction in SaO_2 <u>cause</u> a reduction in max $\dot{V}O_2$ has not been resolved. Classically, one would predict that a reduced CaO_2 would simply limit the extent to which the arterial-venous O_2 content could be widened. Thus, at reduced CaO_2, skeletal muscle capillary PO_2 would be reduced to less than a "critical" level at a lower $\dot{V}O_2$, so that the capillary-to-mitochondrial diffusion gradient would be insufficient; but this has not been proven at least under any other conditions listed above. To the contrary, when exercise load is held constant and blood flow to working skeletal muscles in heavily exercising humans at high altitude is high, femoral venous PO_2 can be reduced to <5 mmHg with no evidence of changes in lactate concentration across the working muscle or changes in metabolic rate (Dempsey et al., 1975). An additional point here is that the arterial hypoxemia has the potential of affecting oxygen transport to <u>all</u> active tissues and that its effects could be exerted on cardiac muscle (and, therefore, maximal cardiac output) or on diaphragm (and, therefore, on max $\dot{V}E$).

Pulmonary ventilation may - depending on the specific circumstances - present a significant determinant to max $\dot{V}O_2$ in two ways, its inadequacy and/or its high metabolic cost.

Maintaining a sufficiently high alveolar PO_2 to ensure sufficient alveolar-to-arterial oxygen transfer is not a problem at maximum exercise in the untrained individual. However, this is a major cause of hypoxemia in many highly trained subjects, in which alveolar PO_2 is not sufficiently high to prevent arterial hypoxemia in the face of marked reductions in mixed venous O_2 and transit time. For example, when untrained and trained subjects are compared, about one-half of the lower arterial PaO_2 in the trained is due to a lower alveolar PO_2 secondary to insufficient hyperventilation. Inadequate ventilation also affects [H+] regulation which may be an important determinant of max $\dot{V}O_2$ and this occurs in all subjects because pH compensation is never complete.

When is the ventilatory response costly enough to provide a significant limitation to max $\dot{V}O_2$? Respiratory muscles will always require some significant amount of $\dot{V}O_2$ and blood flow, which, of course, detracts from those available to working skeletal muscle. The problem here is that this required quantity has really not been measured under experimental conditions which truly mimic the ventilatory work and muscle recruitment which occur during maximum exercise. The efficiency of breathing and the mechanical work required are critically dependent on a number of factors including muscle length, velocity of shortening, duty cycle of contraction, and the use of accessory muscles. Commonly used procedures, which voluntarily imitate exercise ventilation or the use of CO_2 to stimulate $\dot{V}E$, do not mimic the true energy cost of the respiratory muscles that are used during physiologic exercise (Dempsey, 1986). We would expect that the greatest fraction of max $\dot{V}O_2$ - and of maximal cardiac output - diverted to respiratory muscles would occur during exercise at high altitude and in the untrained at sea level, but these estimates range from 5 to 25% of the total body $\dot{V}O_2$ (or $\dot{Q}$) (Pardy et al., 1984; Milic-Emili et al., 1962). In addition to these metabolic and flow requirements, excessive pressure development by the respiratory muscles is eventually manifested in the high CNS; these "sensations" may be especially debilitating and performance-limiting in conditions of prolonged heavy exercise and exercise in hypoxia. These comments are merely more speculation; the _real_ metabolic cost of ventilation during exercise in humans needs quantitation.

So, do imperfections in the pulmonary system form a significant contribution to the limitation of $\dot{V}O_2$ max? Most certainly some significant effect does occur under all conditions - even the apparently ideal conditions of heavy exercise in the untrained working at sea level; and these effects become more significant the higher the work loads achievable and the lower the ambient PO_2.

ACKNOWLEDGMENTS

I thank Michele Miller for manuscript preparation. Original work reported here was supported by NHLBI grant number 15469.

REFERENCES

Arias-Stella, J. and J. Valcancel (1973). The human carotid body at high altitudes. Pathologia and Microbiologia 39:292.

Berssenbrugge, A., J.A. Dempsey and J.B. Skatrud (1984). Effects of sleep state on ventilatory acclimatization to chronic hypoxia. J. Appl. Physiol. 57:1089-1096.

Cerny, F.C., J.A. Dempsey and W.G. Reddan (1973). Pulmonary gas exchange in non-native residents of high altitude. J. Clin. Invest. 52:2993-2999.

Dempsey, J.A. (1986). Is the lung built for exercise? Med. Sci. Sports 18:143-155.

Dempsey, J.A. and J.B. Skatrud (1986). A sleep-induced apneic threshold and its consequences. Am. Rev. Respir. Dis. 133:1163-1170.

Dempsey, J.A., E.H. Vidruk and G.S. Mitchell (1985). Pulmonary control systems in exercise: update. Fed. Proc. 44:2260-2270.

Dempsey, J.A., P. Hanson and K. Henderson (1984). Exercise-induced arterial hypoxemia in healthy human subjects at sea-level. J. Physiol. (Lond.) 355:161-175.

Dempsey, J.A.. N. Gledhill, W.G. Reddan, H.V. Forster, P.G. Hanson and A.D. Claremont (1977). Pulmonary adaptation to exercise: effects of exercise type and duration, chronic hypoxia, and physical training. Ann. NY Acad. Sci. 301:243.

Dempsey, J.A., J.M. Thomson, H.V. Forster, F.C. Cerny and L.W. Chosy (1975). HbO_2 dissociation in man during prolonged work in chronic hypoxia. J. Appl. Physiol. 38:1022-1029.

Dempsey, J.A., J.M. Thomson, S.C. Alexander, H.V. Forster and L.W. Chosy (1975). Respiratory influences on acid-base status and their effects on O_2 transport during prolonged muscular work. In: Metabolic Adaptation to Prolonged Physical Exercise. Proceedings of the 2nd International Symposium on Biochemistry of Exercise. H. Howald and J.R. Poortmans (Eds). Magglingen, Switzerland: Number 7, pp. 56-64.

Dempsey, J.A., W.G. Reddan, J. Rankin, M.L. Birnbaum, H.V. Forster, J.S. Thoden and R.F. Grover (1971). Effects of acute through life-long hypoxic exposure on exercise pulmonary gas exchange. Respir. Physiol. 13:62-89.

Fregosi, R.F. and J.A. Dempsey (1986). Effects of exercise in normoxia and acute hypoxia on respiratory muscle metabolites. J. Appl. Physiol. 60(4):1274-1283.

Garfinkel, R. and R.S. Fitzgerald (1978). The effect of hypoxemia, hypoxia and hypercapnia on FRC and occlusion pressure in human subjects. Resp. Physiol. 33:241-250.

Grimby, G., B. Saltin and L.W. Helmsen (1971). Pulmonary flow-volume and pressure-volume relationships during submaximal and maximal exercise in young well-trained men. Bull. Physiol-Pathol. Respir. 7:157-168.

Grimby, G., M. Goldman and J. Mead (1976). Respiratory muscle actions inferred from rib cage and abdominal V-P partitioning. J. Appl. Physiol. 41:739-751.

Holtgren A. and R.F. Grover (1971). Abnormal circulatory responses to high altitude in subjects with a previous history of high altitude pulmonary edema. Circulation 44:759-770.

Henke, K., M. Sharratt, D. Pegelow and J. Dempsey (1987). Regulation of end-expiratory lung volume during exercise. Fed. Proc. 46(3):1018 (abs).

Lahiri, S., K. Maret and M. Sherpa (1983). Dependence of high altitude sleep apnea on ventilatory sensitivity of hypoxia. Respir. Physiol. 52:281-288.

Milic-Emili, J. Petit and R. Deroanne (1962). Mechanical work of breathing during exercise in trained and untrained subjects. J. Appl. Physiol. 17:43-46.

Pardy, R.L., S.N. Hussain and P.T. Macklem (1984). The ventilatory pump in exercise. In: Clinics in Chest Medicine. J. Loke (Ed). 5:35-49.

Smith, C.A., D. Ainsworth, K. Henderson and J. Dempsey (1987). Differential recruitment of expiratory muscles. Fed. Proc. 46(3):1968 (abs).

Thoden, J.S., J.A. Dempsey, W.G. Reddan, M.L. Birnbaum, H.V. Forster, R.F. Grover and J. Rankin (1969). Ventilatory work during steady-state response to exercise. Fed. Proc. 28:1316-132.

STRUCTURAL VS FUNCTIONAL LIMITATIONS TO OXYGEN TRANSPORT:

IS THERE A DIFFERENCE?

James H. Jones and Richard H. Karas

Department of Physiological Sciences
School of Veterinary Medicine
University of California
Davis, CA 95616
and
Concord Field Station, Museum of Comparative Zoology
Harvard University, Old Causeway Road
Bedford, MA 01730

INTRODUCTION AND BACKGROUND

The flow of O_2 through the mammalian respiratory system has been described as occurring down a cascade of transport steps (Taylor and Weibel, 1981): pulmonary ventilation, pulmonary diffusion, circulatory convection, peripheral tissue diffusion, and reduction of the O_2 by cytochrome oxidases on the inner cristae of mitochondria. When rates of O_2 consumption ($\dot{V}O_2$) are at their maximum ($\dot{V}O_2$max), over 90% of the O_2 flowing through the respiratory system is consumed by mitochondria in cardiac and skeletal muscle tissues (Mitchell and Blomqvist 1971).

The question as to what limits $\dot{V}O_2$max in mammals has been pondered since the early part of this century (Verzar, 1912; Hill and Lupton, 1923). The steps that constitute the O_2 transport cascade and could potentially limit O_2 flux through it can be considered to be of two types: structural and functional. This is an arbitrary grouping as there is ultimately a structural basis for every function; however, in terms of O_2 transport, some steps represent resistances (or their reciprocals, conductances) for diffusion of O_2 down a partial pressure gradient, while other steps involve flow of a medium (air or blood) with a specific capacitance for O_2 (βO_2). The conductances of structural steps of the cascade are largely determined by surface areas and barrier thicknesses for diffusion (Weibel, 1984). Changes in functional transport steps occur with changes in pump frequency or volume, and are affected by the concentration of gas or the quantity of hemoglobin (Hb) present, which changes βO_2.

A number of experimental perturbations of the O_2 transport system have been made in order to determine their effects on $\dot{V}O_2$max (di Prampero, 1985) in attempts to elucidate the limiting factor(s) for $\dot{V}O_2$max. The fraction of O_2 in inspired gas (FIO_2) has been raised and lowered, the concentration of Hb available to bind with O_2 has been increased and decreased, and activities or concentrations of biochemical markers of oxidative metabolism have been measured in muscle biopsy samples taken before and after training. These perturbations have typically been performed and interpreted independently of their effects on other O_2 transport steps in the system.

Most studies attempting to identify a limitation for aerobic metabolism have sought to identify a single step in the O_2 transport cascade as being the limiting factor, and, for humans, cardiovascular transport of O_2 has been regarded as being the limitation for $\dot{V}O_2$max (Saltin and Gollnick, 1983). In recent years, several authors have proposed that regulation of substrate flux through systems of enzymes is not vested in any single "rate-limiting" step, but rather, by the interaction of elements in the system taken as a whole (Heinrich and Rapoport 1973, 1974; Kacser and Burns, 1973, 1979; Kacser, 1983; Porteous, 1985). In many ways, flux of substrate through a system of enzymes is analogous to the flux of O_2 through the respiratory system. The major consideration to date of the possibility that $\dot{V}O_2$max is determined by the integrated action of multiple steps in the O_2 transport cascade has been di Prampero's (1985) analysis and allometric studies of respiratory system design (Taylor and Weibel, 1981; Taylor et al., 1987).

Identification of a limiting step for any process requires a clear definition of what constitutes a limitation. We operationally define a limiting factor as any structural or functional variable that decreases $\dot{V}O_2$ when its value is decreased; we consider an increase in $\dot{V}O_2$ when such a variable is increased to be evidence that excess capacity existed elsewhere in the system when it had previously been limited. The ratio of the fractional change in $\dot{V}O_2$ to the fractional change in the variable is an index of the degree to which that variable limits $\dot{V}O_2$, analogous to the sensitivity coefficient of Kacser and Burns (1973, 1979), the control strength of Heinrich and Rapoport (1973, 1974), the flux control coefficient of Porteous (1985) and the fractional limitation coefficient of di Prampero (1985). The sum of all of the fractional limitations (F) must be standardized so that their sum equals 1.0 for the system:

$$F = \frac{\dot{V}O_2\text{max}/\Delta\text{variable}}{\Sigma(\Delta\dot{V}O_2\text{max}/\Delta\text{variable})}$$

We wished to determine the degree to which individual structural and functional O_2 transport steps limit $\dot{V}O_2$max when considered individually, as well as when functioning within the respiratory system. Changes in PO_2 and $\dot{V}O_2$ elicited by changing a given variable in the O_2 transport cascade could potentially affect other steps in the system. To assess these interactions, we have utilized a Bohr integration model developed for O_2 transfer in the lungs (Karas et al. 1987a), and applied it to diffusion of O_2 in both lung and peripheral (i.e., skeletal muscle) tissues at $\dot{V}O_2$max. We have solved the Bohr integration equations using morphometric estimates of pulmonary diffusing capacity (DLO_2) and capillary volume (Vc) (Weibel 1963), skeletal muscle Vc (Hoppeler et al., 1973), and average values for cardiac output ($\dot{Q}$T) and [Hb] in humans.

RESULTS AND DISCUSSION

We performed the Bohr integration for the lung alone using typical physiological variables and morphometric estimates of structural variables for an average human exercising at $\dot{V}O_2$max: $PAO_2 = PaO_2 = 100$ Torr; DLO_2/Mb = 0.4 ml $O_2 \cdot s^{-1} \cdot Torr^{-1} \cdot kg^{-1}$, where Mb is body mass; Vc/Mb = 3.04 ml·kg^{-1}; $\dot{Q}$T/Mb = 6 ml·s^{-1}·kg^{-1}; [Hb] = 15 g·dl^{-1}; $P\bar{v}O_2$ (mixed venous PO_2) = 21 Torr; arterial and venous pH = 7.40 and 7.25 (Fig. 1A). For the same individual the procedure was repeated with $\dot{Q}$T doubled (Fig. 1B) and with DLO_2 quartered (Fig. 1C). The solid curves in Fig. 1 show the time course for the change in PO_2 and the dashed lines the changes in O_2 saturation (SO_2) of the Hb as the red cells traversed the pulmonary capillaries.

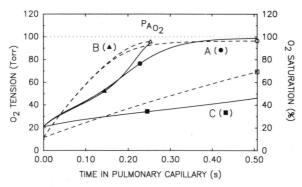

Fig. 1. Bohr integration estimate of the time course of changes in PO_2 (solid lines) and SO_2 (dashed lines) of blood as it traverses the pulmonary capillary. Solid symbols represent $\bar{P}cO_2$ for each curve, open symbols represent arterial SO_2 for the same Bohr integration. A (circles) is average athlete, B (triangles) is elite athlete with $\dot{Q}T$ doubled, C (squares) is pulmonary-limited athlete with DLO_2 quartered.

The average human appears to have ample DLO_2 relative to its $\dot{V}O_2$, as both PO_2 and SO_2 equilibrate long before the red cells exit the lung. There is a slight decrease in SO_2 when $\dot{Q}T$ is doubled, due to the time available for diffusion being halved, and PO_2 and SO_2 are greatly reduced when DLO_2 is quartered. Although arterial SO_2 decreases slightly from normal when $\dot{Q}T$ is doubled, the total quantity of O_2 transferred in the lung is nearly doubled because the shorter capillary transit time (tc) causes the integrated mean capillary PO_2 ($\bar{P}cO_2$) solid symbols in Fig. 1) to be much lower than it is with normal $\dot{Q}T$. The Bohr equation ($\dot{V}O_2 = DLO_2 \cdot [PAO_2 - \bar{P}cO_2]$, (where PAO_2 is alveolar PO_2) indicates that $\dot{V}O_2$ must increase when the integrated pressure head for diffusion increases while DLO_2 remains constant.

These patterns are summarized in Fig. 2, in which the maximum flux of O_2 across the lung ($\dot{V}O_2$max) is plotted for different $\dot{Q}T$ values as DLO_2 varies (with fixed pulmonary Vc). An increase in $\dot{Q}T$ will increase the O_2 flux at any DLO_2, but the increase in $\dot{V}O_2$ will be maximized only if DLO_2 is sufficiently enlarged to allow full saturation of Hb for that $\dot{Q}T$. An increase in [Hb] will have the same effect as increasing $\dot{Q}T$, because the increased βO_2 in the blood will also maintain a lower $\bar{P}cO_2$ as the red cells pass through the pulmonary capillaries.

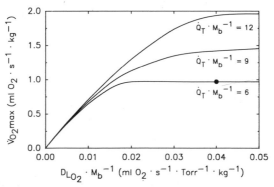

Fig. 2. Maximum $\dot{V}O_2$ across the lung at different combinations of $\dot{Q}T$ and DLO_2, calculated using a Bohr integration. The solid circle represents the average athlete's control value.

These relationships between O_2 flux at a single transport step and changes in a single structural (DLO_2) or functional ($\dot{Q}T$) variable are straightforward. It is not as obvious how changes in a single component of the system will affect flux at other steps in the system, because alterations in variables elicited by changes at one step may affect the boundary conditions of O_2 transport in adjacent steps. For example, when $\dot{Q}T$ is increased and PaO_2 is concomitantly decreased, there are three effects: 1) the alveolar-capillary PO_2 gradient in the lung is enlarged, enhancing $\dot{V}O_2$ in the lung; 2) SO_2 of the blood being circulated is decreased slightly, while bulk flow is increased; 3) the capillary-mitochondrial PO_2 gradient in the muscle is increased, despite the lower SO_2 of the blood, because the shorter tc of the red cells, even with their lower PaO_2, results in a higher integrated $\overline{P}cO_2$. As a result, the increase in $\dot{V}O_2$ seen with an increase in $\dot{Q}T$ could be due to a "cardiovascular limitation" that was overcome by increased flow or capacitance, or a "peripheral tissue limitation" or "pulmonary limitation" that was overcome by the larger integrated pressure gradient in the peripheral or pulmonary capillaries.

To evaluate the effects of changes in specific structural and functional components on O_2 flux through the entire O_2 transport cascade, we employed the Bohr integration model used above, but with two modifications. First, we performed a Bohr integration for the lung (as above, $DLO_2/Mb = 0.4$ ml $O_2 \cdot s^{-1} \cdot Torr^{-1} \cdot kg^{-1}$; $\dot{Q}T/Mb = 6$ ml $\cdot s^{-1} \cdot kg^{-1}$) and used the resulting arterial values as the input for a Bohr integration for the muscle. We assumed that mitochondrial $PO_2 = 0$ Torr and that whole body skeletal muscle Vc was equal to pulmonary Vc (Conley et al. 1987; Weibel et al. 1987); then varied the tissue diffusing capacity for O_2 (DTO_2) in the Bohr integration until the PO_2 of venous blood leaving the muscle matched $P\bar{v}O_2$ entering the lung. We utilized the Fick principle to calculate whole animal $\dot{V}O_2$ from $\dot{Q}T$ and arterial and venous O_2 contents, calculated from the PO_2 and pH.

The results of these calculations yielded a control $\dot{V}O_2max$ against which the magnitude of changes in $\dot{V}O_2$ resulting from perturbations of individual structural and functional steps in the system could be compared. The steps that we altered in the analysis were: PAO_2 (corresponding to a change in minute ventilation or FIO_2) DLO_2, $\dot{Q}T$ and DTO_2. In each case, after the value of one variable was altered, the Bohr integration was run alternately in muscle and lung with all boundary conditions other than the altered variable remaining constant. The output from one compartment was used as the input to the other and the PO_2 was allowed to change until the system reached steady-state, at which time the new $\dot{V}O_2max$ of the system was calculated from the Fick equation. We performed the analyses for control

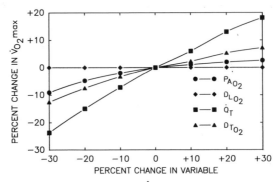

Fig. 3. Percent changes in $\dot{V}O_2max$ in an average athlete resulting from changes of individual variables in the O_2 transport cascade, calculated using a steady-state two-compartment Bohr integration.

Table I. Fractional limitations to $\dot{V}O_2$max.

Fractional Limitation	Average Athlete		Elite Athlete	
	−	+	−	+
F_{PA}	0.21	0.09	0.35	0.17
F_{DL}	0	0	0.18	0.11
F_{QT}	0.52	0.63	0.22	0.22
F_{DT}	0.27	0.28	0.24	0.50
Total	1.00	1.00	0.99	1.00

Fractional limitations for PAO_2 (F_{PA}), DLO_2 (F_{DL}), $\dot{Q}T$ (F_{QT}), and DTO_2 (F_{DT}) for average $\dot{Q}T/Mb = 6$ $ml \cdot s^{-1} \cdot kg^{-1}$) and elite ($\dot{Q}T/Mb = 12$ $ml \cdot s^{-1}$) athletes. The "−" column is fractional response to a decrease in the variable (−30%), the "+" column is fractional response to an increase in the variable (+30%).

$$F = \frac{\dot{V}O_2max/\,\Delta variable}{\Sigma\,(\Delta\dot{V}O_2max/\Delta variable)}$$

conditions, with $\dot{Q}T/Mb = 6$ $ml \cdot s^{-1} \cdot kg^{-1}$, then repeated the analyses with $\dot{Q}T/Mb = 12$ $ml \cdot s^{-1} \cdot kg^{-1}$. The latter value is a remarkably high $\dot{Q}T$ for a human, but not for non-human "elite" athletes, e.g., dog (Karas et al. 1987b) and horse (unpublished data). We used the higher $\dot{Q}T$ as an extreme example to determine if the responses of the system to changes in individual variables might be different for athletes with cardiovascular systems that may have evolved to "match" other components of the system (e.g., DLO_2), as opposed to the average human athlete in which it is lower. In either case, our primary interest was not in the magnitudes of the particular values chosen, but rather, the way in which the system responded to changes in different variables.

The analyses for average athletes (Fig. 3) and for elite athletes (Fig. 4) for changes of ±30% in O_2 transport variables yield different results. In both systems, changes in individual variables yield larger changes in $\dot{V}O_2$max when the variables are decreased than when they are increased, due to the sigmoid shape of the O_2 equilibrium curve causing larger decreases in SO_2 with falling PO_2 than increases with rising PO_2. If we now consider the fractional limitations imposed by individual variables, i.e. the change in $\dot{V}O_2$max due to a change in a given variable compared to the changes in $\dot{V}O_2$max elicited by changes in other variables, we see that the average athlete (Fig. 3) appears to be more sensitive to changes in $\dot{Q}T$ than any other variable (Table I). The fractional limitation due to $\dot{Q}T$ (F_{QT}) is −0.52 for decreases and +0.63 for increases. DTO_2 imposes the next largest fractional limitation ($F_{DT} = -0.27$ and ± 0.28), followed by PAO_2 ($F_{PA} = -0.21$ and +0.09). Changes of ±30% in DLO_2 have no effect on $\dot{V}O_2$max in the average athlete.

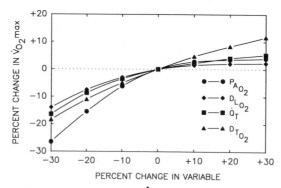

Fig. 4. Percent changes in $\dot{V}O_2$max in an elite athlete resulting from changes of individual variables in the O_2 transport cascade, calculated using a steady-state two-compartment Bohr integration.

In the elite athlete (Fig. 4) $\dot{V}O_2$max is most sensitive to decreases in PAO_2 (F_{PA} = -0.35) and to increases in DTO_2 (F_{DT} = +0.50). Changes in $\dot{Q}T$ have less than half the effect on $\dot{V}O_2$max that they have in the average athlete ($F_{\dot{Q}T}$ = -0.22 and +0.22), and changes in DLO_2 have significant effects on $\dot{V}O_2$max (F_{DL} = -0.18 and +0.22).

These results suggest that 50-60% of the limitation to $\dot{V}O_2$max in average human athletes is due to $\dot{Q}T$, which is similar to the 78% fractional limitation that di Prampero (1985) ascribed to $\dot{Q}T$. Some degree of limitation to peripheral tissue diffusion (or mitochondrial oxidative capacity) must exist at $\dot{V}O_2$max, as $P\bar{v}O_2$ never falls to 0 Torr. Our analysis indicates that this diffusion limitation may account for 28% of the limitation at $\dot{V}O_2$max, again similar to di Prampero's (1985) estimate, although he did not consider what role PAO_2 might play in limiting $\dot{V}O_2$max.

Elite athletes (Fig. 4) appear to be limited at $\dot{V}O_2$max differently than average athletes. Decreases in PAO_2 cause the largest change in $\dot{V}O_2$max, and decreases in DTO_2, $\dot{Q}T$ and DLO_2 all have quantitatively similar limiting effects. Increases in DTO_2 overcome approximately 50% of the limitation when variables are increased, and $\dot{Q}T$, PAO_2 and DLO_2 combined exert the same degree of limitation. Elite athletes are more sensitive to change in FIO_2 or ventilation than are average athletes, and elite athletes are sensitive to changes in DLO_2, which average athletes are not.

Our analysis suggests that for an average athlete, the greatest increase in $\dot{V}O_2$max per unit increase of a single variable in the O_2 transport cascade would occur with an increase in either the functional variable $\dot{Q}T$, or the structurally related component, [Hb]; for an elite athlete, the largest increase would occur with an incrase in the structural variable DTO_2. This change in limiting factors with increased aerobic capacity is consistent with the hypothesis that structures and functional capacities of O_2 transport steps in elite athletes may be better matched and come closer to achieving simultaneous limitations (i.e., all steps with similar fractional limitations) to O_2 flux than in averge athletes. We used the same DTO_2 for elite and average athletes in this analysis, whereas in reality both $\dot{Q}T$ and DTO_2 are larger in elite athletes than in average athletes (see Taylor et al. and Weibel et al. in this volume). Although we have ignored the effects of differences in mitochondrial volumes and oxidative capacities between average and elite athletes, these would presumably increase the effective DTO_2 in elite athletes by reducing back pressure during diffusion into the peripheral tissues. The major findings of this analysis in terms of the response of the system to perturbations are:

1. the limitation for maximal O_2 flux in a series of transport steps is shared by multiple steps in the system;

2. the degree to which any individual step limits flux through the entire system depends on the relative magnitude of its structural or functional features compared to other elements in the system, and the major limitation may be either structural or functional in nature; the $\dot{V}O_2$ is determined by the interactions of both structural and functional fractional limitations;

3. changes in the relative magnitudes of individual steps in the O_2 transport cascade with training or evolution might lead to equivalence of fractional limitations among all steps in the system. Such a finding would be in conformance with Taylor and Weibel's (1981) principle of symmorphosis, although it does not appear to be applicable to average athletes, e.g., humans.

REFERENCES

Conley, K.E., S.R. Kayar, K., Rösler, H., Hoppeler, E.R. Weibel, C.R. Taylor and O. Mathieu-Costello (1987). Adaptive variation in the mammalian respiratory system. III. Capillaries and their relationship to oxidative capacity. Respir. Physiol. 69:47-64.

di Prampero, P.E. (1985). Metabolic and circulatory limitations to $\dot{V}O_2$max at the whole animal level. J. Exp. Biol. 115:319-331.

Heinrich, R. and T.A. Rapoport (1973). Linear theory of enzymatic chains; its application for the analysis of the cross-over theorem and of glycolysis of human erythrocytes. Acta Biol. Med. Ger. 31:479-494.

Heinrich, R. and T.A. Rapoport (1974). A linear steady-state treatment of enzymatic chains. Eur. J. Biochem. 42:89-95.

Hill, A.V. and H. Lupton (1923). Muscular exercise, lactic acid, and the supply and utilization of oxygen. Quart. J. Med. 16:135-171.

Hoppeler, H., P. Luthi, H.Claassen, E.R. Weibel and H. Howald (1973). The ultrastructure of the normal human skeletal muscle. Pflugers Arch. 344:217-232.

Kacser, H. (1983). The control of enzymes in vivo: elasticity analysis of the steady state. Biochem. Soc. Trans. 11:35-40.

Kacser, H. and J. Burns (1973). The control of flux. Symp. Soc. Exp. Biol. 27:65-104.

Kacser, H. and J. Burns (1979). Molecular democracy: who shares the controls? Biochem. Soc. Trans. 7:1149-1160.

Karas, R.H., C.R. Taylor, J.H. Jones, S.L. Lindstedt, R.B. Reeves and E.R. Weibel (1987a). Adaptive variation in the mammalian respiratory system. VI. Flow of oxygen across the pulmonary gas exchanger. Respir. Physiol. 69:101-115.

Karas, R.H., C.R. Taylor, K., Roseler and H. Hoppeler (1987b). Adaptive variation in the mammalian respiratory system. IV. Limits to oxygen transport by the circulation. Respir. Physiol. 69:65-79.

Mitchell, J.H. and G. Blomqvist (1971). Maximal oxygen uptake. N. Engl. J. Med. 284:1018-1022.

Porteous, J.W. (1985). Enzyme catalysed fluxes in metabolic systems. Why control of such fluxes is shared among all components of the system. In: Gilles, R., ed. Circulation, Respiration and Metabolism. Springer-Verlag, New York, pp. 263-277.

Saltin, B. and P.D. Gollnick (1983). Skeletal muscle adaptability: significance for metabolism and performance. In: Handbook of Physiology. Skeletal Muscle, pp. 555-631. Am. Physiol. Society.

Taylor, C.R. and E.R. Weibel (1981). Design of the mammalian respiratory system. I. Problem and strategy. Respir. Physiol. 44:1-10.

Taylor, C.R., E.R. Weibel, R.H. Karas and H. Hoppeler (1987). Adaptive variation in the mammalian respiratory system. VIII. Structural and functional design principles determining the limits to oxidative metabolism. Respir. Physiol. 69:117-127.

Verzar, F. (1912). The gaseous metabolism of striated muscle in warm-blooded animals. J. Physiol. (Lond.) 44:243-258.

Weibel, E.R. (1963). Morphometry of the human lung. Springer, Berlin.

Weibel, E.R. (1984). The Pathway for Oxygen: Structure and Function in the Mammalian Respiratory System. Harvard Univ. Press: Cambridge.

Weibel, E.R., L.B. Marques, M. Constantinopol, F. Doffey, P. Gehr and C.R. Taylor (1987). Adaptive variation in the mammalian respiratory system. V. The pulmonary gas exchanger. Respir. Physiol. 69:81-100.

REGULATION OF THE ACID-BASE BALANCE DURING PROLONGED HYPOXIA: EFFECTS OF

RESPIRATORY AND NON-RESPIRATORY ACIDOSIS

N.C. Gonzalez, J. Pauly, G. Widener, L.P. Sullivan and
R.L. Clancy

Department of Physiology
University of Kansas Medical Center
Kansas City, KS   66103

INTRODUCTION

Prolonged hypoxia results in adaptive changes in the oxygen transfer system which appear to be designed to insure adequate oxygen supply to the cells (for review, see Bouverot, 1985). Many of these adaptations, however, produce other functional changes not directly concerned with oxygen transfer. Examples of this are the acid-base changes of high altitude hypoxia. Fig. 1 shows the time course of changes in plasma acid-base balance in male conscious rats exposed to a barometric pressure of 370-380 torr (Gonzalez and Clancy, 1986a). The hyperventilation initially results in respiratory alkalosis; this is eventually corrected by a proportionate decrease in plasma $[HCO_3^-]$, which is largely due, in turn, to the renal compensatory mechanisms put into motion by the alkalosis. After three weeks of hypoxia, plasma pH has returned to normal, but at the expense of depletion of HCO_3^-.

Fig. 2 shows that the HCO_3^- depletion associated with prolonged hypoxia is not limited to plasma: intracellular $[HCO_3^-]$ of cardiac and skeletal muscles of the hypoxic rats decreases in proportion to the PCO_2, and cell pH is not different from that of normoxic controls in any of the tissues studied (Gonzalez and Clancy, 1986b). The low cellular $[HCO_3^-]$ observed by us confirms the original observation of Freeman and Fenn (1953) that prolonged hypoxia results in depletion of tissue total CO_2 stores in rats.

These acid-base features of prolonged hypoxia have been known for some time (for review, see Bouverot, 1985). In contrast, little is known about the ability of hypoxic animals to withstand challenges to the acid-base balance. These challenges frequently are the result of alterations in the balance between production and elimination of acids or bases, and they may occur as a consequence of disease or of increased physical activity. The mechanisms employed by the body to correct these alterations include chemical buffering and changes in metabolic, respiratory and renal functions, upon which the effect of prolonged hypoxia is incompletely understood. The objective of the studies reviewed here was to determine if prolonged exposure to hypoxia influences the ability of animals to regulate their acid-base balance when they are challenged with an acid load.

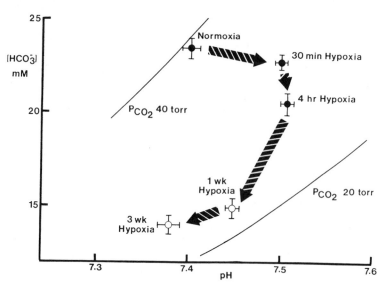

Fig. 1. Plasma acid-base balance at various times of exposure to hypobaric hypoxia. Intact, conscious rats were kept at a PB of 370-380 torr (PIO$_2$ 68-70 torr). PaO$_2$ varied between 33 and 37 torr. From Gonzalez and Clancy (1986a). Reprinted with permission of Respiration Physiology.

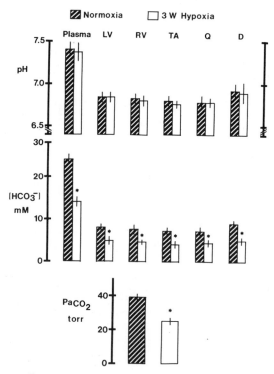

Fig. 2. pH and [HCO$_3^-$] of plasma, and of left ventricle (LV), right ventricle (RV), tibialis anterior (TA), quadriceps (Q) and diaphragm (D) muscle cells in normoxic and three-week hypoxic rats. Intracellular pH was calculated from the steady-state distribution of 5,5-dimethyl-2,4-oxazolidinedione (DMO). Cellular [HCO$_3^-$] was calculated from cell pH and PaCO$_2$. Stars indicate a significant difference between normoxic and hypoxic. Redrawn from data of Gonzalez and Clancy, 1986b.

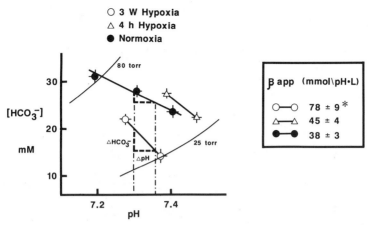

Fig. 3. Plasma acid-base composition of normoxic, 4h-hypoxic and three-week hypoxic rats exposed to various levels of inspired PCO_2 for 4h. The slope of the lines connecting the various groups ($\Delta[HCO_3^-]/\Delta pH$) is the negative of the apparent non-bicarbonate buffer value of plasma (βapp). Redrawn from data of Gonzalez and Clancy (1986a).

Sprague-Dawley rats were exposed for three weeks to a barometric pressure of 370-380 torr, which results in a PO_2 of moist inspired air of 68-70 torr (Gonzalez and Clancy, 1986a, b; Widener et al., 1986). The animals were fitted with indwelling arterial, venous and, occasionally, urinary bladder catheters and were studied in the conscious state in a chamber where PIO_2 was maintained at 68-70 torr by mixing air and N_2 at ambient barometric pressure (740-750 torr). Controls were normoxic rats of the same age and weight maintained at ambient PO_2. The capacity of hypoxic rats to regulate the acid-base balance was tested by introducing acid loads in the form of an increase in inspired PCO_2 (respiratory acidosis) or by the infusion of non-carbonic acid (non-respiratory acidosis).

RESPIRATORY ACIDOSIS

Fig. 3 illustrates the effects of 4 hours of hypercapnia on the plasma acid-base composition of three groups of rats: normoxic controls, rats exposed to hypoxia for three weeks, and rats exposed to the same level of hypoxia for only 4 hours. The straight lines connect points obtained at different $PICO_2$ in a given group. Their slopes, $\Delta[HCO_3^-]/\Delta pH$, represent the apparent non-bicarbonate buffer value of plasma (βapp). βapp of three-week hypoxic rats is approximately twice that of either normoxic or 4h-hypoxic rats. This indicates an improved pH regulation by the prolonged hypoxic rats, since a given level of hypercapnia resulted in a larger increase in plasma $[HCO_3^-]$ and a smaller decrease in pH than in the other groups. Since PaO_2 of the 4h hypoxic rats was lower than that of the three-week hypoxic rats, it seems that the increased pH regulation observed in the latter is not due to low PO_2 per se, but to some other mechanism associated with prolonged exposure to hypoxia.

The value of plasma βapp observed at any given time of hypercapnia in vivo is the result of the interaction of several factors. These are principally chemical buffering by intra- and extracellular non-bicarbonate buffers, increase in the rate of excretion of proton equivalents by the kidneys and by the lungs, and exchange of proton equivalents between plasma, interstitial, and intracellular fluids. Although these mechanisms are set

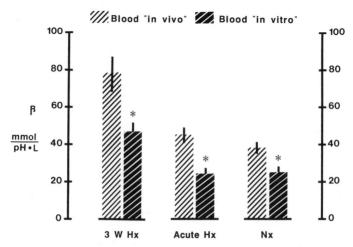

Fig. 4. Non-bicarbonate buffer value of plasma obtained after in vivo equilibration of animals (light columns) or of blood in vitro (dark columns) with various CO_2 tensions. Stars indicate a significant difference between in vivo and in vitro values. Redrawn from data of Gonzalez and Clancy, 1986a.

in motion simultaneously, their time course is different. This characteristic has been used to determine the mechanisms underlying the changes in plasma βapp observed during hypercapnia (Brown, 1971; Davenport, 1964; Gonzalez, 1986; Lai et al., 1973; Pryss-Roberts, 1968).

The increased βapp of hypoxic rats could have been produced by one of more of the following mechanisms:

1. Chemical buffering

An estimate of the contribution of blood buffers to the in vivo plasma βapp can be obtained by determining the plasma $\Delta[HCO_3^-]/\Delta pH$ after in vitro CO_2 equilibration of blood. This in vitro plasma buffer value, β in vitro, is the result of the titration of plasma and red cell non-bicarbonate buffers with CO_2, and of the redistribution of the newly generated bicarbonate between plasma and red cells. Fig. 4 shows β in vitro of normoxic, short-term, and prolonged hypoxic rats. Also represented are the corresponding values of in vivo plasma βapp shown in Fig. 3. β in vitro is approximately twice as high in the three-week hypoxic as in the normoxic or in the 4h-hypoxic rats. Since β in vitro largely reflects the buffer value of hemoglobin, it is not surprising to see that this parameter was increased in the prolonged hypoxic rats which show a large increase in Hb concentration. In addition, we observed that, in three week hypoxic rats the volume of distribution of inulin (extracellular fluid volume) is decreased, plasma volume is unchanged, and red blood cell mass is increased (Gonzalez and Clancy, unpublished observations). These changes result in an increased intravascular-to-interstitial volume ratio which, for any given Hb concentration, will tend to increase the plasma βapp observed in vivo (Dell and Winters, 1970; Heisler, 1986).

We do not know whether chemical buffering of fluids other than plasma and red blood cells contributed to the high in vivo βapp of plasma of hypoxic rats. If the cellular chemical buffer value were increased in prolonged hypoxia, this could contribute to a higher plasma βapp. Souhrada and Bullard (1972) reported a low myocardial total chemical buffer value in prolonged hypoxic rats. We are not aware of determinations of chemical buffer value in other tissues of prolonged hypoxic rats.

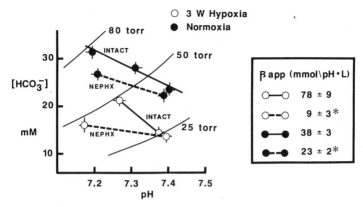

Fig. 5. Effect of nephrectomy on plasma βapp. Intact and bilaterally nephrectomized rats were exposed for 4h to different levels of CO_2 in the inspired air. Stars indicate a significant difference between intact and nephrectomized. Redrawn from data of Gonzalez and Clancy, 1986a.

2. Renal compensation

The higher βapp of hypoxic rats could be due, at least in part, to a more effective renal compensation in response to hypercapnia in hypoxic rats. To test this possibility we exposed hypoxic and normoxic rats to hypercapnia immediately after nephrectomy. Four hours of nephrectomy in normocapnic conditions did not result in significant plasma acid-base changes in either normoxic or hypoxic rats (Fig. 5). On the other hand, 4h of nephrectomy plus hypercapnia resulted in lower plasma pH and $[HCO_3^-]$ than those of intact rats exposed to comparable CO_2 levels. Therefore, plasma βapp was lower in nephrectomized than in intact rats. This was true for both normoxic and hypoxic rats; however, the effect of nephrectomy was more marked in the three week hypoxic rats, in which βapp decreased to a value below that of the normoxic rats. Thus, elimination of the renal compensation has a more profound effect in the prolonged hypoxic than in the normoxic rats.

A second approach to this problem was to determine the rate of renal acid excretion during hypercapnia in intact rats. Fig. 6 shows the total renal acid excretion (titratable acid plus NH_4^+ urinary excretion) of normoxic and three-week hypoxic rats during a two-hour normocapnic period and during two consecutive two-hour hypercapnic periods (1 and 2 in Fig. 6). Hypercapnia elicited an increase in total renal acid excretion above control levels in both groups. This increase, however, was approximately 70% larger in the hypoxic than in the normoxic rats.

The higher rate of renal acid excretion during hypercapnia in the hypoxic rats is consistent with the data in the nephrectomized rats and supports the idea that one of the causes of the high βapp of prolonged hypoxia is a more effective renal compensation of the acidosis in these animals. The mechanism responsible for this is not clear; a possible explanation is related to the acid-base composition of the extracellular fluid of prolonged hypoxia. Total acid excretion by the kidney during hypercapnia is not a linear function of the PCO_2: a given increase in PCO_2 elicits a larger increase in acid excretion at lower (less than 40 torr) than at higher PCO_2 values. Furthermore, for any given PCO_2, acid excretion is highest at low extracellular $[HCO_3^-]$ (Cohn and Steinmetz, 1980; Sasaki et al., 1983). Therefore, hypercapnia should result in a larger increase in total renal acid excretion in the prolonged hypoxic rats which show a lower initial PCO_2 and plasma $[HCO_3^-]$ than normoxic, or short term hypoxic rats.

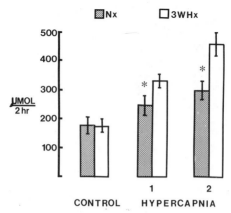

Fig. 6. Effect of hypercapnia on the rate of total acid excretion (titratable acid plus NH_4^+) by the kidneys of normoxic and three-week hypoxic rats. The height of the bars represents the number of micromoles of proton equivalents excreted per rat in a 2h period. In the control, CO_2 was absent from the inspirate. 1 and 2 represent two successive 2h periods where $PICO_2$ was elevated to 35-45 torr. Stars indicate a significant difference between normoxic and hypoxic. Redrawn from data of Widener et al., 1986.

In spite of a larger increase in acid excretion during hypercapnia, urine pH decreased less in the hypoxic rats, suggesting a high urine buffer value (Widener et al., 1986). Although the nature of the buffer(s) is not known, a high urine buffer value would favour a higher acid excretion in acidosis.

3. Transfer of acid-base equivalents between intra- and extracellular fluids

Hypercapnia in intact, normoxic rats is accompanied by a net transfer of base equivalents from cells into plasma (Lai et al., 1973). The source of this base is largely skeletal muscle; this exchange contributes to elevate plasma βapp, and to lower skeletal muscle βapp (Cechetto and Main-wood, 1978; Heisler and Piiper, 1972; Heisler, 1986). A higher net base equivalent efflux from skeletal muscle cells could contribute to the high plasma βapp of hypoxic rats. Indirect evidence, however, suggests that this is not the case. First, after nephrectomy in normoxic rats, plasma βapp increases as a function of duration of hypercapnia (Fig. 7). In the absence of renal compensation, this increase has been attributed to transfer of base equivalents from muscle to plasma (Lai et al., 1973). In the hypoxic, nephrectomized rats, on the other hand, we observed that plasma βapp actual-ly decreased as a function of duration of hypercapnia. Although other mechanisms could have participated, a decrease in plasma βapp with time is not consistent with a net movement of base equivalents from extrarenal sources into plasma during hypercapnia; rather, it would support a net base equivalent flux out of plasma. The second line of evidence stems from the skeletal muscle cell acid-base changes of hypercapnia. Under in vivo normoxic conditions, the net intra- to extracellular transfer of base equivalents that occurs in hypercapnia, results in a skeletal muscle cell βapp which is considerably lower than the chemical buffer value obtained by in vitro CO_2 equilibration of skeletal muscle cells (Cecheto and Mainwood, 1978; Heisler and Piiper, 1972; Heisler, 1986). If the transfer of base equivalents from cells to plasma were higher in hypoxia than in normoxia, it would be expected that skeletal muscle cell βapp would be lower in hypoxic than in normoxic rats. The opposite, however, is true: hypercapnia resulted

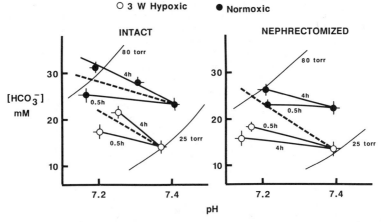

Fig. 7. Plasma acid-base changes at various times after elevating $PICO_2$ in intact and nephrectomized rats. The slope of dashed lines indicates the plasma non-bicarbonate buffer value obtained equilibrating blood _in vitro_. In the case of nephrectomized normoxic rats, the _in vitro_ buffer value coincided with that obtained _in vivo_ after 4h of hypercapnia. With the exception of the hypoxic, nephrectomized rats (right panel, open circles), _in vivo_ βapp increases with time of exposure to high CO_2. Redrawn from data of Gonzalez and Clancy (1986a).

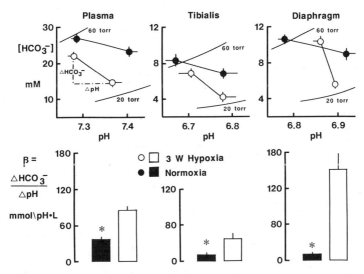

Fig. 8. Top: Acid-base composition of plasma, and of intracellular fluid of tibialis anterior and diaphragm muscles of normoxic and three week hypoxic rats exposed to different $PICO_2$ levels for 4h. The slopes of the lines connecting points at different PCO_2 values ($\Delta[HCO_3^-]/\Delta pH$) are represented in bar form in the bottom portion. Stars indicate a significant difference between normoxic and hypoxic. Redrawn from data of Gonzalez and Clancy, 1986a and 1986b.

in a higher, not lower, skeletal muscle cell βapp than that of normoxic rats (Fig. 8). Other mechanisms, including a higher chemical buffer value of hypoxic cells, could contribute to this result; however, the direction of the change is not consistent with a larger net transfer of base from cells to plasma in hypoxia than in normoxia. Furthermore, nephrectomy, which lowers plasma βapp more in hypoxia than in normoxia, also has a larger effect on skeletal muscle βapp of the hypoxic rats (Gonzalez and Clancy, 1986a,b). Perhaps the high chemical buffer value of the extracellular fluid combined with the lower interstitial fluid volume, plus the more effective renal compensation of hypoxic rats result in either a smaller loss of base by the skeletal muscle during hypercapnia, or in an actual uptake of base from the extracellular fluid. Alternatively, other cellular sources may contribute base equivalents to the extracellular fluid during hypercapnia.

The observation of an increased cell pH regulation in skeletal muscle in hypoxia is interesting, since it indicates that the acid-base changes are not restricted to the extracellular fluid. The underlying mechanism is not clear: the fact that the increased pH regulation is either abolished or limited by nephrectomy (Gonzalez and Clancy, 1986b) suggests a dependence of cell pH on extracellular acid-base composition (Boron et al., 1979).

In summary, prolonged hypoxia results in an improved ability to regulate plasma and skeletal muscle pH during hypercapnia. This seems to be the result of at least two factors: a higher concentration of blood buffers, largely hemoglobin - possibly combined with a smaller volume of distribution of the HCO_3^- generated by titration of non-bicarbonate buffers - and a more effective renal compensation in the hypoxic rats.

NON-RESPIRATORY ACIDOSIS

Although hypoxic rats show an increased capacity to regulate their plasma pH during hypercapnia, their response to non-respiratory acidosis may be different. First, bicarbonate, a very effective buffer for non-carbonic acids in open systems, is decreased in plasma and cellular fluids. Second, the ventilatory and renal responses of hypoxic rats to non-respiratory acidosis could be different from those to respiratory acidosis. Observations in humans exercising at high altitude suggest a lower capability to regulate plasma pH after acclimatization (Cerreteli, 1980).

Fig. 9 shows the results of experiments where 9 mmol/kg of a 0.3M solution of gluconic acid (pKa $\cong$ 3.7) were infused i.v. into intact, conscious, three-week hypoxic and normoxic rats over a period of 15 min. Acid infusion initially produced decreases in plasma pH, PCO_2 and $[HCO_3^-]$ in both groups; these decreases were followed by a recovery towards control values. The magnitude of the initial changes as well as the extent of the recovery were different in both groups. pH decreased more in the hypoxic rats after the infusion of acid. Also, recovery of pH was incomplete in this group: 4h after infusion, pH was still significantly lower than control. In contrast, pH had reached control values 3h after infusion in the normoxic rats.

The pattern of pH recovery was different in the hypoxic and in the normoxic rats. The slopes of the straight lines drawn through the control (A) and the 0.5h samples (B) are -7 and -18 mmol/pH.L for the normoxic and hypoxic rats, respectively. These are the plasma βapp obtained previously by us in nephrectomized normoxic and hypoxic rats made hypercapnic for 0.5h (Fig. 7). In these conditions - nephrectomy and short term hypercapnia -the acid-base regulatory mechanisms, with the exception of chemical buffering, are either absent or, due to the short CO_2 exposure, their effect is small. Since CO_2 equilibrates rapidly across body fluids, and transfer of proton equivalents across red cell and capillary membranes is faster than across cell membranes, plasma βapp obtained under these conditions is thought to represent largely the combined chemical non-bicarbonate buffer value of the red cells, plasma, and interstitial fluid (Brown and Clancy, 1965; Dell and Winters, 1970; Heisler, 1986; Lai et al., 1973).

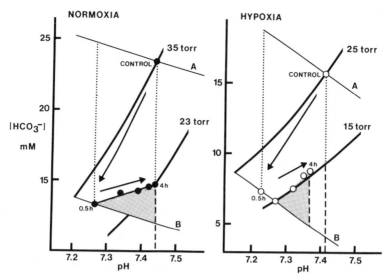

Fig. 9. Plasma acid-base changes after infusion of gluconic acid to normoxic and three week hypoxic rats. Each point represents the mean of 8 rats. Arterial blood samples were taken immediately before and at 0.5, 1, 2, 3 and 4 hours after acid infusion was finished. Infusion was completed in 15 min. The slopes of the lines labelled A and B were obtained by exposing nephrectomized normoxic and three-week hypoxic rats to 0.5h of hypercapnia (Fig. 7). The shaded areas illustrate the contribution of non-respiratory mechanisms to the correction of plasma pH after acid infusion. (Unpublished data.)

Line B represents the plasma acid base composition that would have occurred if the recovery of pH after the 0.5h sample would have been effected solely through a decrease in PCO_2 (respiratory compensation), and no renal compensation or exchange of acid-base equivalents with tissue cells had occurred. The shaded areas between the actual samples and line B reflect the contribution of non-respiratory mechanisms to the changes in plasma acid-base composition observed after the 0.5h sample. It is apparent that in normoxic rats both respiratory and non-respiratory components played a role in the return of plasma pH to normal: PCO_2 decreased continuously and plasma $[HCO_3^-]$ increased above what would have been predicted from a pure respiratory compensation. In the hypoxic group, on the other hand, after a marked decrease in PCO_2 in the 0.5 and 1h samples, plasma pH was corrected almost exclusively by non-respiratory mechanisms, since PCO_2 remained essentially constant between the 1 and 4h samples. This suggests that the hypoxic rats reached a maximal level of alveolar ventilation approximately one hour after the infusion of acid was terminated and this level remained constant thereafter. This contrasts with the data obtained in the normoxic rats, in which PCO_2 decreased until plasma pH reached control levels at the 3h sample. Failure to increase ventilation further in the hypoxic rats is probably responsible for the incomplete compensation of plasma pH observed in this group. The reason for the difference in the ventilatory response to metabolic acidosis is not clear. It is possible that the hypoxic rats may be unable to increase their ventilation further due to the already high level of ventilation secondary to the hypoxic stimulus. On the other hand it could be that the site responsible for the hyperventilatory stimulus during acidosis reached control or near-control pH values earlier in the hypoxic than in the normoxic rats, and, therefore, the stimulus for hyperventilation was reduced in the hypoxic rats.

The contribution of the non-respiratory component to the acid-base changes occurring after the 0.5h sample is represented by the vertical distance between any given sample and line B. This distance represents the mmol of base equivalent added to each liter of plasma by non-respiratory mechanisms of acid-base regulation after the 0.5h sample. The contribution is larger in the hypoxic rats; for the last sample it was 3.9±0.2 mM in the hypoxic and 2.7±0.3 mM in the normoxic rats. The mechanisms responsible for the non-respiratory compensation probably include a higher rate of renal acid excretion in the hypoxic rats, as suggested by the observations in respiratory acidosis reported above. Whether other mechanisms participate should be the subject of further study.

CONCLUSION

Our data indicate that prolonged hypoxia markedly affects the ability of intact rats to regulate their acid-base balance when challenged with an acid load. The patterns of acid-base regulation of short term hypoxia investigated by us were not strikingly different from those of normoxia. This suggests that the effect of prolonged hypoxia on the acid-base balance is not a direct effect of low PO_2 per se, but of the adaptive changes that result from prolonged hypoxic exposure. At least two acid-base regulatory mechanisms are affected by prolonged hypoxia: the concentration of chemical buffers, and the rate of renal acid excretion. The effect of hypoxia on the former is such that the net result depends on the type of acid load introduced. The elevated non-bicarbonate buffer value results in an increased buffer capacity for carbonic acid; the decrease in plasma and cell HCO_3 concentration, on the other hand, represents a limitation when the acid load is in the form of a non-carbonic acid. This may explain the larger initial pH decrease in the hypoxic rats after infusion of gluconic acid. The more effective renal compensation should result in faster excretion of excess acid, and, therefore, in an earlier correction of the acid-base disturbance in hypoxic rats. The mechanism of the increased renal compensation of acidosis, as well as the contribution of other possible factors to the changes in acid-base regulation in hypoxia, should be the subject of further study.

ACKNOWLEDGMENTS

The research reviewed here was supported by grants from the American Heat Association, Kansas Affiliate.

REFERENCES

Bouverot, P. (1985). Zoophysiology. Vol. 16. Adaptation to altitude hypoxia in vertebrates. Berlin, Springer-Verlag.

Boron, W.F., W.C. McCormick and A. Roos (1979). pH regulation in muscle barnacle fibers: dependency on intracellular and extracellular pH. Am. J. Physiol. 237:C185-C193.

Brown, E.B., Jr. (1971). Whole body buffer capacity. In: Ion Homeostasis in Brain. Ed. B.K. Siesjo and S.C. Sorensen, Copenhagen, Munskgaard.

Brown, E.B., Jr. and R.L. Clancy (1965). In vivo and in vitro CO_2 blood buffer curves. J. Appl. Physiol. 20:885-889.

Cechetto, D. and G.W. Mainwood (1978). Carbon dioxide and acid-base balance in isolated rat diaphragm. Pfluegers Arch. 376:251-258.

Cerretelli, P. (1980). Gas exchange in high altitude. In: Pulmonary Gas Exchange, Vol. II: Organism and Environment. Ed. J.B. West, New York, Academic Press, pp 97-147.

Cohn, L.H. and P.R. Steinmetz (1980). Control of active proton transport in the turtle urinary bladder by cell pH. J. Gen. Physiol. 76:381-393.

Davenport, H.T., P.A. Auld, P. Sekelj, W. Jegier and M. McGregor (1964). Hypercarbia during halothane anesthesia with neuromuscular block. Anesthesiology 25:307-311.

Dell, R.B. and R.W. Winters (1970). A model for the in vivo CO_2 equilibration curve. Am. J. Physiol. 219:37-44.

Freeman, F.H. and W.O. Fenn (1953). Changes in carbon dioxide stores of rats due to atmospheres low in oxygen or high in carbon dioxide. Am. J. Physiol. 174:422-430.

Gonzalez, N.C. (1986). Acid-base regulation in mammals. In: Acid-base Regulation in Animals. Ed. N. Heisler. Amsterdam, Elsevier, pp. 175-202.

Gonzalez, N.C. and R.L. Clancy (1986a). Acid-base regulation in prolonged hypoxia: effect of increased PCO_2. Respir. Physiol. 64:213-227.

Gonzalez, N.C. and R.L. Clancy (1986b). Intracellular pH regulation during prolonged hypoxia in rats. Respir. Physiol. 65:331-339.

Heisler, N. and J. Piiper (1972). Determination of intracellular buffering properties in rat diaphragm muscle. Am. J. Physiol. 222:747-753.

Heisler, N. (1986). Buffering and transmembrane ion transfer processes. In: Acid-Base Regulation in Animals, Ed. N. Heisler, Amsterdam, Elsevier, pp. 3-47.

Lai, Y.L., E. Martin, B. Attebery and E.B. Brown, Jr. (1973). Mechanisms of extracellular pH adjustment during hypercapnia. Respir. Physiol. 19:107-114.

Pryss-Roberts, C., G. Kelman and J.F. Nunn (1968). Determination of the "in vivo" carbon dioxide titration curve of anesthetized man. Br. J. Anesthesiol. 38:500-509.

Sasaki, S., C.A. Berry and F.C. Rector (1983). Effect of luminal and peritubular HCO_3^- concentration and PCO_2 on HCO_3^- reabsorption in rabbit proximal convoluted tubuli perfused in vitro. J. Clin. Invest. 70:639-649.

Souhrada, J.F. and R.W. Bullard (1972). Adaptation to hypobarism: sensitivity of myocardial tissue to carbon dioxide. J. Appl. Physiol. 32:501-505.

Widener, G., L.P. Sullivan, R.L. Clancy and N.C. Gonzalez. (1986). Renal compensation of hypercapnia in prolonged hypoxia. Respir. Physiol. 65:341-350.

THE EFFECT OF PROLONGED HYPOXIA ON STRIATED MUSCLE pH REGULATION

R.L. Clancy, C. Gernon, D.P. Valenzeno and N.C. Gonzalez

Department of Physiology
University of Kansas Medical Center
Kansas City, KS 66103

INTRODUCTION

The extracellular acid-base status of mammals exposed to low inspired PO_2 for prolonged periods of time has been studied extensively (Bouverot, 1985). However, less is known regarding the intracellular acid-base status during prolonged hypoxia. In a previous study we observed that the intracellular pH of striated muscles of rats subjected to 3 weeks of hypobaric hypoxia was not significantly different from that of normoxic rats (Gonzalez and Clancy, 1986b). However, when the cells were acid-loaded by rendering the animals hypercapnic, the decreases in intracellular pH of tibialis anterior, quadriceps and diaphragm were significantly less than in hypercapnic normoxic rats. The magnitude of cell pH regulation, as measured by the apparent non-bicarbonate buffer value, ranged from 3 (tibialis anterior) to 12 (diaphragm) times greater in the hypobaric hypoxic than the normoxic animals. This augmentation of striated muscle pH regulation during hypercapnia in hypobaric hypoxic rats could result from increased physicochemical buffering and/or increased efflux of proton equivalents. The question arises as to whether augmentation of these cellular pH-regulating mechanisms resulted from intrinsic changes in the cell interior and/or the sarcolemma engendered by the prolonged hypoxia or whether extra-cellular factors which enhance these cellular pH-regulating mechanisms were increased to a greater extent in the hypercapnic hypoxic rats than in the hypercapnic normoxic rats. To resolve this question cell pH regulation of diaphragms from prolonged hypoxic and normoxic rats was studied under in vitro conditions. Under these conditions the contribution of extracellular factors to cell pH regulation was the same in diaphragms from hypoxic and normoxic rats. Therefore, any difference in cell pH regulation should be attributable to intrinsic changes in one or more of the cell pH-regulating mechanisms.

METHODS

Male Sprague-Dawley rats (200-225 g) were used for this study. Hypobaric hypoxia was produced by placing rats in a chamber maintained at 370-380 torr. Normoxic rats were housed in a similar chamber maintained at ambient barometric pressure (740-750 torr). Normoxic rats were given the same amount of standard rat chow as consumed by the hypoxic rats to produce equal weight gains in the two groups. After three weeks of hypobaric

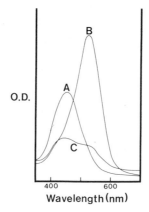

O.D.

400   600

Wavelength(nm)

Fig. 1. Optical density of neutral red dye solution as a
function of wavelength and pH. A:pH 7.7; B:pH 7.0, C:pH 7.4.
Decreasing the pH results in a decrease in the ratio of the
optical density at 453 to 526.5 nm.

exposure the hypoxic and normoxic rats were anesthetized with methoxy-
flurane and the diaphragms excised. One half of each diaphragm was
incubated in a modified Ringer solution while the other half was incubated
in modified Ringer solution containing 25 μM neutral red, a pH sensitive
dye. Following incubation, both hemidiaphragms were mounted in a lucite
chamber through which Ringer solution maintained at 37°C was perfused at 25
ml/min. These solutions did not contain neutral red dye. The muscle
chamber was positioned in a spectrophotometer and optical densities of the
hemidiaphragm at 526.5 and 453 nm were determined at 40 sec intervals. The
spectrophotometer was programmed to subtract the optical densities at these
two wavelengths of the undyed muscle from those of the dyed muscle. The
resulting signal was largely a measure of the optical densities of neutral
red dye in the diaphragm. The spectrophotometer was interfaced with a
computer programmed to determine the ratio of the optical densities at 453
and 526.5 nm. The optical density ratio was displayed as a function of time
on a video monitor and stored for subsequent printed display. Fig. 1 shows
the absorption spectra of a 25 μM solution of neutral red at pH 7.4 (C), and
at acid (B) and alkaline (A) pH. At acid pH the maximum absorption peak is
526.5 nm, while at alkaline pH it is at 453 nm. As pH increases there is a
progressive fall of the absorbance at 526.5 nm and a concommitant increase
in absorbance at 453 nm. Consequently, increasing the pH results in an
increase in the optical density ratio at 453 to 526.5 nm. Thus, although
the absolute optical density at any given wavelength is the result of
multiple interactions of the incident light with the preparation, the ratio
of optical densities at 453 and 526.5 nm is a function of pH only.

The experimental protocol for hemidiaphragms from both hypoxic and
normoxic rats is illustrated in Fig. 2. Shown are the optical density
ratios measured at 40 second intervals. Initially the hemidiaphragm was
superfused with a Ringer solution containing 12.6 mM HCO_3^- and equilibrated
with 3% CO_2 - 97% O_2. The resulting pH was 7.39. This was followed by
superfusion with a 48 mM HCO_3^- Ringer solution equilibrated with 10% CO_2 -90%
O_2. The pH of this solution was also 7.39. Increasing the CO_2 concentration results in acid loading of the cells as a result of intracellular
generation of carbonic acid. After 30 min of superfusion at the increased
CO_2 concentration, the perfusate was changed to that used before acid
loading. Since the pH of the perfusate was maintained constant throughout
the experiment, changes in the optical density ratio should result primarily from changes in the absorbance of neutral red dye located in the cells
and hence reflect changes in intracellular pH.

314

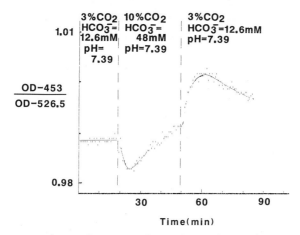

Fig. 2. Experimental protocol. The acid-base conditions of the perfusing solutions are shown. Hemidiaphragms were acid loaded for 30 minutes by increasing the CO_2 concentration of the perfusate from 3 to 10%. Optical density ratios were recorded at 40 second intervals.

In other experiments, the Ringer solutions contained 1 mM amiloride or 1mM amiloride and 1mM SITS (4-acetamido-4-isothiocyanatostilbene-2-2 disulfonic acid). These agents were used to attenuate transmembrane Na^+/H^+ exchange (amiloride) and HCO_3^-/Cl exchange (SITS).

Calibration of the optical density ratio in terms of cell pH was carried out in the following manner. Following completion of the above experimental protocol the hemidiaphragms were superfused with a modified Ringer solution containing 112 mM K^+ and 5 mg/L nigericin. Nigericin is an ionophore which mediates H^+/K^+ exchange; under these conditions the cell membrane becomes highly permeable to protons resulting in equilibration of intracellular and extracellular pH (Chaillet and Boron, 1985). The hemidiaphragms were superfused with high K^+-nigericin solutions having pHs ranging from 6.8 to 7.6. The data from a representative experiment are shown in Fig. 3. The relationship between the optical density ratio and perfusate pH was found to be linear by least square analysis. The changes in the OD ratio during and following acid loading (Fig. 2) were divided by slope of the OD ratio vs pH relationship (Fig. 3); this provided a measure of the cell pH changes occurring during and following acid loading. The computer was programmed to determine the OD-pH relationship and convert the OD ratios obtained during the experimental procedures to pH_i values.

RESULTS

Effect of acid loading on cell pH

Data obtained from a hemidiaphragm from a normoxic rat are shown in Fig. 4. In the left and right panels, labeled C, the hemidiaphragm was superfused with a 12.6 mM HCO_3^- Ringer solution equilibrated with 3% CO_2 97% O_2. The pH of the solution was 7.39. In the center panel, the perfusate HCO_3^- concentration was 48 mM. The perfusate was equilibrated with 10% CO_2 – 90% O_2 resulting in a pH of 7.39. Superfusion with the 10% CO_2 solution resulted in a decrease of cell pH of 0.15 pH units in approximately 4.5 min. This decrease was followed by a return toward the control cell pH at an initial rate (designated by the dashed lines) of 0.009 pH units/min as determined by least square analysis. Cell pH continued to increase at a lower rate and at the end of 30 min of acid loading was almost equal to the

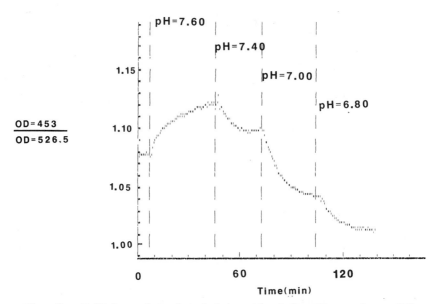

Fig. 3. Cellular neutral red dye optical density ratio at 453
to 526.5 nm as a function of cell pH. The hemidiaphragm was
superfused with a modified Ringer solution containing 112 mM
K^+ and 5 mg/L nigericin. Under these conditions cell pH is
equal to the pH of the Ringer solution.   pH's of the Ringer
solution are shown in each panel. The slope of pH vs optical
density ratio was determined using linear regression analyses.

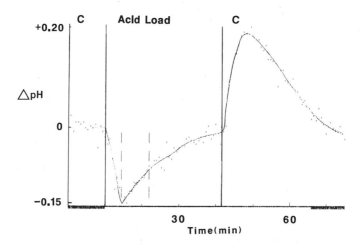

Fig. 4. Cell pH changes in a hemidiaphragm from a normoxic
rat. See Fig. 2 for description of experimental protocol. pH
of the perfusion solutions was 7.39.

control cell pH. Decreasing the CO_2 concentration (C) was accompanied by a
rapid increase in cell pH which became 0.19 pH units higher than that prior
to acid loading. Cell pH then decreased and 20 minutes later was equal to
that prior to acid loading.

Data from a hemidiaphragm excised from a 3 week hypoxic rat are summarized in Fig. 5. The composition of the Ringer solution was the same as that
in Fig. 4. During the period of acid loading, cell pH decreased 0.24 pH
units over a period of 7 minutes. Cell pH then increased, the initial rate

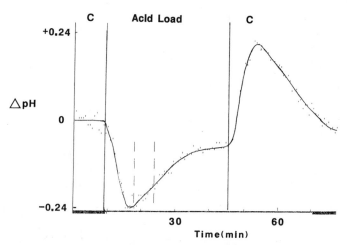

Fig. 5. Cell pH changes in a hemidiaphragm from a 3 week hypobaric hypoxic rat. See Fig. 2 for description of experimental protocol. pH of the perfusion solutions was 7.38.

	1		2		3	
	H	N	H	N	H	N
MEAN	−0.13	−0.11	0.009	0.009	+0.12	+0.11
SEM	±0.03	±0.02	±0.002	±0.002	±0.03	±0.02
N	11	13	11	13	11	13

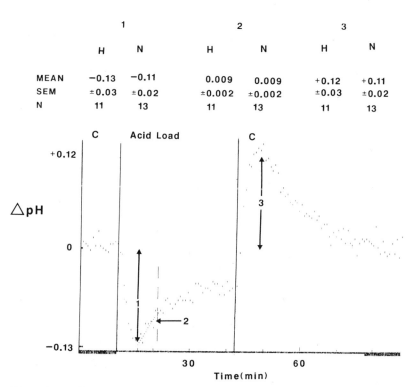

Fig. 6. Effect of acid-loading rat hemidiaphragms. 1 = Initial cell pH change, 2 = Maximum rate of cell pH recovery (pH units/min), 3 = Maximum cell pH change upon removal of the acid load. H = hypobaric hypoxic hemidiaphragms. N = normoxic hemidiaphragms, SEM = Standard Error of the Mean, N = Number of hemidiaphragms.

of increase being 0.008 pH units/min. Following 30 min of acid loading cell
pH was only 0.06 pH units lower than before acid loading. Removal of the
acid load, by decreasing the CO_2 concentration, was accompanied by a rapid
increase in cell pH, the maximum value being 0.23 units greater than the
control cell pH. Cell pH then decreased becoming equal to the pre-acid
loading pH approximately 22 minutes later.

The results from 13 three week hypobaric hypoxic and 11 normoxic rat
hemidiaphragms are shown in the upper portion of Fig. 6. The columns of
data correspond to the numbered cell pH changes shown in the lower portion
of the figure. The experimental conditions were the same as those in Figs.
4 and 5. The initial decreases in cell pH accompanying acid loading (Column
1) in the hemidiaphragms from hypoxic (H) and normoxic (N) rats were −0.13
and −0.11 pH units respectively. These decreases were not significantly
different. The initial rate of cell pH recovery during the acid load was
the same in the hypoxic and normoxic diaphragms (column 2). Removal of the
acid load was accompanied by peak increases in cell pH of +0.12 and +0.11 pH
units in the hypoxic and normoxic diaphragms (column 3). These increases
were not significantly different.

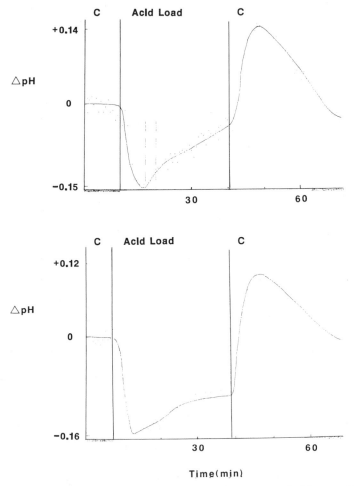

Fig. 7. Effect of amiloride on cell pH regulation. Upper
panel: Cell pH changes during and following acid-loading in a
hemidiaphragm from a 3 week hypoxic rat. Lower panel: Cell pH
changes in the same hemidiaphragm superfused with Ringer
solution containing 1 mM amiloride.

318

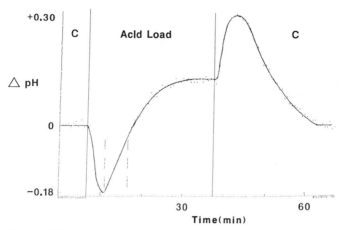

Fig. 8. Cell pH changes in 3 week hypobaric hypoxic hemi-diaphragm. See Fig. 2 for description of experimental protocol. pH of the perfusing solutions was 7.37.

Mechanisms of cell pH recovery during acid loading

The recovery of cell pH following the initial decrease in cell pH during acid loading could be the result of an efflux of H^+ coupled to Na^+ influx and/or an exchange of extracellular HCO_3^- for intracellular Cl^- (Roos and Boron, 1981). The contribution of the Na^+/H^+ antiporter to cell pH recovery during acid loading was determined by superfusing hemidiaphragms with Ringer solutions containing 1 mM amiloride. Hemidiaphragms were superfused with Ringer-amiloride solution for 1 hour before acid loading as well as during and following acid loading. The data from an experiment using a hemidiaphragm from a hypoxic rat are shown in Fig. 7. The upper graph shows the cell pH changes during and following acid loading in the absence of amiloride. The lower graph illustrates the cell pH changes in the presence of 1 mM amiloride. The initial decrease in cell pH during acid loading was not affected by amiloride. However, the initial rate of cell pH recovery was decreased from 0.011 to 0.002 pH units/min by amiloride. In 4 hypoxic and 3 normoxic hemidiaphragms, amiloride decreased the initial rate of cell pH recovery by 50 and 40%, respectively. Amiloride treatment also attenuated the increase in cell pH following removal of the acid load by 44% in hypoxic and 49% in normoxic hemidiaphragms.

The combined contribution of Na^+/H^+ plus HCO_3^-/Cl^- exchange to cell pH recovery was determined by superfusing hemidiaphragms with Ringer solutions containing 1 mM amiloride and 1 mM SITS. Cell pH changes in a hemidiaphragm from a hypoxic rat before amiloride and SITS treatment are presented in Fig. 8. Fig. 9 illustrates the cell pH changes in the same hemidiaphragm superfused with amiloride-SITS Ringer for 1.5 hour before acid loading as well as during and following acid loading. Prior to superfusion with amiloride and SITS (Fig. 8) the initial rate of cell pH recovery during acid loading was 0.035 pH units/min. In the presence of amiloride and SITS, the initial rate of cell pH recovery was only 0.007 pH units/min. The extent of cell pH recovery following 30 min of acid loading was decreased 77% in the presence of amiloride and SITS. In 3 hypoxic preparations these blocking agents decreased the initial rate of cell pH recovery by 83% and the overshoot in cell pH following acid loading by 66%. In 3 normoxic preparations, these decreases were 67 and 51%, respectively. Thus the degree of attenuation of cell pH recovery by amiloride and SITS during acid loading appeared to be comparable in hypoxic and normoxic hemidiaphragms.

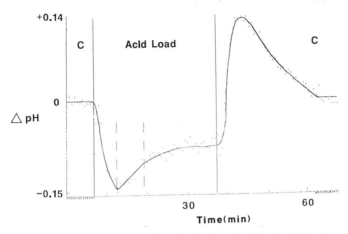

Fig. 9. Cell pH changes in 3 wk hypobaric hypoxic hemidiaphragm superfused with Ringer solution containing 1 mM
amiloride and 1 mM SITS. The effect of acid loading on this
hemidiaphragm preparation before adding amiloride and SITS is
shown in Fig. 8.

DISCUSSION

The present study was undertaken to determine if the enhanced cell pH
regulation accompanying prolonged hypoxia in vivo is due to intrinsic
changes in chemical buffering and/or H^+ and $\overline{HCO_3}$ fluxes or if it is due to
the differences in the extracellular acid-base status during acid loading
in hypoxia. This was accomplished by making the contribution of the extracellular fluid to cell pH regulation during acid loading the same in hypoxic
and normoxic diaphragms.

The magnitude of the initial decrease in cell pH accompanying acid
loading is principally determined by the type and quantity of non-
bicarbonate buffers, i.e., primarily proteins and organo-phosphates. This
initial decrease in cell pH was not different in normoxic (-0.11) and
hypoxic (-0.13) hemidiaphragms. These results indicate that prolonged
hypoxia does not result in an increase in chemical non-HCO_3 buffer value.
Thus the enhanced cell pH regulation of hypoxic diaphragms in vivo does not
appear to be due to an increase in chemical non-HCO_3 buffer value.
Souhradra and Bullard (1972) reported that the total (HCO$_3$ plus non-HCO$_3$)
buffer value was decreased in cardiac muscle during prolonged hypoxia. The
decrease in total buffer value could have resulted from a decrease in HCO_3,
without any change in non-HCO_3 buffer value. To our knowledge, there are no
published reports of the effect of prolonged hypoxia on the chemical non-
HCO_3 buffer value of skeletal muscle.

In both normoxic and hypoxic hemidiaphragms, the initial decrease in
cell pH accompanying acid loading was followed by a partial recovery. This
partial recovery of cell pH during acid loading has been observed in a
variety of tissues (Roos and Boron, 1981). One mechanism accounting for
cell pH recovery is an increase in the activity of the Na^+/H^+ antiporter
resulting from an increase in the intracellular proton concentration
(Grinstein and Rothstein, 1986). A second possible mechanism is an influx
of HCO_3 in exchange for Cl^- resulting from an increase in the transmembrane
HCO_3 concentration gradient during acid loading. A measure of the maximum
rate at which these ionic fluxes occurred was obtained by determining the
maximum rate of cell pH recovery. This was found to be the same (0.009 pH
units/min) in the hypoxic and normoxic hemidiaphragms.

The overshoot in cell pH above the initial control value, observed after removing the acid load, is a function of the extent of pH recovery during acid loading. The overshoot values for the hypoxic and normoxic diaphragms were 0.12 and 0.11 pH units, respectively. These data indicate that the extent of pH recovery during the 30 minute period of acid loading was not different in the hypoxic and normoxic hemidiaphragms.

The experiments employing the Na^+/H^+ antiporter blocking agent, amiloride, indicate that a significant portion of cell pH recovery during acid loading was attributable to this ion exchange mechanism. The observation that the attenuation of cell pH recovery by amiloride was comparable in both hypoxic and normoxic hemidiaphragms suggests that the activity of the Na^+/H^+ antiporter is not affected by prolonged hypoxia.

Superfusion of hemidiaphragms with both SITS and amiloride attenuated the degree of cell pH recovery to a somewhat greater extent than amiloride alone. These results indicate that a HCO_3^-/Cl^- exchange contributes to cell pH recovery during acid loading. However, if these agents completely blocked Na^+/H^+ and HCO_3^-/Cl^- exchange and these were the only ionic fluxes affecting cell pH there would be no pH recovery during acid loading, or "overshoot" in cell pH upon removing the acid load. The observation that cell pH recovery and overshoot were still evident suggests that other factors were operative. One possibility is a change in transmembrane HCO_3^- flux during acid loading whch is not inhibitable by SITS. If during acid loading the transmembrane electrical potential did not change appreciably, the electrochemical gradient for HCO_3^- efflux would be decreased as a result of increasing the extracellular HCO_3^- concentration from 12.6 to 48 mM. This could result in a HCO_3^- influx, which is not inhibited by SITS.

In previous experiments we observed that in vivo cell pH regulation during acid loading is greater in diaphragms of hypoxic than normoxic rats. It is possible that in vivo acid loading by rendering the animals hypercapnic is accompanied by an increase in one or more hormones which augment Na^+/H^+ and/or HCO_3^-/Cl^- exchange. If the hormonal increases are greater in hypoxic than normoxic rats, cell pH regulation would be greater in hypoxic rats. It is also possible that the in vivo observations are attributable to the relative changes in the extracellular acid-base status of hypoxic and normoxic rats. In vivo the increment in extracellular H^+ concentration accompanying hypercapnia is less in hypoxic than normoxic rats, while the increment in extracellular HCO_3^- is greater in hypoxic than normoxic rats. These differences between hypoxic and normoxic rats are a result of 1) an increased chemical buffering of the blood associated with an increased hemoglobin concentration in hypoxic rats and 2) a greater renal compensation during hypercapnia in hypoxic rats (Gonzalez and Clancy, 1986a; Widener et al., 1986). The lower extracellular H^+ concentration in the hypoxic rats could result in a smaller decrease in cell pH by virtue of a greater H^+ efflux via the Na^+/H^+ exchanger or a smaller H^+ influx resulting from a smaller increase in the electrochemical gradient for H^+ influx. Similarly, the higher extracellular HCO_3^- concentration in the hypoxic rats could result in a greater HCO_3^- influx via the HCO_3^-/Cl^- exchanger or a smaller HCO_3^- efflux accompanying a decrease in the electrochemical gradient for HCO_3^- efflux.

Support for the greater cell pH regulation during hypoxia under in vivo conditions being due to changes in the extracellular acid base status was obtained in experiments where nephrectomized rats were made hypercapnic (Gonzalez and Clancy, 1986b). Following nephrectomy, cell pH regulation in hypoxic diaphragms during hypercapnia was markedly decreased. Concomitantly, the increase in extracelluar H^+ concentration was much greater than in non-nephrectomized hypoxic rats. Likewise the increase in extracellular HCO_3^- concentration was much less in the nephrectomized than the non-nephrectomized rats. Thus it appears that the degree of cell pH regulation during acid loading in vivo is directly related to the degree of acid base regulation of the extracellular fluid.

In summary, the results of this study indicate that prolonged hypobaric hypoxia does not directly produce intrinsic changes in the cell interior and/or the sarcolemma mechanisms involved in cell pH regulation. Rather, it appears that changes in the extracellular environment augment intrinsic cell pH regulation during prolonged hypobaric hypoxia.

REFERENCES

Bouvert, P. (1985). Adaptation to altitude hypoxia in vertebrates. Zoophysiology, Vol. 16, Berlin, Springer-Verlag, 176 p.

Chaillet, J.R. and W.F. Boron (1985). Intracellular calibration of a pH-sensitive dye in isolated perfused salamander proximal tubules. J. Gen. Physiol., 86:765-794.

Gonzalez, N.C. and R.L. Clancy (1986a). Acid-base regulation in prolonged hypoxia: effect of increased PCO_2. Respir. Physiol., 64:213-227.

Gonzalez, N.C. and R.L. Clancy (1986b). Intracellular pH regulation during prolonged hypoxia in rats. Respir. Physiol., 65:331-339.

Grinstein, S. and A. Rothstein (1986). Mechanisms of regulation of the Na^+/H^+ exchanger. J. Membrane Biol., 90:1-12.

Roos, A. and W.F. Boron (1982). Intracellular pH. Physiol. Rev., 61:297-434.

Souhrada, J.F. and R.W. Bullard (1972). Adaptation to hypobarium: sensitivity of myocardial tissue to carbon dioxide. J. Appl. Physiol. 32:501-505.

Widener, G., L.P. Sullivan, R.L. Clancy and N.C. Gonzalez (1986). Renal compensation of hypercapnia in prolonged hypoxia. Respir. Physiol. 65:341-350.

HCl INFUSION STIMULATES THE RELEASE OF A SUBSTANCE FROM THE BLOOD

WHICH ALTERS BREATHING AND BLOOD PRESSURE

James A. Orr, Hashim Shams*, M. Roger Fedde** and P. Scheid*

University of Kansas, Lawrence, KS
*Ruhr-Universitat Bochum, Federal Republic of Germany
**Kansas State University, Manhattan, KS

INTRODUCTION

Intravenous infusion of strong acids, such as hydrochloric acid, has been previously used by various investigators as a means to lower blood pH in the study of respiratory or cardiovascular responses to acidosis. A previous report from our laboratory has demonstrated that infusion of HCl at a constant circulating blood pH causes the release of a substance from the blood that affects breathing and blood pressure (Orr et al., 1987). Repeated infusion of HCl in the same animal failed to elicit the response, either because the causative agent was depleted during the initial infusion or because the animal became refractory to the agent.

This previous report has revealed, for the first time, that respiratory and blood pressure changes during an acidosis induced by the infusion of HCl are not solely attributable to lowering of circulating blood pH and stimulation of chemoreceptors, but are in part due to the release of a substance from the blood at the site of the acid infusion. The experiments described below are a continuation of the investigation of this strong acid-blood response. We tested the hypothesis that pre-treatment of the animal with a cyclooxygenase inhibitor (indomethacin) would attentuate the respiratory and hemodynamic response to HCl infusion. This hypothesis is based on the premise that the infusion of a strong acid, such as HCl, into the blood stream stimulates platelets to release thromboxane, perhaps at the tip of the infusion catheter where the strong acid meets the blood. Thromboxane is an arachidonic acid metabolite with known vasoconstrictor and platelet-aggregating properties (Moncada et al., 1985)

METHODS

Five cats of either sex were anesthetized with a mixture of chloralose and urethane, and an extracorporeal loop was installed from a femoral artery to a femoral vein. Into this loop, acid (0.25 M HCl), and 10 cm downstream, base (0.25 M NaOH) were infused for 10 min at a rate of 0.75 ml/min resulting in a dose of about 1.0 mmol of salt/kg of body weight (Fig. 1). Right ventricular systolic blood pressure (RVBP), mean systemic arterial blood pressure (SAP), respiratory rate and tidal volume were measured continuously.

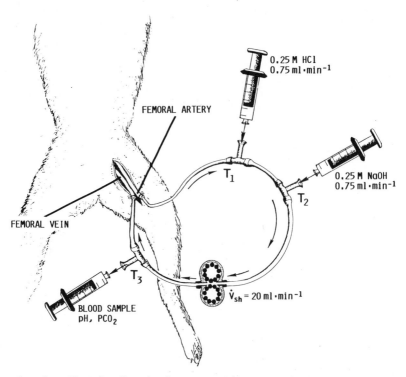

Fig. 1. Sketch of animal preparation. Ports T_1 and T_2 were used for infusion of acid and base respectively. Port T_3 served as a sampling site for shunt loop blood. $\dot{V}_{sh}$, shunt loop blood flow. (Figure is taken from Orr et al., 1987 with permission of the American Physiological Society.)

The results of these experiments were compared to data collected from a second group of cats (n = 5) that were pretreated with the cyclooxygenase inhibitor, indomethacin. Indomethacin was administered intravenously at a dose of 2.5 mg/kg 30 minutes prior to the extracorporeal acid-base infusion.

RESULTS

In the control (non-treated) animals, a dramatic increase in RVBP and fall in mean systemic arterial blood pressure was observed 2-3 minutes after the start of the acid-base infusion. On the average, RVBP increased from a control value of 25 mmHg to 53 mmHg and SAP fell from a control value of 136 mmHg to 88 mmHg. These alterations in blood pressure were accompanied by increases in respiratory rate and decreases in tidal volume (Table I).

These responses in blood pressure and respiration were substantially attenuated in the animals pre-treated with indomethacin. In these animals, the increase in RVBP and fall in SAP were only 20% of the change recorded in the control animals. The diminution of the respiratory changes was even more dramatic, in that respiratory rate was unchanged in the indomethacin treated animals and tidal volume increased slightly (as opposed to decreasing) during acid-base infusion (Table I).

Table I. Cardiopulmonary response to acid-base infusion with and without inhibition of prostaglandin/thromboxane synthesis

| | Group I (Control) | | Group II (Indomethacin-treated) | |
	Before Infusion	During Infusion	Before Infusion	During Infusion
Right Ventricular blood pressure (mmHg)	25.2±0.8	52.5±4.0	26.7±2.7	33.5±1.8
Systemic arterial blood pressure (mmHg)	136.7±4.2	88.4±8.6	133.0±11	119.0±16
Tidal Volume (ml)	34.3±2.9	13.8±3.4	41.5±6.7	50.2±8.7
Respiratory rate (min^{-1})	18.5±3.2	52.2±11.4	21.6±4.0	22.9±4.0

Values are mean ± standard error of the mean for 5 animals in each group. The cats in Group II were pretreated with indomethacin (2.5 mg kg^{-1}) 30 minutes prior to the acid-base infusion.

CONCLUSION

These results are consistent with the hypothesis that acid-base infusion stimulates platelets to release an arachidonic acid metabolite, probably thromboxane, that alters blood pressure and affects respiration. The data provide further evidence that infusion of strong acids, such as HCl, leads to respiratory and hemodynamic responses that are not entirely due to reductions in circulating blood pH and consequent stimulation of known chemoreceptors. It is unknown from these data whether the arachidonic acid metabolite released from the blood is thromboxane, but this seems a likely candidate since platelets are known to release this agent when stimulated (Moncada et al., 1985). It is also unknown from these experiments whether other vasoactive substances known to be present in platelets (e.g., serotonin) are also released during this phenomenon. Answers to these questions await further investigation.

REFERENCES

Orr, J.A., H. Shams, M.R. Fedde and P. Scheid. (1987). Cardiorespiratory changes during HCl infusion unrelated to decreases in circulating blood pH. J. Appl. Physiol. 62:2362-2370.
Moncada, S., R.J. Flower and J.R. Vane. (1985). Prostaglandins, prostacyclin, thromboxane A$_2$ and leukotrienes. In: Goodman and Gilman's The Pharmacological Basis of Therapeutics. Eds: A.G. Gilman, L.S. Goodman, T.W. Rall and F. Murad. Seventh Ed. MacMillan Pub. Co.: New York, pp. 660-673.

OXYGEN TRANSFER IN THE TRAINED AND UNTRAINED QUARTER HORSE

H.H. Erickson, W.L. Sexton, B.K. Erickson and J.R. Coffman

College of Veterinary Medicine
Kansas State University
Manhattan, KS 66506

INTRODUCTION

The recognition of the horse as a natural athlete has led to increased interest in equine sports medicine, exercise physiology, new training methods, and factors improving the performance of horses (Engelhardt, 1977; Milne et al., 1977; Bayly et al., 1983b; Persson et al., 1983). A considerable amount of research has been performed in recent years in order to describe the physiological responses of the horse to exercise and training (for reviews see Engelhardt, 1977; Physick-Sheard, 1985; Rose, 1985). Most of these studies, however, have examined the effects of acute exercise training rather than long-term, strenuous exercise training similar to that of a competitive performance horse. Furthermore, there are very few studies that describe the equine athlete's response to detraining, how soon it starts, how much of it occurs, and how rapidly a horse can be retrained and placed back in competition. The purpose of this study was to: (1) evaluate selected cardiopulmonary and metabolic responses of the Quarter Horse to a standard exercise (treadmill) test during a 25-week program of endurance and interval training and (2) to examine the effects of an extended period of detraining and subsequent retraining on exercise performance.

METHODS

Five 2-year old Quarter Horses were used in the study. A standard exercise test (SET) was performed by each horse prior to training and periodically during training (Sexton et al., 1987), detraining, and retraining (Erickson et al., 1987). The SET consisted of four 5-minute increments of increasing running speed (1.0 to 2.8 m/sec) on a 12% grade. The SET was performed by each horse prior to the start of training, following endurance training (13 weeks), after 6 and 12 weeks of combined endurance and interval training, at 6-week intervals during detraining, and after 10 and 18 weeks of retraining. During the exercise test, arterial pressure, right ventricular pressure and dP/dt, pulmonary artery pressure and flow velocity, electrocardiogram, blood temperature, and heart rate were recorded continuously. Arterial blood samples were taken at the end of each work level for blood gas, pH, lactate, hemoglobin, and hematocrit analyses.

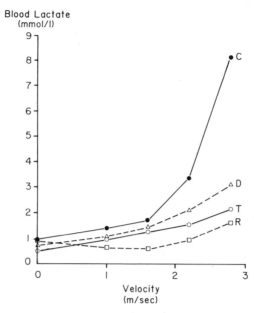

Fig. 1. Effects of initial training (T), detraining (D) and retraining (R) on blood lactate concentration in Quarter Horses during a standard exercise test on a treadmill at a 12% grade. (C) represents the control response before training.

The total training program lasted 174 days, including 90 days of long, slow, distance (LSD) endurance work followed by 84 days of combined LSD and interval training. Running speed was regulated by a rider using an on-board heart rate computer (Equine Biomechanics, Unionville, PA) to keep heart rate within a prescribed range. Endurance training was performed at a slow gallop with heart rate maintained at 140-150 b/min, beginning with 10 min/day and increasing to 20 min/day by day 70. The first 42 days of interval training consisted of gradual acceleration to a heart rate of 180-200 b/min over 20-30 sec, followed by a decrease in running speed to a heart rate of 100-110 b/min. Accelerations were performed 2 days/week. On the other 3 training days, the horses performed 20 min of endurance training. During the final 42 days of interval training, the horses performed sustained high-speed intervals 1 day a week. The horses were accelerated until a heart rate of 190-200 b/min was reached, maintained at this rate for 20-30 sec, and then slowed to a trot until the heart rate recovered to 100-110 b/min. Detraining consisted of 6 month's rest in a paddock, followed by 18 weeks retraining on a similar program of LSD and interval exercise.

RESULTS

Running speed, at a heart rate of 140 b/min, increased for all horses during endurance training. At 2.8 m/sec (12% grade) following 90 days of endurance training, the average heart rate decreased from 180±7 to 158±6 b/min, right ventricular dP/dt from 2210±49 to 1740±93 mm Hg/sec, pulmonary artery blood temperature from 40.3±0.1 to 39.7±0.2 C, blood lactate from 7.6±0.9 to 2.2±0.8 mmol/l(Sexton, et al., 1987). Fig. 1 shows the effects of training on blood lactate during the exercise test.

Following 6 and 12 weeks of combined endurance and interval training, the arterial PO_2 remained at resting levels throughout the treadmill exercise test, rather than falling as in previous tests (Fig. 2). The heart

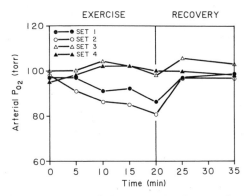

Fig. 2. Effects of endurance and interval training on arterial oxygen tension in Quarter Horses during a standard exercise test on a treadmill at a 12% grade. SET 1 = Control; SET 2 = Endurance trained; SET 3 = 6 weeks interval training; SET 4 = 12 weeks interval training.

rate versus pulmonary artery mean flow velocity relationship suggested an increase in stroke volume associated with more intense interval training.

Following detraining, there were no statistically significant differences in the measured variables either at rest, during the exercise test, or during recovery (Erickson et al., 1987). However, the performance of the exercise tests on the treadmill was more difficult for the horses as detraining progressed, evidenced by increased sweating, more labored breathing, and prolonged recovery time. There was a small upward shift in the blood lactate curve, suggesting an increased reliance on anaerobic metabolism (Fig. 1). Exercise training and detraining had no effect on the significant increases in hemoglobin concentration and hematocrit observed during the exercise tests. Furthermore, no significant changes were observed in arterial, right ventricular, and pulmonary arterial pressures measured during the exercise tests.

DISCUSSION

The results of this study suggest that 25 weeks of combined low-intensity, endurance training and high-intensity, interval exercise significantly enhance cardiopulmonary and metabolic efficiency and improve oxygen transfer during exercise. These adaptations are probably responsible, in part, for the improved exercise performance observed in the horses during the exercise training and testing sessions.

The lower heart rates measured during submaximal exercise after training are well documented in recent reviews (Evans, 1985; Physick-Sheard, 1985). The lack of change in the heart rate response to the exercise tests following combined endurance and interval training is probably attributable to the low intensity of the treadmill test. Adaptive changes that may have resulted from high intensity interval training were probably not taxed and, therefore, not reflected in the submaximal heart rate responses. The apparent increase in stroke volume following combined endurance and interval training, concommitant with a decrease in heart rate at a given workload, would maintain cardiac output while being more energetically favorable (Stone, 1977).

The increases in hemoglobin concentration and hematocrit that occur during exercise in the horse are attributed to increased sympathetic activity, which results in splenic contraction and release of additional red blood cells into the circulation, and is well-documented in the literature (Persson, 1967). Training had no effect on the resting hemoglobin concentration or the increase in hemoglobin concentration during exercise, which markedly increases blood oxygen transfer capacity.

The fall in PaO_2 observed during the first two exercise tests on the treadmill after endurance training is similar to that described by Bayly et al., (1983a) and Thornton et al., (1983). However, the horses could not be considered to be hypoxemic, since the lowest PaO_2 values measured were well above 75 torr. After combined endurance and interval training, the PaO_2 measured during exercise was significantly higher. This is in contrast to a report by Thornton et al., (1983), who observed that PaO_2 fell during an exercise test following a 5-week training program of endurance and interval exercise. The difference in these results may be related to the longer duration of our training program and the submaximal exercise test that we used. The maintenance of PaO_2 during exercise after more intense interval training may be related to improvements in ventilation to pefusion matching and/or the abatement of diffusion limitations (Robinson, 1985).

Blood lactate concentration increases during exercise, when the rate of lactate production exceeds metabolism by the muscle and liver, and is indicative of a work intensity that requires the recruitment of muscle fibers dependent on anaerobic metabolism (Brooks, 1985). Exercise training increases oxidative capacity of skeletal muscle and enhances the capacity of muscle to utilize lactate during exercise. The net effect of these adaptations is a delay in the onset of blood lactate accumulation. Detraining resulted in a small, nonsignificant increase in blood lactate during the highest level of exercise. The changes in lactate accumulation might have been larger during exercise after detraining, if the horses could have been subjected to a more strenuous test. Similarly, Thornton et al., (1983) observed a significant decrease in blood lactate during exercise after 5 weeks of strenuous training in Standardbred trotters and no change with 5 weeks of detraining. These authors plotted blood lactate vs. running speed and observed no significant differences in the slope of the curve before and after training and detraining. The intercept increased with training, but did not change with detraining.

Pulmonary artery blood temperature progressively increased during the standard exercise test as the work load increased, declined rapidly during the first minute of recovery, and then slowly returned toward resting values during the next 20 minutes of recovery. The decline in blood temperature following exercise training suggests more efficient cooling, probably by the respiratory system and heat exchange through the skin. Following 6 weeks of detraining, there was a small increase in the peak blood temperature during the exercise tests, suggesting that cooling efficiency was reduced. These results indicate that the well-trained horse is better able to dissipate heat during exercise and can probably run farther and work harder before hyperthermia begins to limit performance.

In summary, endurance and interval training appear to enhance oxygen transfer in the Quarter Horse during exercise. Endurance training resulted in enhanced exercise performance associated with improving cardiac and metabolic efficiency, as seen by lower exercise heart rates, blood lactate concentrations, and blood temperatures. High-intensity, interval training appeared to enhance gas exchange during exercise through maintenance of the arterial oxygen tension and to further augment cardiac function by apparent increases in stroke volume.

ACKNOWLEDGEMENTS

This study was supported by grants from the American Quarter Horse Association, the Grayson Foundation, and USDA Animal Health and Disease Research Funds. Contribution No. 87-409-A from the Kansas Agricultural Experiment Station.

REFERENCES

Bayly, W.M., B.D. Grant, R.G. Breeze and J.W. Kramer (1983a). The effects of maximal exercise on acid-base balance and arterial blood gas tension in Thoroughbred horses. In: Equine Exercise Physiology, edited by D.H. Snow, S.G.B. Persson and R.J. Rose. Granta Editions, Cambridge, p. 400-407.

Bayly, W.M., A.A. Gabel and S.A. Barr (1983b). Cardiovascular effects of submaximal aerobic training on a treadmill in Standardbred horses, using a standardized exercise test. Am. J. Vet. Res. 44:544-553.

Brooks, G.A. (1985). Anaerobic threshold: review of the concept and directions for future research. Med. Sci. Sports. Exerc. 17:22-31.

Engelhardt, W.V. (1977). Cardiovascular effects of exercise and training in horses. Adv. Vet. Sci. Comp. Med. 21:173-205.

Erickson, H.H., W.L. Sexton, B.K. Erickson, and J.R. Coffman (1987). Cardiopulmonary response to exercise and detraining in the Quarter Horse. In: Equine Exercise Physiology 2, edited by J.R. Gillespie and N.E. Robinson (in press).

Evans, D.L. (1985). Cardiovascular adaptations to exercise and training. In: Veterinary Clinics of North America: Equine Practice, edited by R.J. Rose, W.B. Saunders, Philadelphia, PA, pp. 513-531.

Milne, D.W., A.A. Gabel, W.W. Muir, and R.T. Skarda (1977). Effects of training on heart rates, cardiac output, and lactic acid in Standardbred horses, using a standardized exercise test. J. Am. Vet. Med. Assoc. 1:131-135.

Persson, S.G.B. (1967). On blood volume and working capacity in horses. Acta. Vet. Scand. (Suppl) 19:1-188.

Persson, S.G.B., B. Essen-Gustavsson, A. Lindholm, D. McMiken, and J.R. Thornton (1983). Cardiorespiratory and metabolic effects of training of Standardbred yearlings. In: Equine Exercise Physiology, edited by D.H. Snow, S.G.B. Persson and R.J. Rose. Granta Editions, Cambridge, pp. 458-469.

Physick-Ehard, P.W. (1985). Cardiovascular response to exercise and training in the horse. In: Veterinary Clinics of North America: Equine Practice, edited by J.D. Bonagura. W.B. Saunders, Philadelphia, PA, 1:383-417.

Robinson, N.E. (1985). Respiratory adaptations to exercise. In: Veterinary Clinics of North America: Equine Practice, edited by J.R. Rose, W.B. Saunders, Philadelphia, PA, 1:497-512.

Rose, R.J., Editor (1985). Symposium on Exercise Physiology. In: Veterinary Clinics of North America: Equine Practice, W.B. Saunders, Philadelphia, PA, Vol. 1, No. 3.

Sexton, W.L., H.H. Erickson and J.R. Coffman (1987). Cardiopulmonary and metabolic responses to exercise in the Quarter Horse: effects of training. In: Equine Exercise Physiology 2. edited by J.R. Gillespie and N.E. Robinson (in press).

Stone, H.L. (1977). Cardiac function and exercise training in conscious dogs. J. Appl. Physiol. 42:824-832.

Thornton, J., B. Essen-Gustavsson, A. Lindholm, D. McMiken and S.G.B. Persson (1983). Effects of training and detraining on oxygen uptake, cardiac output, blood gas tensions, pH and lactate concentrations during and after exercise in the horse. In: Equine Exercise Physiology, edited by D.H. Snow, S.G.B. Persson and R.J. Rose. Granta Editions, Cambridge, pp. 470-486.

O_2 TRANSPORT IN THE HORSE DURING REST AND EXERCISE

Gail L. Landgren[1], Jerry R. Gillespie[1], M. Roger Fedde[2],
Bryon W. Jones[3], Richard L. Pieschl[2], and Peter D. Wagner[4]

Departments of Surgery & Medicine[1] and
Anatomy & Physiology[2]
College of Veterinary Medicine and
Institute for Environmental Research[3]
Kansas State University, Manhattan, KS 66506
and
Section of Physiology[4], Department of Medicine,
University of California-San Diego, La Jolla, CA 92093

ABSTRACT

We studied mechanisms of O_2 transport in 6 adult (2-5 year old) horses at rest and during steady-state exercise on a treadmill (0% slope) at 12 m/s (a submaximal gallop). Oxygen consumption was measured using an open-flow system. Arterial and mixed venous blood samples were simultaneously obtained for measurement of O_2 content and hemoglobin concentration. $\dot{V}O_2$ increased from 1.5±0.2 L/min at rest to 46.2±4.8 L/min during exercise. HR increased from a resting value of 36.9±2.5 bpm to 196.5±10.9 bpm and the arterio-venous O_2 content difference (a-$\bar{v}$ O_2) increased from 4.2±0.8 ml O_2/100 ml blood to 20.3±1.6 ml O_2/100 ml blood.

The 30.4-fold increase in oxygen consumption in the horse at submaximal $\dot{V}O_2$ versus only a 10-fold increase in man at $\dot{V}O_2$max demonstrates the marked ability of the horse to transfer O_2 at each step in the O_2 transport pathway.

INTRODUCTION

The equine athlete has more than 3 times the capacity to increase its oxygen consumption ($\dot{V}O_2$) during exercise than does man. $\dot{V}O_2$ has been measured in ponies (Seeherman et al., 1981; Forster et al., 1984; Powers et al., 1987) and horses (Karlsen and Nadaljak, 1964; Thomas and Fregin, 1981; Thornton et al., 1983; Bayly et al., 1987; Evans et al., 1987; Hoppeler et al., 1987; Thornton et al., 1987) at rest and during several levels of exercise. We extended these measurements in horses while they were running on a treadmill (0% slope) and at a higher velocity than in previous studies to determine the capacity of the cardiopulmonary system to transport O_2 to the working muscles and the muscle's ability to extract O_2 from the blood.

METHODS

Animals

We used one Thoroughbred and five Quarter Horse geldings ranging in age from 2 to 5 years (average body weight of 521.7±34.2 Kg).

Instrumentation

To measure $\dot{V}O_2$, a bias flow of room air was passed through a loose-fitting mask. The expired gas and bias flow air were mixed in baffled mixing barrels and then analyzed by a polarographic analyzer (Beckman model OM-11). Total flow through the system was measured with a Meriam flow straightener.

Anaerobic samples of arterial and mixed venous blood were simultaneously obtained from the carotid and pulmonary arteries and analyzed for oxygen content and hemoglobin concentration (Hb) using a Lex-O_2-Con (Lexington Instruments) and a hemoglobinometer, respectively.

Heart rate (HR) was determined from an ECG recording.

Exercise Protocol

We tested the horses while at rest and during 3 minutes of galloping at 12 m/sec on a Sato treadmill (0% slope). A 1 minute warmup trot at 4 m/s was allowed before the exercise test. We took blood samples and other measurements at rest and every 30 seconds during the exercise test.

RESULTS

Oxygen Consumption, Cardiac Output and Heart Rate

$\dot{V}O_2$ increased from 1.5±0.2 L/min at rest to 46.2±4.8 L/min at 12 m/sec. The 30.4-fold increase in $\dot{V}O_2$ was accompanied by an estimated 6-fold increase in cardiac output ($\dot{Q}$) (36.6±4.1 L/min at rest to 228.1±26.1 L/min during exercise) and a 4.8-fold increase in the arterio-venous O_2 difference (a-$\bar{v}$ O_2) (4.2±0.8 vol. % at rest to 20.3±1.6 vol. % during exercise). The increase in $\dot{Q}$ was primarily due to a 5.3-fold increase in HR (36.9±2.5 bpm to 196.5±10.9 bpm).

O_2 Transport Capacity of the Blood

Hb concentration in the blood was increased by exercise (10.5±2.3 gm/100 ml blood to 17.6±1.4 gm/100 ml blood); this was associated with an increase in arterial O_2 content (CaO$_2$) from 14.2±1.7 vol. % to 23.0±2.5 vol. %.

O_2 Extraction by the Muscle

Mixed venous O_2 content (C$\bar{v}$O$_2$) decreased during exercise from 10.0±1.2 vol. % to 2.6±1.6 vol. %.

DISCUSSION

Oxygen Consumption

In man, $\dot{V}O_2$ increases approximately 10-fold over rest during maximal exercise (Brooks and Fahey, 1984), while in the horse, $\dot{V}O_2$ increases more than 30 times (Karlsen and Nadaljak, 1964; Thomas and Fregin, 1981; Thornton et al., 1983). We measured a 30.4 times increase in $\dot{V}O_2$ (2.9 ± 0.4 to 88.5 ± 7.6 ml/min·kg) in horses during submaximal galloping at 12 m/sec for 3 minutes on a treadmill (0% slope), while Thomas and Fregin (1981) found an even greater increase (36-fold) while running at only 3.9 m/sec, but on an 11.5% slope. Thornton et al., (1983) reported an increase in $\dot{V}O_2$ of 37.6 times that of rest in Standardbreds trotting for one minute at 8.5 m/sec on a 10% slope. Hoppeler et al., (1987) measured a "$\dot{V}O_2max$" of 136 ± 3.9 ml/min·kg (a 45-fold increase over rest) in Standardbreds on an inclined treadmill. The larger $\dot{V}O_2$ increase in the horse is accounted for by a larger increase in a-$\bar{v}$ O_2 and in $\dot{Q}$.

Cardiac Output and Heart Rate

The horse also has greater ability than man to increase its $\dot{Q}$ (6 times [Thomas and Fregin, 1981] vs 3.3 times [Brooks and Fahey, 1984]) during exercise. This is brought about by a much larger fractional increase in HR (8 times [Engelhardt, 1977] vs 2.6 times [Brooks and Fahey, 1984]). Horses have a lower resting HR (30 bpm) and a higher maximum HR (240 bpm) (Engelhardt, 1977) than man (70 to 185 bpm) (Brooks and Fahey, 1984). This allows more oxygenated blood to be delivered per unit time to the working muscles during exercise.

O_2 Transport Capacity of the Blood

In the horse, the large increase in O_2 carrying capacity of blood and high O_2 extraction by the working muscles appears to play a large role in its aerobic capacity. With the onset of exercise in the horse, the spleen, which stores approximately 1/3 to 1/2 of the horse's total red blood cells, contracts and there is a dramatic rise in the hematocrit and hemoglobin concentration (Persson, 1967). We measured a 1.7-fold increase in Hb and a 1.6-fold increase in CaO_2 from rest to 12 m/sec, illustrating the large increase O_2 carrying capacity of blood. The spleen in man does not contract during exercise, although there is a small hemoconcentration of the blood due to fluid loss to the cells and interstial fluid (Astrand and Rodahl, 1986). Astrand et al., (1964) found in man that the CaO_2 was only 3% higher during heavy exercise than at rest.

O_2 Extraction by the Muscle

The $C\bar{v}O_2$ decreases similarly in horse and man. We measured $C\bar{v}O_2$ in the horse at 10.0 ± 1.2 vol. % at rest and 2.6 ± 1.6 vol. % during submaximal exercise. In man, the $C\bar{v}O_2$ decreased from about 12 vol. % at rest to 2 vol. % during maximal exercise (Astrand et al., 1964). Therefore, horse muscle does not appear to be superior in O_2 extraction.

We measured a 4.8-fold increase in the a-$\bar{v}$ O_2 in the submaximally running horse, while in the maximally exercised man, the a-$\bar{v}$ O_2 increases only 2.9 times the resting value. This larger increase in the a-$\bar{v}$ O_2 in the horse is associated with the larger increase in the CaO_2, resulting from the dramatic increase in Hb concentration.

SUMMARY

The superior oxygen utilization during exercise by the horse as compared to man is aided by an enhanced O_2 transport from the lungs to the tissues. One primary factor in this process is the control of the splenic reservoir of erythrocytes.

ACKNOWLEDGMENTS

Supported by The Dean's Fund, College of Veterinary Medicine, Kansas State University; The Grayson Foundation; The American Quarter Horse Association; HL 17703; and American Lung Association of California.

REFERENCES

Astrand, P.O., T.E. Cuddy, B. Saltin and J. Stenberg (1964). Cardiac output during submaximal and maximal work. J. Appl. Physiol. 19:268.

Astrand, P.O. and K. Rodahl (1986). Textbook of Work Physiology. New York, McGraw-Hill Book Co., pp. 181-185.

Bayly, W.M., R.B. Schultz, D.R. Hodgson and P.D. Gollnick (1987). Diffusion impairment as a cause of exercise-induced hypoxemia. In: Equine Exercise Physiology 2, eds. by J.R. Gillespie & N.E. Robinson. Davis, CA, ICEEP Public. (in press)

Brooks, G.A. and T.D. Fahey (1984). Exercise Physiology: Human Bioenergetics and Its Applications. New York, John Wiley & Sons, pp. 332-333.

Engelhardt, W. von (1977). Cardiovascular effects of exercise and training in horses. Adv. in Vet. Sci. and Comp. Med. 21:173-205.

Evans, D.L. and R.J. Rose (1987). Maximal aerobic capacity in the racehorse: prediction and correlation. In: Equine Exercise Physiology 2, ed J.R. Gillespie and N.E. Robinson. Davis, CA, ICEEP Public. (in press)

Forster, H.V., L.G. Pan, G.E. Bisgard, S.M. Dorsey, and M.S. Britton (1984). Temporal pattern of pulmonary gas exchange during exercise in ponies. J. Appl. Physiol. 57(3):760-767.

Hoppeler, H., J. Jones, S.L. Lindstedt, C.R. Taylor, E.R. Weibel, and A. Lindholm (1987). Relating $\dot{V}O_2$max to skeletal muscle mitochondria in horses. In: Equine Exercise Physiology 2, eds. J.R. Gillespie and N.E. Robinson. Davis, CA, ICEEP Public. (in press)

Karlsen, G.G. and E.A. Nadaljak (1964). Konevodstvo I Konnyisport 34(11):27-31.

Persson, S.G.B. (1967). On blood volume and working capacity in horses. Acta Vet. Scand. Suppl. 19:1-189.

Powers, S., R. Beadle, J. Lawler and D. Thompson (1987). Respiratory gas exchange kinetics in transitions from rest or prior exercise in ponies. In: Equine Exercise Physiology 2, eds. J.R. Gillespie and N.E. Robinson. Davis, CA, ICEEP Public. (in press)

Seeherman, J.J., C.R. Taylor, G.M.O. Maloiy and R.B. Armstrong (1981). Design of the mammalian respiratory system. II. Measuring maximum aerobic capacity. Respir. Physiol. 44:11-23.

Thomas, D.P. and G.F. Fregin (1981). Cardiorespiratory and metabolic responses to treadmill exercise in the horse. J. Appl. Physiol. 50(4):864-868.

Thornton, J., G. Essen-Gustavsson, A. Lindholm, D. McMiken, S. Persson (1983). Effects of training and detraining on oxygen uptake, cardiac output, blood gas tensions, pH, and lactate concentrations during and after exercise in the horse. In: Equine Exercise Physiology, eds. D.H. Snow, S.G.B. Persson, and R.J. Rose. Cambridge, Granta Editions, pp. 470-486.

Thornton, J., J. Pagan and S. Persson (1987). The oxygen cost of weight loading and uphill treadmill exercise in the horse. In: Equine Exercise Physiology 2, eds. J.R. Gillespie and N.E. Robinson. Davis, CA, ICEEP Public. (in press)

MORPHOMETRY OF RIGHT VENTRICULAR PAPILLARY MUSCLE IN RAT DURING

DEVELOPMENT AND REGRESSION OF HYPOXIA-INDUCED HYPERTENSION

Kuen-Shan Hung, Henry Pacheco, Dianna Lessin,
Kerry Jordan, and Leone Mattioli

Departments of Anatomy and Pediatrics
University of Kansas Medical Center
Kansas City, KS   66103

ABSTRACT

Morphometric analyses of the right ventricular papillary muscle, as well as measurements of right ventricular pressure and weight, were carried out in the rat during the development and recovery of hypoxic pulmonary hypertension. Animals were divided into hypoxic and normobaric control groups. The hypoxic rats were placed in hypobaric chambers for 1, 2, and 3 wks; and after 3 wks exposure, subgroups of hypoxic rats were allowed to recover in normoxia for 1 to 9 wks. Hematocrit (HCT) and right ventricular systolic pressure (RVSP) were measured prior to sacrifice. The heart was perfused, and the right ventricle (RV) was separated from the left ventricle and septum (LV+S) and weighed. The papillary muscles were dissected and processed for ultrastructural morphometry. Results showed that HCT, RVSP, and RV weight increased in the rats during the hypoxic exposure and then gradually returned to control levels after 3 to 4 wks of normobaric recovery. The papillary muscle of the hypoxic rats showed increased volume density of interstitium, increased diameter and cross sectional area of the cardiac myocytes, reduced volume density of mitochondria, and reduced mitochondria to myofilament ratio. During normoxic recovery, these morphometric indices returned toward control values at various periods of time ranging from less than 3 wks to 9 wks. The results indicate that the adaptive ultrastructural changes of the papillary muscle in RV hypertrophy paralleled the RVSP changes, and also demonstrate the reversibility of these changes in ambient oxygen.

INTRODUCTION

It has been repeatedly shown in various experimental models that chronic hypoxia leads to the development of pulmonary arterial hypertension and right ventricular hypertrophy (RVH). Both the pulmonary arterial pressure and RV mass return to near normal levels upon removal of the hypoxic stimulus (Ressl et al., 1974; Herget et al., 1978; Kay, 1980; Kentera and Susic, 1980; Rabinovitch et al., 1981; Sobin et al., 1983; Kentera et al., 1985). However, the ultrastructural changes underlying the development and regression of hypoxia-induced RVH have received little

337

attention. Previously, morphometric studies by Herbener et al., (1973) have shown no changes in the volume density of mitochondria in the right ventricular myocytes of mice exposed to hypobaric hypoxia.

The purpose of this study is to correlate the ultrastructural changes of the RV papillary muscle with indices of hypoxic pulmonary hypertension (hematocrit and RV systolic pressure and weight) in rats during the progression and regression of hypoxic pulmonary hypertension.

METHODS

Male Sprague-Dawley rats (n = 93) with initial body weights (BW) of 150 to 175g were used. The hypoxic rats (n = 51) were placed in hypobaric chambers for 1, 2, and 3 wks; the controls (n = 42) were pair-fed in the same room in a normobaric environment (Hung et al., 1986). Subgroups of hypoxic rats (n = 32) were removed from the chambers after 3 wks of hypoxic exposure, and allowed to recover in normoxia for 1, 2, 3, 4, 5, 7 and 9 wks. Each hypobaric chamber has 0.25 m^3 capacity; and the chamber pressure was adjusted to approximately 370 torr. At the end of each exposure period, the animals were anesthetized with pentobarbital sodium (60 mg/kg) and weighed. Hematocrit (HCT) was measured from a tail blood sample. A 3.5 French catheter was placed in the right ventricle via the right jugular vein (Stinger et al., 1981), and the systolic pressure (RVSP) was measured by a Statham 23DB transducer and recorded on a EFM recorder while the animal was breathing spontaneously in ambient oxygen.

The chest was opened, and the heart was arrested in diastole by injecting 1 ml of 1 N KCl into the RV through the catheter and perfused with 2% glutaraldehyde in phosphate buffer, pH 7.2, under 100 cm H_2O pressure for two minutes. After further fixation in fresh 2% glutaraldehyde for at least 2 hours the RV was separated from the left ventricle and septum (LV+S), and weighed.

The papillary muscles from at least 3 animals from each of the following subgroups were obtained for quantitative ultrastructural morphometric analysis: 1, 2, and 3 wks in hypoxia, and their controls under 3 wks normoxia; 3 wks hypoxia followed by 3 wks normoxic recovery and their controls; 3 wks hypoxia followed by 7 wks recovery and their controls; and 3 wks hypoxia followed by 9 wks recovery and their controls. After perfusion fixation, two RV papillary muscles from each rat were removed and rinsed in 0.1M phosphate buffer, postfixed in 1% osmium tetroxide, dehydrated in graded ethanols and embedded in Araldite.

The muscle was oriented for exact cross sections and viewed under a JEOL 100S transmission electron microscope. Low power micrographs (total magnification, 2,500X) were used to determine the volume densities of the myocytes, capillaries (including endothelium and lumen), and interstitial tissue using a point counting technique (Marino et al., 1986). The myocyte diameters and cross sectional areas were measured with a Bioquant Image Analysis System consisting of a Hipad digitizer interfaced with an Apple IIe microcomputer. Intermediate power micrographs (total magnification, 10,000X) were used to determine the volume densities of mitochondria and myofilaments by the same point counting technique. The mitochondria-to-myofilaments ratios were calculated by their volume densities.

Certain tissue sampling criteria were established to insure consistency regarding the areas photographed. A random grid opening was selected and micrographs were taken of the cells lying at each corner of the grid opening as well as two cells from the center of the grid. Only the cells cut in cross sections were used, and cells in the border region of the papillary muscle were disregarded.

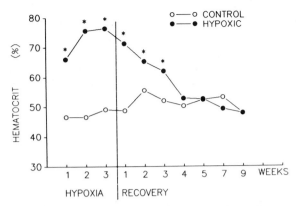

Fig. 1. Changes in hematocrit during hypoxic exposure
and subsequent normoxic recovery. * indicates
significant difference (p < 0.05) between hypoxic and
control groups.

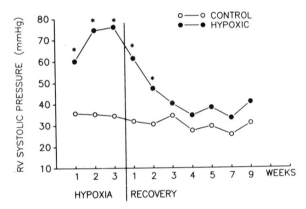

Fig. 2. Changes in right ventricular systolic pressure
during hypoxic exposure and subsequent normoxic recovery.
* indicates significant difference (p < 0.05) between
hypoxic and control groups.

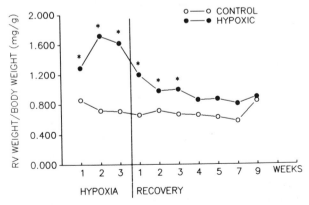

Fig. 3. Changes in right ventricle to body weight ratios
during hypoxic exposure and subsequent normoxic recovery.
* indicates significant difference (p < 0.05) between
hypoxic and control groups.

Data were expressed as mean ± SE. Statistical significance for the values of HCT, RVSP, and RV/BW and RV/LV+S weight ratios comparing the control and experimental groups were determined by using analysis of variance and Duncan's multiple range test with a "P" value of 0.05 (Modrak, 1983). For morphometric analyses, data for an individual rat were represented by calculated averages derived from the photographs of the two papillary muscles. The values of the hypoxic and normoxic recovery groups were compared with the value of one combined control group, because among the control subgroups no statistically significant differences were found in the initial morphometric analyses.

RESULTS

There were no differences in the body weights between all hypoxic and control groups. As expected, HCT (Fig. 1), RVSP (Fig. 2), and RV/BW (Fig. 3) rose during the 3 wks of hypoxic exposure, and gradually returned to near normal values after 3 to 4 wks of normoxic recovery. Although statistically nonsignificant, the values of RVSP and RV/BW were consistently higher in the recovery groups even at 5, 7, and 9 wks. The hypoxic exposure did not affect the left ventricular weight (Fig. 4).

The myocyte volume densities of the hypoxic rats and those during normoxic recovery were generally lower than that of the controls, although the differences did not reach statistical significance (Fig. 5). The volume density of the interstitium (excluding capillaries) of the hypoxic rats was slightly higher than that of the controls, but the difference was statistically significant only at 2 wks of hypoxic exposure. During the recovery period, there was a tendency toward normalization of the volume density of the interstitium (Fig. 6). No significant differences were found in the volume density of the capillaries in all groups (Fig. 7).

The myocyte diameter of the hypoxic rats was significantly increased after 3 wks of hypoxia, and did not return to normal until the 7th wk of normoxic recovery (Fig. 8). The cross sectional area of the myocytes after 3 wks hypoxia was greater than that of the controls, and the difference was still significant at the 7th wk of normoxic recovery (Fig. 9).

The volume density of mitochondria was significantly reduced in hypoxic rats by the second wk, and returned to the control value by the third wk of normoxic recovery (Fig. 10). The slight increase in the volume density of myofilaments in the hypoxic rats was not significantly different from that of the controls (Fig. 11). The mitochondria-to-myofilaments ratio in all hypoxic groups was significantly lower than that of the controls, and the ratio returned to normal during recovery (Fig. 12).

DISCUSSION

Our studies showed that the ultrastructural changes in the right ventricular myocardium during the progression and regression of hypoxic pulmonary hypertension and RVH followed the same general pattern of changes in HCT, RVSP, and RV weight. Thus, during the development of RVH, while HCT, RVSP and RV/BW increased in the hypoxic rats, the volume densities of the myocardial interstitium, myocyte diameters and areas also increased, and the volume density of mitochondria, and the mitochondria-to-myofilaments ratio decreased. Additionally, the volume density of myocytes was reduced and the volume density of myofilaments was increased, but the differences did not reach statistical significance. The volume density of the capillaries in the myocardium of the hyoxic rats did not differ significantly from the controls during hypoxia and normoxic recovery periods. This

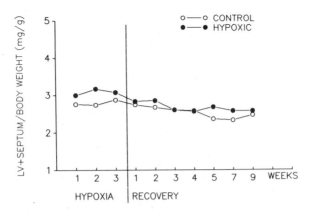

Fig. 4. Changes in the ratio of left ventricle plus septum to body weight during hypoxic exposure and subsequent normoxic recovery.

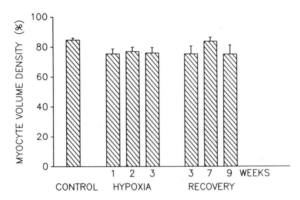

Fig. 5. Changes in the myocyte volume density during hypoxic exposure and subsequent normoxic recovery.

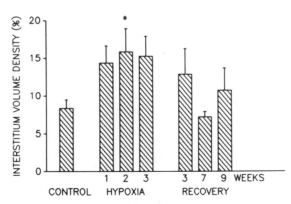

Fig. 6. Changes in the interstitial volume density during hypoxic exposure and subsequent normoxic recovery. * indicates significant difference (p < 0.05) between hypoxic and control groups.

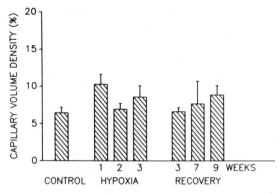

Fig. 7. Changes in the capillary volume density during hypoxic exposure and subsequent normoxic recovery.

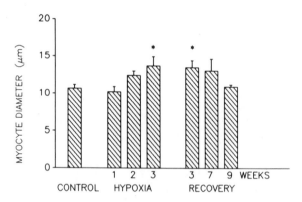

Fig. 8. Changes in the myocyte diameter during hypoxic exposure and subsequent normoxic recovery. * indicates significant difference ($p < 0.05$) between hypoxic and control groups.

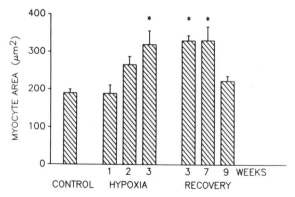

Fig. 9. Changes in the myocyte cross sectional area during hypoxic exposure and subsequent normoxic recovery. * indicates significant difference ($p < 0.05$) between hypoxic and control groups.

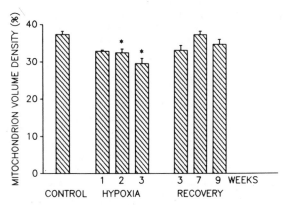

Fig. 10. Changes in the mitochondrion volume density during hypoxic exposure and subsequent normoxic recovery. * indicates significant difference (p < 0.05) between hypoxic and control groups.

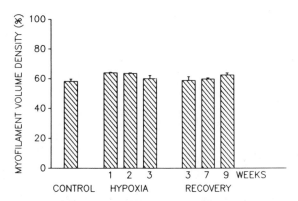

Fig. 11. Changes in the myofilament volume density during hypoxic exposure and subsequent normoxic recovery.

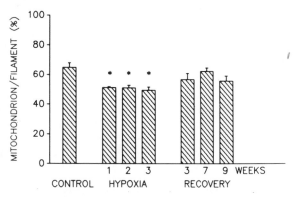

Fig. 12. Changes in the mitochondria to myofilaments ratio during hypoxic exposure and subsequent normoxic recovery. * indicates significant difference (p < 0.05) between hypoxic and control groups.

finding suggests a proportional growth of the capillaries during the development of RVH as well as a proportional regression during normoxic recovery. In addition, our results indicate that hypobaric hypoxia did not produce left ventricular hypertrophy.

During recovery, HCT, RVSP, and RV/BW returned toward control values within 4 wks. Complete reversal of HCT was shown. However, despite nonsignificant statistical differences, RVSP and RV/BW of the hypoxic groups were still higher than those of the controls at 9 wks. This finding was consistent with previous reports showing incomplete reversal of RVH in hypoxic pulmonary hypertension (Kay, 1980; Kentera and Susic, 1980; Rabinovitch et al., 1981; Kentera et al., 1985). Morphometric data indicated the same general pattern of recovery although the measured parameters returned to normal values at various periods of time. Specifically, the volume densities of the interstitium and mitochondria, and the mitochondria-to-myofilaments ratio, were quickly restored within the first 3 wks of normoxic recovery, while the myocyte diameter and area were reversed more slowly at 7 and 9 wks, respectively.

In comparison with other studies, some differences and similarities were noted. When exposed to hypobaric hypoxia, the mouse (Herbener et al., 1973) and rat (Tomanek, 1979a) developed left ventricular hypertrophy; a finding which is not supported by our data. This difference is very likely due to our pair-feeding methods which equalized the body weight in both hypoxic and control rats. During induction of RVH we have shown a reduction in the volume density of mitochondria as well as a reduction in the mitochondria-to-myofilaments ratio in the papillary muscle of the RV. These results are consistent with those of Bishop and Cole (1969) who found a reduced mitochondrial volume per myocyte in the RV of dogs with experimental pulmonic stenosis, and those of Lund and Tomanek (1978), Tomanek (1979b) and Breisch et al. (1984) who observed a decreased mitochondria-to-myofilaments ratio in the hypertrophic left ventricular myocytes. In contrast, no changes in either the volume density of mitochondria or the mitochondria-to-myofilaments ratio were shown by Herbener et al. (1973) who induced RVH by hypobaric hypoxia in the mouse, by Anversa et al. (1984) who produced RVH through experimental myocardial infarction of the left ventricle in rat, and by Marino et al. (1983, 1986) who induced RVH by banding of the pulmonary artery in cat. The discrepancies in these results may be due to species variations and/or experimental methods used to produce RVH. In our experiments, RVH could result from pressure overload due to increased pulmonary vascular resistance, as well as increased blood viscosity and reduced alveolar oxygen tension.

Marino et al. (1986) have correlated morphometric data with physiological measurements in their studies on the recovery of RVH in the cat. They have stressed the importance in the restoration of the volume density of interstitium (connective tissue), as well as the volume density and size of the cardiac myocytes when the pressure overload was removed. While our data tend to support their conclusion, we have also shown normalization of mitochondria-to-myofilaments ratio upon returning to normoxia. Our finding indicates that there is a restoration of balance between the energy producing and energy consuming organelles in our experimental model of RVH. The differences in these results may be accounted for by species variation and experimental methods used. Specifically, Marino et al. (1986) were dealing with a simple pressure unloading of the RV, while the recovery in our model was more complex involving not only the reduction of pressure overload, but also normalization of blood viscosity and alveolar oxygen tension. In conclusion, our studies demonstrate that the complex adaptive ultrastructural changes of the papillary muscle in hypoxia-induced RVH are reversible. Whether RVH produced by longer hypoxic exposure is readily reversible as well remains to be demonstrated.

ACKNOWLEDGMENTS

This project was supported in part by a grant-in-aid from the American Heart Association, Kansas Affiliate. We acknowledge the use of the KUMC EM Research Center, which is supported in part by the J.W. & E.E. Speas Trust, BRSG S07 RR 05373, the Shared Biomedical Equipment Fund No. 244301, and NIH Instrumentation Grant S10 RR 01582.

REFERENCES

Anversa, P., C. Beghi, S.L. McDonald, V. Levicky, Y. Kikkawa and G. Olivetti (1984). Morphometry of right ventricular hypertrophy induced by myocardial infarction in the rat. Am. J. Pathol. 116:504-13.

Bishop, S.P. and C.R. Cole (1969). Ultrastructural changes in the canine myocardium with right ventricular hypertrophy and congestive heart failure. Lab. Invest. 20:219-229.

Breisch, E.A., F.C. White and C.M. Bloor (1984). Myocardial characteristics of pressure overload hypertrophy. A structural and functional study. Lab. Invest. 51:333-342.

Herbener, G.H., R.H. Swigart and C.A. Lang (1973). Morphometric comparison of the mitochondrial populations of normal and hypertrophic hearts. Lab. Invest. 28:96-103.

Herget, J., A.J. Suggett, E. Leach and G.R. Barber (1978). Resolution of pulmonary hypertension and other features induced by chronic hypoxia in rats during complete and intermittent normoxia. Thorax 33:468-473.

Hung, K.-S., J.C. McKenzie, L. Mattioli, R.M. Klein, C.D. Menon and A.K. Poulose (1986). Scanning electron microscopy of pulmonary vascular endothelium in rats with hypoxia-induced hypertension. Acta Anat. 126:13-20.

Kay, J.M. (1980). Effect of intermittent normoxia on chronic hypoxic pulmonary hypertension, right ventricular hypertrophy, and polycythemia in rats. Am. Rev. Resp. Dis. 121:993-1001.

Kentera, D. and D. Susic (1980). Dynamics of regression of right ventricular hypertrophy in rats with hypoxic pulmonary hypertension. Respiration 39:272-275.

Kentera, D., D. Susic and V. Kanjuh (1985). Experimental evidence that hypoxic pulmonary hypertension can be altered by drugs. Prog. Resp. Res. 20:26-30.

Lund, D.D. and R.J. Tomanek (1978). Myocardial morphology in spontaneously hypertensive and aortic-constricted rats. Am. J. Anat. 152:141-152.

Marino, T.A., S.R. Houser and G. Cooper IV (1983). Early morphological alterations of pressure-overloaded cat right ventricular myocardium. Anat. Rec. 207:417-426.

Marino, T.A., E. Brody, I.K. Lauva, R.L. Kent and G. Cooper IV (1986). Reversibility of the structural effects of pressure overload hypertrophy of cat right ventricular myocardium. Anat. Rec. 214:1431-1437.

Modrak, J.B. (1983). A computer program for multigroup comparisons. Trends Pharmacol. Sci. 4:490-492.

Rabinovitch, M., W.J. Gamble, O.S. Miettinen and L. Reid (1981). Age and sex influence on pulmonary hypertension of chronic hypoxia and on recovery. Am. J. Physiol. 240:H62-H72.

Ressl, J., D. Urbanova, J. Widimsky, B. Ostadal, V. Pelouch and J. Prochazka (1974). Reversibility of pulmonary hypertension and right ventricular hypertrophy induced by intermittent high altitude hypoxia in rats. Respiration 31:38-46.

Sobin, S.S., H.M. Tremer, J.D. Hardy and H.P. Chioidi (1983). Changes in arteriole in acute and chronic hypoxic pulmonary hypertension and recovery in rat. J. Appl. Physiol.: Respirat. Environ. Exercise Physiol. 55:1445-1455.

Stinger, R.G., V.J. Iacopino, I. Alter, T.M. Fitzpatrick, J.C. Rose and P.A. Kot (1981). Catheterization of the pulmonary artery in the closed-

chest rat. J. Appl. Physiol.: Respirat. Environ. Exercise Physiol.
    51:1047-1050.
Tomanek, R.J. (1979a).   Quantitative ultrastructural aspects of cardiac
    hypertrophy. Texas Rep. Biol. Med. 39:111-122.
Tomanek, R.J. (1979b). The role of prevention or relief of pressure over-
    load on the myocardial cell of the spontaneously hypertensive rat. A
    morphometric and stereologic study. Lab. Invest. 40:83-91.

THERMOREGULATION AND METABOLISM IN HYPOXIC ANIMALS

R.K. Dupre, A.M. Romero and S.C. Wood*

University of Nevada, Las Vegas, NV
and
Lovelace Medical Foundation*
Albuquerque, New Mexico

INTRODUCTION

It is well established that altered body temperature (Tb) has an exponential effect on the oxygen consumption ($\dot{V}O_2$) of resting animals (Krogh, 1914). Alterations of Tb over a range larger than $20^{\circ}C$ are routine for ectothermic vertebrates. The range of altered core Tb is smaller for homeotherms but is still significant in hyperthermia and hypothermia. In general, a $1^{\circ}C$ change in Tb causes an 11% change in metabolic rate (Q_{10} = 2.5). Consequently, fever or exercise-induced hyperthermia, will elevate oxygen demand and amplify the hypoxic stress of a mammal at high altitude or with cardiopulmonary disease. Conversely, hypothermia could be beneficial to any animal faced with a limited oxygen supply.

Many lower vertebrates (reptiles, fish, amphibians) regulate body temperature by behavioral as well as physiological means. When placed in an experimental temperature gradient they will move to an ambient temperature that provides their "preferred" body temperature. The phenomenon of hypoxia-induced reduction of selected body temperature was hypothesized by Wood (1984) and verified with experimental data for lizards (Hicks and Wood, 1985) and a variety of other ectothermic animals (Wood et al., 1985).

Mammals also utilize behavioral as well as physiological means to regulate body temperature. In contrast to the lower vertebrates, the regulation of core temperature is usually precise and within narrow limits. However, a number of previous studies have shown that acute hypoxia results in a lowering of Tb (Gellhorn and Janus, 1936; Miller and Miller, 1966; Lister, 1984; Wood et al., 1985). Knowledge of hypoxic hypothermia is not new; however, a majority of studies have focused on the phenomenon rather than ascribing any physiological significance, either positive or negative, to the phenomenon. Only a few studies have attempted to understand the mechanisms underlying hypoxic hypothermia (Blatteis and Lutherer, 1973, 1974; Horstman and Banderet, 1977; Schnakenberg and Hoffman, 1972) or the implications of such a hypothermia (Hicks and Wood, 1985; Miller and Miller, 1966). Despite the benefits to oxygen transport and utilization accrued with hypothermia which we have discussed above, the decrease in body temperature associated with hypoxia has sometimes been attributed to a failure of the thermoregulatory system of the animal (Moore, 1959).

The mechanism of hypothermia in lower vertebrates is primarily behavioral (the animals seek out a cooler environment) with a physiological component of altered peripheral blood flow. We showed that hypoxia-induced hypothermia in lizards is a controlled response; i.e., the animals, when placed in a "shuttle box" where they must choose between an ambient temperature of either 10 or 50° C, show a reduction in both the upper and lower exit temperatures (Hicks and Wood, 1985). The temperature selected in a thermal gradient is below the upper exit temperature during normoxia but above the upper exit temperature during hypoxia. Therefore, the animal moves to attain a lower body temperature that is above the hypoxic lower exit temperature.

In mammals, the mechanism of hypothermia is also not completely understood. This is partly due to the difficulty of separating metabolism from Tb regulation and determining if reduced Tb is a cause or effect of reduced metabolic rate. In mammals, the interactions between hypoxia and hypothermia are also complicated by thermogenic responses to decreasing body temperature. These responses, shivering and non-shivering thermogenesis, increase $\dot{V}O_2$ and heat production providing partial homeostasis of normal body temperature in neonates and complete homeostasis in adults. In this classical scheme, metabolism is maintained at a basal level over a range of ambient temperatures called the thermoneutral zone (TNZ). In adults, body temperature is maintained at ambient temperatures below the TNZ by thermogenic responses.

This study tested the hypothesis that the classical scheme of thermal neutrality does not necessarily apply to hypoxia. We examined the thermoneutral zone of growing Wistar rats during acute hypoxic hypoxia to determine if the thermoregulatory system of this endothermy is reset or fails in response to the hypoxic stress.

METHODS

Growing Wistar rats ranging in weight from 110-140g were obtained from a breeding colony maintained at the School of Medicine of the University of New Mexico. The animals of the colony were housed after weaning with littermates (4-6 animals per cage) in a windowless room at 25°C on a 12L:12D light cycle. Water and commercial laboratory chow were provided ad libitum.

$\dot{V}O_2$ of rats fasted for 6 hours was measured by open-flow indirect calorimetry over an ambient temperature range of 1 to 37°C. Thermocouples (20 gauge copper-constantan) for body temperature measurement were inserted approximately 5 cm into the rat's colon and secured to the tail with adhesive tape. The animals were then placed in an air-tight chamber through which air was flowed at a rate of 740 ml/min. The oxygen concentration of the incurrent gas to the chamber was regulated by a Wosthoff 231 a/F gas-mixing pump. One of two oxygen concentrations, 21% or 10%, balance N_2, was delivered to 22 individual Wistar rats. $\dot{V}O_2$ was calculated as the product of the air flow and the oxygen concentration difference between the gas entering and exiting the respiratory chamber and corrected to STPD.

Ambient temperature was modulated by placing the respiratory chamber in an environmental chamber with an effective range of temperature control from 0-50°C. The temperature of the chamber was monitored by a 40 gauge chromel-alumel thermocouple located in the respiratory chamber. After each change in ambient temperature, data were not collected until steady-state measurements of respiratory chamber temperature, animal body temperature, and $\dot{V}O_2$ were obtained (a period typically of 1.5 to 2 hours).

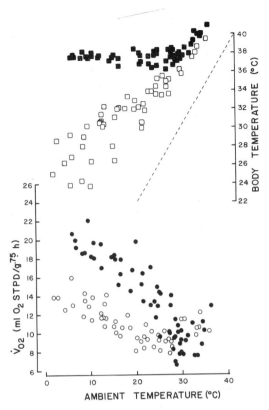

Fig. 1. Body temperature and oxygen consumption of normoxic
(closed symbols) and hypoxic (open symbols) Wistar rats over a
range of ambient temperatures (N=11 for each treatment).

The limits of the thermoneutral zone and the estimation of thermal
conductance were statistically determined by stepwise linear regression
analysis of the oxygen consumption-ambient temperature relationship for
each inspired oxygen concentration.

RESULTS

The results of this study are illustrated in Fig. 1. The thermoneutral
zone of the normoxic rats ranged from about $29^{O}C$ to $34^{O}C$. Below the lower
critical temperature of $29^{O}C$, the increase in oxygen consumption provides
an estimate of thermal conductance of -0.498 ml O_2 $g^{0.75}$/hr $\cdot$ ^{O}C.

When the rats were placed in an environment containing only 10% O_2
there was no significant change in the basal metabolic rate. However, there
was a substantial difference in the thermoneutral zone. Both the upper and
lower critical temperatures of the thermoneutral zone decreased during
hypoxic exposure to $30^{O}C$ and $22^{O}C$, respectively. This represents a broadening of the thermoneutral zone as well. Finally, the slope of the increase
in VO_2 below the lower critical temperature of the hypoxic rats was significantly lower than that of the normoxic rats, suggesting a thermal conductance of -0.275 ml O_2/g$^{0.75}$/hr $\cdot$ ^{O}C.

DISCUSSION

The phenomenon of hypoxic hypothermia appears to be accompanied, at
least in Wistar rats, by both a broadening of the ambient temperature range

over which $\dot{V}O_2$ is independent of ambient temperature, and a significant decrease in the thermal limits of that range. This shift in the relationship of metabolic heat production and ambient temperature during hypoxic hypoxia is certainly supportive of a shift in thermoregulatory operational points. Although these data do not definitively address changes in the thermal reference upon which the thermoregulatory system cues, the data clearly suggest that the endothermic thermoregulatory system has not failed. This is evident in two features of Fig. 1. First, at temperatures below $22^{\circ}C$, the hypoxic rats increased oxygen consumption in what appears to be a response similar to the increase in metabolic heat production at temperatures below the thermoneutral zone of normoxic animals. Secondly, and perhaps more telling, although the body temperature was not maintained at a single temperature over the range of ambient temperatures under hypoxic conditions, the slope of the relationship between body temperature and ambient temperature (0.39) is significantly different than that of the isothermal line (unity). However, the data are currently insufficient to ascertain the relative performance of the thermoregulatory system (at the system level) during hypoxia.

The decrease in thermoregulatory operational points is consistent with the observations of Hicks and Wood, (1985), who found that hypoxic lizards in a two-temperature shuttle box would continue to thermoregulate during hypoxia but at lower upper and lower thresholds for behavioral thermoregulation. In addition, Dupre et al., (1986) noted that the thermal thresholds for the onset of panting also decreased during hypoxia.

Such shifts in the thermoregulatory patterns of ectotherms were predicted based on a two-compartment model of oxygen transport in organisms with significant shunts (Wood, 1985). Based on the position of oxyhemoglobin dissociation curves at an ectotherm's "preferred" body temperature, as the ambient oxygen concentration was reduced, a point would be reached – the "crossover" point – at which the amount of oxygen loading would not be sufficient to meet the animal's metabolic needs. However, if the animal were to lower its body temperature and change the temperature-dependent position of the oxyhemoglobin dissociation curve, then the oxygen demands of the animal – lower due to the van't Hoff effect – could be maintained.

Although the significance of hypoxic hypothermia has been largely ignored, there are a few studies which suggest that this response is not only theoretically interesting, but a very practical solution to the problems of matching oxygen transport to oxygen demand during periods of limited oxygen availability. Hypothermia has been found to increase the time of survival of hypoxic neonates from several mammalian species (Miller and Miller, 1966). Dunn and Miller, (1969) established the potential clinical importance of hypothermia during hypoxia when they made an experimental group of asphyxic human neonates hypothermic in addition to the generally accepted treatment. This treatment resulted in a decrease in the mortality of the experimental group to about 11% as compared to an international average of about 44%. Carlsson et al., (1976) have also shown that hypothermia provides a protective effect to the brain during hypoxic hypoxia.

Despite these studies, the typical clinical treatment for infants and children with hypoxemia due to cardio-pulmonary disease is to externally regulate their temperature such that a core temperature of $37^{\circ}C$ is maintained. If the hypothermia associated with hypoxia is a regulated phenomenon, then a more appropriate treatment might be to allow the body temperature to decrease to their new controlled level. Or, if a thermostat is still deemed necessary, then temperature might be maintained at $2-5^{\circ}C$ below $37^{\circ}C$, thus providing for a more favorable matching between oxygen transport and oxygen demand. The focus of this question remains as to

350

whether hypoxic hypothermia is a regulated response. While the data presented herein suggest that some aspects of thermoregulation change during hypoxia, further studies will be required to conclude that this is due to a resetting of the set-point of the thermoregulatory system.

REFERENCES

Blatteis, C.M. and L.O. Lutherer (1973). Cold-induced thermogenesis in dogs: its reduction by moderate hypoxia. J. Appl. Physiol. 35:608-612.

Blatteis, C.M. and L.O. Lutherer (1974). Reduction by moderate hypoxia of the calorigenic action of catecholamines in dogs. J. Appl. Physiol. 36:337-339.

Carlsson, C., M. Hagerdal and B.K. Siesjo (1976). Protective effect of hypothermia in cerebral oxygen deficiency caused by arterial hypoxia. Anesthesiol. 44:27-35.

Dunn, J.M. and J.A. Miller, Jr. (1969). Hypothermia combined with positive pressure ventilation in resuscitation of the asphyxiated neonate. Am. J. Obstet. Gynecol. 104:58-67.

Dupre, R.K., J.W. Hicks and S.C. Wood (1986). The effect of hypoxia on evaporative cooling thresholds of lizards. J. Therm. Biol. 11:223-227.

Gellhorn, E. and A. Janus (1936). The influence of partial pressures of O_2 on body temperature. Am. J. Physiol. 116:327-329.

Hicks, J.W., R.K. Dupre and S.C. Wood (1984). Blood oxygen alters preferred body temperature in lizards. Physiologist 27:284.

Hicks, J.W. and S.C. Wood (1985). Temperature regulation in lizards: effects of hypoxia. Am. J. Physiol. 248:R595-R600.

Horstman, D.H. and L.E. Banderet (1977). Hypoxia-induced metabolic and core temperature changes in the squirrel monkey. J. Appl. Physiol. 42:273-278.

Krogh, A. (1914). The quantitative relation between temperature and standard metabolism in animals. Intern. Z. Physik-chem. Biol. 1:491-508.

Lister, G. (1984). Oxygen transport in the intact hypoxic newborn lamb: acute effects of increasing P_{50}. Ped. Res. 18:172-177.

Miller, J.A., Jr. and F.S. Miller (1966). Interactions between hypothermia and hypoxia-hypercapnia in neonates. Federation Proc. 25:1338-1341.

Moore, R.E. (1959). Oxygen consumption and body temperature in newborn kittens subjected to hypoxia and reoxygenation. J. Physiol. (Lond.) 149:500-518.

Schnakenberg, D.D. and R.A. Hoffman (1972). Hypoxic inactivation of cold-induced brown fat thermogenesis. Experientia 28:1172-1173.

Wood, S.C. (1984). Cardiovascular shunts and oxygen transport in lower vertebrates. Am. J. Physiol. 247:R3-R14.

Wood, S.C., R.K. Dupre and J.W. Hicks (1985). Voluntary hypothermia in hypoxic animals. Acta Physiol. Scand. 124:46.

Wood, S.C. and J.W. Hicks (1985). Oxygen transport in vertebrates with cardiovascular shunts. In: Alfred Benzon Symposium 21. Cardiovascular shunts: Phylogenetic, Ontogenetic, and Clinical Aspects, ed. by K. Johansen and W. Burggren, Munksgaard, Copenhagen, pp. 354-362.

PARTICIPANTS

Thomas Albrecht, Department of Physiology, University of Kansas Medical Center, Kansas City, KS 66103

Marvin Bernstein, Department of Biology, New Mexico State University, Las Cruces, NM 88003

Toni Bray, Department of Physiology, University of Kansas Medical Center, Kansas City, KS 66103

G. Brice, Department of Physiology, Medical College of Wisconsin, Milwaukee, WI 53226

Steven M. Cain, Department of Physiology and Biophysics, University of Alabama at Birmingham, University Station, Birmingham, AL 35294

H.K. Chang, Department of Biomedical Engineering, University of Southern California, Los Angeles, CA 90089

C.K. Chapler, Department of Physiology, Queen's University, Kingston, Canada K7L 3N6

Asita Chatterjee, Department of Anatomy and Physiology, Kansas State University, Manhattan, KS 66506

Richard L. Clancy, Department of Physiology, University of Kansas Medical Center, Kansas City, KS 66103

Richard J. Connett, Department of Physiology, University of Rochester, School of Medicine and Dentistry, Rochester, NY 14642

Jerome A. Dempsey, Department of Preventive Medicine, University of Wisconsin Medical School, Madison, WI 53705

Howard H. Erickson, Department of Anatomy and Physiology, College of Veterinary Medicine, Kansas State University, Manhattan, KS 66506

Kipp Erickson, Department of Anatomy and Physiology, College of Veterinary Medicine, Kansas State University, Manhattan, KS 66506

Mary Ernst, Department of Physiology and Cell Biology, The University of Kansas, Lawrence, KS 66045

M. Roger Fedde, Department of Anatomy and Physiology, College of Veterinary Medicine, Kansas State University, Manhattan, KS 66506

Lisa Flood, Department of Physiology and Cell Biology, The University of Kansas, Lawrence, KS 66045

Hubert V. Forster, Department of Physiology, Medical College of Wisconsin, Milwaukee, WI 53226

Thomas Gayeski, Department of Anesthesiology, University of Rochester Medical Center, Rochester, NY 14642

Jerry R. Gillespie, Department of Surgery and Medicine, College of Veterinary Medicine, Kansas State University, Manhattan, KS 66506

Norberto C. Gonzalez, Department of Physiology, University of Kansas Medical Center, Kansas City, KS 66103

Guillermo Gutierrez, Department of Internal Medicine, University of Texas Health Sciences Center, Houston, TX 77025

Mike D. Hammond, Division of Pulmonary and Critical Care Medicine, James Haley Veterans Hospital, Tampa, FL 33612

George J. Heigenhauser, Department of Medicine, McMaster University Medical Center, Hamilton, Ontario, Canada, L8N 3Z5

Esther P. Hill, Section of Physiology, Department of Medicine, University of California at San Diego, La Jolla, CA 92093

Michael Hlastala, Division of Respiratory Diseases, Department of Medicine, University of Washington, School of Medicine, Seattle, WA 98195

Mike Hogan, Section of Physiology M-023A, Department of Medicine, University of California at San Diego, La Jolla, CA 92093

Kuen-Sahn Hung, Department of Anatomy, University of Kansas Medical Center, Kansas City, KS 66103

Russell E. Isaacks, Veterans Administration Medical Center, Miami, FL 33125

James H. Jones, Department of Physiological Sciences, University of California, School of Veterinary Medicine, Davis, CA 95616

Robert Klemm, Department of Anatomy and Physiology, Kansas State University, Manhattan, KS 66506

Paul Kubes, Department of Physiology, Queen's University, Kingston, Canada K7L 3N6

Gail Landgren, Department of Anatomy and Physiology, College of Veterinary Medicine, Kansas State University, Manhattan, KS 66506

M. Harold Laughlin, Department of Biomedical Sciences, College of Veterinary Medicine, University of Missouri, Columbia, MO 65211

Stan L. Lindstedt, Department of Zoology and Physiology, University of Wyoming, Laramie, WY 82071

Jack A. Loeppky, Department of Bioengineering and Physiology, Lovelace Medical Foundation, Albuquerque, NM 87108

Tim Martin, Sickle Cell Trait Project, William Beaumont Army Medical Center, El Paso, Texas 79920

Odile Mathieu-Costelo, Section of Physiology M-023A, Department of Medicine, University of California at San Diego, La Jolla, CA 92093

Mat Melinyshyn, Department of Physiology, Queen's University, Kingston, Canada K7L 3N6

James A. Orr, Department of Physiology and Cell Biology, The University of Kansas, Lawrence, KS 66045

Janet Pauley, Department of Physiology, University of Kansas Medical Center, Kansas City, KS 66103

Andrew Peacock, Cardiovascular and Pulmonary Research Laboratory, University of Colorado Health Science Center, Denver, CO 80262

Johannes Piiper, Abteilung Physiologie, Max-Planck-Institut fur Experimentelle Medizin, D-3400 Gottingen, Federal Republic of Germany

Frank L. Powell, Section of Physiology M-023A, Department of Medicine, University of California at San Diego, La Jolla, CA 92093

Josep Roca, Section of Physiology M-023A, Department of Medicine, University of California at San Diego, La Jolla, CA 92093

Rick Rogers, Department of Anatomy and Physiology, College of Veterinary Medicine, Kansas State University, Manhattan, KS 66506

David Saunders, Department of Anatomy and Physiology, College of Veterinary Medicine, Kansas State University, Manhattan, KS 66506

Kevin Sink, Department of Physiology and Cell Biology, The University of Kansas, Lawrence, KS 66045

Steve Spencer, Department of Physiology and Cell Biology, The University of Kansas, Lawrence, KS 66045

C. Richard Taylor, Harvard University, Concord Field Station, Bedford, MA 01730

Steven P. Thomas, Department of Biological Sciences, Duquesne University, Pittsburgh, PA 15282

Peter Wagner, Section of Physiology M-023A, Department of Medicine, University of California at San Diego, La Jolla, CA 92093

Ewald Weibel, Institute of Anatomy, Universitat Bern, Postfach 139, Buhlstrasse 26, CH-3000 Bern 9, Switzerland

John B. West, Section of Physiology M-023A, Department of Medicine, University of California at San Diego, La Jolla, CA 92093

Jane Westfall, Department of Anatomy and Physiology, College of Veterinary Medicine, Kansas State University, Manhattan, KS 66506

Gail Widener, Department of Physiology, University of Kansas Medical Center, Kansas City, KS 66103

David C. Willford, Section of Physiology M-023A, Department of Medicine, University of California at San Diego, La Jolla, CA 92093

Jeff Wilson, Department of Anatomy and Physiology, College of Veterinary Medicine, Kansas State University, Manhattan, KS 66506

Robert M. Winslow, Division of Blood Diseases, Letterman Army Institute of Research, Presidio, San Francisco, CA 94129

Jonathan B. Wittenberg, Albert Einstein College of Medicine, Bronx, NY 10561

Stephen Wood, Department of Bioengineering, Lovelace Medical Foundation, Albuquerque, NM 87108

Robert Woodson, Department of Medicine, University of Wisconsin School of Medicine, Madison, WI 53792

Symposium on Oxygen Transfer from Atmosphere to Tissues

Kansas City   March 27 & 28, 1987

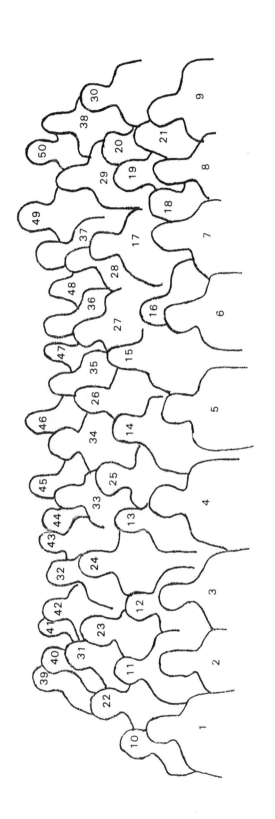

1. M. Roger Fedde
2. Mike D. Hammond
3. Frank L. Powell
4. Johannes Piiper
5. Guillermo Gutierrez
6. Russell E. Isaacks
7. G. Brice
8. Mary Ernst
9. Norberto C. Gonzalez
10. Kipp Erickson
11. Mike Hogan
12. James A. Orr
13. Robert Woodson

14. Steven M. Cain
15. C.K. Chapler
16. Paul Kubes
17. Hubert V. Forster
18. Gail Widener
19. Lisa Flood
20. Tim Martin
21. Toni Bray
22. Mat Melinyshyn
23. Gail Landgren
24. H.K. Chang
25. Odile Mathieu-Costelo
26. C. Richard Taylor

27. John B. West
28. Thomas Albrecht
29. David C. Willford
30. Janet Pauley
31. Rick Rogers
32. Michael Hlastala
33. Robert M. Winslow
34. Jonathan B. Wittenberg
35. Stan L. Lindstedt
36. James H. Jones
37. Ewald Weibel
38. M. Harold Laughlin
39. George J. Heigenhauser

40. Jeff Wilson
41. Richard L. Clancy
42. Jack A. Loeppky
43. Jerome A. Dempsey
44. Richard J. Connett
45. Peter Wagner
46. Andrew Peacock
47. Josep Roca
48. Asita Chatterjee
49. Steven P. Thomas
50. David Saunders

INDEX